DATE DUE

DEMCO 38-296

Physics in the Nineteenth Century

Michael Faraday (1791–1867). *Photograph by Sir R. Hadfield.*

Physics in the Nineteenth Century

ROBERT D. PURRINGTON

Rutgers University Press
New Brunswick, New Jersey, and London

Library of Congress Cataloging-in-Publication Data

Purrington, Robert D.
    Physics in the nineteenth century / Robert D. Purrington.
      p. cm.
    Includes bibliographical references and index.
    ISBN 0–8135–2441–5 (cloth : alk. paper).—ISBN 0–8135–02442–3
  (pbk. : alk. paper)
    1. Physics—History—19th century. I. Title.
QC7.P84 1997   96–49115
530'.90'034—dc21   CIP

British Cataloging-in-Publication information available

Manufactured in the United States of America.

CONTENTS

List of Illustrations   *vii*
Preface   *ix*
Acknowledgments   *xvii*

1 *Prologue: The Century of Science*   *1*

2 *Nineteenth-Century Science in Context*   *9*

3 *Electromagnetism*   *32*

4 *Heat and Thermodynamics*   *75*

5 *Energy and the Energy Principle*   *102*

6 *Atomism*   *113*

7 *The Kinetic Theory of Gases and*
   Statistical Mechanics   *132*

8 *Fin de Siècle*   *148*

9 *Epilogue*   *169*

Notes   *175*
Index   *239*

ILLUSTRATIONS

Plates

Frontispiece. Michael Faraday.

1. William Whewell. 24
2. Title page of Benjamin Franklin's *Experiments and Observations on Electricity.* 35
3. Charles Augustin Coulomb. 36
4. Hans Christian Oersted. 42
5. André-Marie Ampère. 43
6. Faraday's laboratory at the Royal Institution. 51
7. William Thomson. 59
8. George Gabriel Stokes. 60
9. James Clerk Maxwell. 64
10. Newcomen steam engine. 81
11. James Watt. 82
12. Sadi Carnot. 87
13. James Prescott Joule. 90
14. Sir William Thomson. 97
15. W.J.M. Rankine. 99
16. Thompson's illustration of cannon boring. 106
17. Hermann von Helmholtz. 111
18. Bernoulli's illustration of his kinetic theory of gases. 134
19. Rudolph Clausius. 137

Figures

3–1. Oersted's notes on the experiment in which he "discovered electromagnetism." 41

3–2. Ampère's device for measuring the force between two
        current-carrying spirals.   *45*

3–3. Faraday's device that first achieved "magnetic rotation."   *49*

3–4. Maxwell's mechanical model of hexagonal molecular vortices
        with rolling idle wheels between them.   *69*

4–1. Carnot cycle, displayed on a p-V, or indicator, diagram.   *89*

6–1. Dalton's models of spherical atoms.   *121*

8–1. Experimental blackbody curves for different temperatures.   *151*

8–2. Observed temperature dependence of the specific heat of
        hydrogen gas at constant volume (c_v).   *159*

PREFACE

I am sure that many writers have wanted to make the point that William Berkson did in his preface to *Fields of Force*. The idea for the book arose from the fact that I wanted to read such a book, had looked for it, but found there was none.[1] The result is this work, a study of physics in the nineteenth century, which readers could equally well consider a book about physics between the French Revolution (or the Industrial Revolution) and World War I. Above all, it is the story of the flowering of classical physics that is the hallmark of the nineteenth century. The nineteenth century, and especially the last half, saw the creation of physics as an independent and coherent scientific discipline, with a well-developed sense of profession, with vigorous scientific organizations, and with widely disseminated and widely read professional journals. It was the century in which the scope, subject matter, and divisions of classical physics became so clearly defined that they are for the most part unchanged in the twentieth century.

This book offers a scientist's view of a crucial century in the history of physics, something that has become increasingly rare. It is, in a sense, an attempt to regain for scientists some lost ground, namely the prerogative to explore the history and evolution of their discipline. Perhaps the key word here is *evolution,* a perspective that historians often reject as anachronistic. There are sound reasons why this is so, yet to be dogmatic about the issue impoverishes the corpus of historical writing about science. It denies a large audience, consisting of scientifically literate and historically oriented readers for whom narrow, specialized, methodologically correct writing has little to offer, any insight into the connections between modern science and its forebears. Furthermore, this book's evolutionary or developmental perspective is a valid view of scientific change. Endangered and increasingly unfashionable, such an approach emphasizes that the course of scientific discovery in a given field has a progressive character that deserves careful recounting, despite spurts and stops, blind alleys and reversals. An arrow of time exists, whether or not one believes that it moves in the direction of progress.

The scientist's view of the development of nineteenth-century science is often different from the historian's—not necessarily superior, but different. Thus, the history of science should be judged not only as history but also as science, and scientists are uniquely equipped to deal with this dimension. Of course, the historian brings a sensitivity to historiography that few scientists possess and a determination to see the scientific activity of a period without any reference to the present. The validity of both views should be obvious to anyone not blinded by professional rhetoric and dogma. In saying this, I acknowledge that a great deal of what average scientists take to be the "facts" of the history of their discipline are just plain wrong. Often discoveries did not take place in the ways in which even the discoverers themselves have claimed, and the role of historians in setting the record straight has been crucial.

We all know, or should know, that pure unadorned historical fact, devoid of context and interpretation, is rare or nonexistent. Thus, historians of science endeavor to understand the way in which scientific practice was molded by the culture of its time, how it was shaped by economic and societal forces. The scientist-historian, while not ignoring such influences, may focus more strongly on the connections between and among discoveries, on the way in which the work of one scientist influenced the work of contemporaries and successors through correspondence and published papers. It does not follow that the importance of the scientific activity of an earlier era is to be judged from a modern perspective, only that science has its own inner imperatives that operate alongside those of the cultural context. In other words, the substance of the discipline evolves and develops in ways that one could never predict from the social context alone. Of course, the converse is true as well. Thus, an understanding of both dimensions is vital, whether one sees scientific change in terms of the old polar opposites known as *internal* and *external* or in terms of more subtle and complex interactions among forces at work in the culture and factors that are intrinsic to the subject itself, and whether one admits that such a separation is even possible. If I have emphasized intrinsic factors in this book, showing how a discovery was shaped by existing results, both published and unpublished, experimental and theoretical, and by interaction with others working in the field, the reasons are twofold: first, because that is usually how the practitioners themselves saw it; and second, perhaps, because that is how a scientist today sees discovery and change in a field.

Even scientists ought to be able to see the importance of treating the history of their discipline in the same way that the history of any other set of ideas is explored: by learning what scientists were actually doing and saying and how these actions were related to the culture in which they were working. By dismissing the efforts of predecessors who pursued what hindsight has revealed as blind alleys, we not only denigrate their particular search for understanding but also implicitly condemn to irrelevance that portion of today's science, however important it may now seem, that will fail to become be part of the received truth or canon of the next generation. Because we cannot know in advance what will prevail and what will not, the importance of all scientific work is put into doubt or judged not for its own sake but by how future commentators will view it. Thomas Kuhn's advice is worth heeding in this regard: that "the past of science should be approached as an alien culture, one that the historian strives first to understand and then to make accessible to others."[2] In other words, the history of a particular era is not simply the story of a few great figures whose names are household words. As it is today, much of the actual work in the nineteenth century was carried out by scientists or philosophers whose names, for whatever reason, are rarely remembered now.

The message here is that we must take seriously the day-to-day work of our predecessors in spite of the gap between what they knew and what we know, avoiding the easy route of ridiculing their errors or dismissing their methods as unscientific. Yet even while we take pains to understand the work of those who held views we now know to be wrong, we can, in the physical sciences, select from the historical record those beliefs or discoveries on which our current understanding is founded. That this can be done in the sciences (to a degree not possible in other areas of intellectual inquiry) is the result of the substantially linear character of scientific discovery. Granting that much of what we take to be established truth in any scientific discipline is subject to revision and falsification by future developments and is peculiar to a particular time and place, it is nonetheless characteristic of the sciences that they are built on empirical facts, however much those facts may be relative to current theoretical structures. We err in treating scientific data as purely relative, as nothing more than part of a belief system, thus reducing the history of science to a description of what scientists believed as they practiced their professions, just as we err in assuming that empirical facts are unchanging and that the course of science merely progresses from ignorance to enlightenment.

For nearly all practicing scientists—not all, to be sure—realism is an unequivocal commitment, rarely reflected upon very deeply.[3] Science, according to this view, is not merely another cultural activity, not simply fashion or metaphor, not simply an alternative way of viewing the world. The success of science, its efficacy, its law-giving character—indeed the "progress" of science—clearly distinguishes it from other, no less important, areas of human inquiry. Fashions change in science, it is true; theories are found to be wrong; ideas once held are discarded. But like it or not, science moves ponderously but inexorably forward. In some areas, such as the ecological and geological sciences, this statement is self-evident. But even particle physics, operating in a realm beyond the reach of commonsense ideas, is different only in a limited and technical sense. We know that our picture of the fundamental aspects of matter (for example, the microscopic world) will be different in a century or a millennium, that the language itself will be different. But the phenomena survive, however differently they may be described and interpreted. The discoveries are not wasted but incorporated into a new description.

Science has a sociological dimension, of course; and its impact, its speed of progress, its public face, do depend on the time in which it is carried out. No one would be quicker than the scientist to admit that there is no single scientific method that is applicable in all fields and at all times or to both theorists and experimentalists. Similarly, no one operates under the misapprehension that the choice of problems considered to be important, the distribution of research money by funding agencies, the award of prizes, and so on are unrelated to larger social and culture issues. In a deeper sense, it may often be true that the metaphors used by scientists in understanding their discoveries, the models they employ, owe as much to intellectual fashion outside science as within it. But the way that science is done or why it is done is quite separate from *what* is done.

Any work in the history of science lies somewhere between two extremes. One extreme reflects the growth of the history of science as a professional academic discipline and tends to be largely descriptive—or, in the jargon, contextual. It describes the history of a scientific field or the science of a particular era without any reference to the present or future of that discipline and is, as nearly as possible, the history of a field of science as it was actually practiced. The other extreme is sometimes called developmental historiography, which follows the evolution of a discipline or a specific idea from its historical origins to the present or takes special notice of those ideas that have had the greatest impact on current science.[4] It tends to see science less as a cultural or a social activity and more as a search for truth (although in some suitably restricted sense), and it focuses on the substance of the scientific discovery itself, on the connections between ideas and results, rather than on process or context.[5] I hope that this book will fall somewhere between those two extremes. I have made every effort to be factually correct, with the goal of creating for the reader a sense of what it was like to be involved with physics in the nineteenth century.

One does not embark on a history of nineteenth-century science without trepidation. There has been more written about science in that century than in any other (except for our own), meaning that the primary sources have for the most part survived intact. The century combined a predilection for written correspondence with increasingly rapid transportation and communication, so there exist almost innumerable letters among and between the figures who shaped nineteenth-century science.[6] It was also the period in which the professional journal came into being—although a few journals, such as those of the Royal Society, have a much longer history. The result is a plethora of both resources and secondary literature on the century. Yet there have been few attempts at providing a comprehensive account of physical science in the nineteenth century, largely (but not solely) because of the magnitude of the task. René Taton devoted one volume of his noted *Histoire des sciences* to the nineteenth century and had the temerity to review *all* the sciences, but the result could hardly be called successful and certainly not comprehensive.[7] Another reason for the absence of comprehensive histories of nineteenth-century science is a disinclination on the part of historians of science to undertake a work involving significant integration and generalization, especially for such a rich and important period. It demands a hefty amount of intellectual hubris to undertake this kind of synthesis, which simultaneously attempts to describe the details of everyday scientific activity during a particular time while placing them into a larger context. Depth and completeness are inevitably sacrificed to breadth and scope, and one need only look at the size of Crosbie Smith and Norton Wise's biography of Kelvin (William Thomson) to get some sense of what a truly comprehensive theory of nineteenth-century physics might look like.[8] Of course, the goal of a book such as this is not to make the detailed journal articles, biographies, and monographs obsolete but rather to provide connections, which would otherwise be lacking, among these works and among their central ideas and personalities. I hope that the reader will feel that this larger scope provides sufficient justification for the endeavor— that the interconnections among ideas, individuals, and eras will make the history more meaningful, especially to those who cannot or will not consult the specialized journals.

Another reason for the absence of a comprehensive study of physics in the nineteenth century is the reluctance of historians of science to acknowledge the evolutionary or progressive character of science. Indeed, some of the best attempts to write this sort of history, works such as Donald Cardwell's *From Watt to Joule,* are no longer fashionable. As a result, the history of science rarely reaches beyond specialized journals and monographs. Perhaps most slighted are scientists who teach and study these disciplines today, who would profit enormously from contact with the history of their disciplines.

A final reason explaining the dearth of comprehensive studies of nineteenth-century science is the knowledge that our understanding of the topic is incomplete. Nevertheless, our lack of knowledge will remain incomplete even a century from now. Since World War II, our understanding of nineteenth-century science and the figures who helped to shape it has increased enormously. Given the growing professionalism and specialization of the twentieth century, it would be difficult, if not impossible, to write a history of all of nineteenth-century physical science. Thus, I have undertaken the substantially more modest, but nonetheless challenging, task of writing a history of nineteenth-century physics. It is, I repeat, enormously challenging; for the resources are appallingly rich, one's abilities only too finite, and a publisher's forbearance limited.

It is difficult to understand how a subtle person such as Henry Guerlac could have claimed that "the history of science is primarily . . . the history of thought about nature," yet the claim is not entirely off the mark as a description of a great deal of writing in the history of science.[9] I note with interest that David Wilson, while supporting Thomas Kuhn's brilliant thesis concerning Planck and quantum discontinuity, takes Kuhn to task for downgrading the importance of experimental physics.[10] Perhaps this tendency results from the fact that historians of science are often historians of ideas who specialize in scientific thought and have little taste or sympathy for physics as an experimental science.[11] But physics is, after all, an empirical science; the raw material upon which the theorist ruminates is observation and experiment. The theoretical physicist is a late nineteenth-century invention; before that time, most of the great natural philosophers were adept at theory *and* experiment: One thinks of Galileo, Hooke, Huygens, Newton, and even Ampère, Maxwell, and Helmholtz.[12] But while the heritage of d'Alembert, Lagrange, Laplace, Poisson, Euler, and Gauss can be traced through figures such as Maxwell and Thomson to those we consider to be the first real theoretical physicists (Boltzmann, Planck, Lorentz, and Einstein), the "discovery" of modern physics is largely a story of experimental physics or at least one in which experimental physics and experimental physicists played a major role.[13]

Of course, historians and physicists will continue to make competing claims on behalf of theory and experiment. At one time or another, experimental physics may be strongly guided by theory; but such guidance is the sine qua non of experimental investigation, something we should remember even as we argue for giving more attention to the role of empiricism. Some scientists, such as Arthur Eddington, believed that experiment played a minor role in understanding nature. The contrary view of Lenard and Stark—that experiment alone could be counted upon—was partly responsible for their hostility to Einstein's ideas and indeed to much "non-Aryan" science. Max Born, a great theorist himself, who was of Jewish origin, was sufficiently disturbed by Eddington's ideas, which he called "a considerable danger to the sound development of science," that he answered them in his speech "Experiment and Theory in Physics," addressed to the Durham Philosophical Society in 1943.[14] In recent times, Derek de Solla Price has strongly defended the autonomous role of experiment or observation in the process of scientific discovery.[15]

The fact is that such generalizations are hazardous and that the balance between theory and experiment in one time or in one discovery may be very different from that in another. Largely empirical discoveries often dominate an emerging area of physics, where the phenomena themselves are not even known. Such was the case from the 1880s into the early years of the twentieth century. On the other hand, at least in microphysics, experimental results are no more "facts" than theoretical entities are; they have no meaning except as they are viewed in some theoretical structure, and they can only be connected through theory, which is the only way to generalize from them.[16] For Paul Dirac, one of the founders of modern quantum theory, the mathematics preceded everything else, followed by theoretical interpretation, and finally by a connection with experiment. In the end, many mathematical devices that are fundamental to some of the most successful physical theories have no experimental counterpart at all.

In the late nineteenth century, experimental physics was driven as much by technological innovation as by theoretical imperative. The legacy of the Industrial Revolution, which began about 1750, and the "second Industrial Revolution" 125 years later, which brought widespread electrification and the internal combustion engine, both greatly influenced research in physics. If space permitted, I would explore at length the impact of new technology and the resulting expansion of experimental physics into new realms (small physical scale, high and low temperatures, hard vacua, higher time resolution, and so on), leading to the discovery of phenomena that ultimately could not be absorbed into the framework of classical physics.[17]

Centuries are hardly less arbitrary than generations; and in speaking of nineteenth-century physics, one might be justly accused of imposing boundaries onto the continuous development of science that are totally unrelated to the course of scientific discovery. In the case of the nineteenth century, however, it is difficult to make the charge stick. Certainly the period from the Industrial Revolution to the beginning of World War I is a self-contained and coherent period in the history of physics; and it is essentially the period I deal with in this book, although it generously exceeds the span of a century. A less ambitious view of the epoch would include the French Revolution up to 1900, the year in which the quantum revolution began.

The literature on physics in the nineteenth century is enormous and grows almost faster than one can read it. Not all of it is important or of uniform quality. The notes to each chapter in this book provide citations to works that the reader may want to consult to verify my statements of fact or opinion or to explore any issue more fully, and they include a selection of important monographs and articles that may elucidate issues I have raised. Currently, there is no other comprehensive study of nineteenth-century physics; and if this book does not quite claim to have accomplished that goal, it nonetheless comes closer than others. I have given some fields short shrift for a variety of reasons, including constraints of space and time (as in the case of fluid dynamics and elasticity) or because treating them in detail would take us too far afield. Astronomy remained separate from many of the theoretical and technological developments in the other sciences during the century, and except for celestial mechanics, began to shed its descriptive character and merge with physics only after 1850.

When possible, the reader should consult the primary sources: the *Diary* and *Experimental Researches* of Faraday; the *Scientific Papers* and *Treatise* of Maxwell; the papers written by Joule, Thomson (Lord Kelvin) and Stokes, Tyndall, and so on in the *Philosophical Magazine* or the various publications of the Royal Society and the corresponding documents in *Annalen der Physik* or *Comptes Rendus;* or translations of Carnot, Helmholtz, and Clausius. There is no better way to understand scientific practice as it actually took place. Although access to correspondence is usually only available to specialists, there has been a recent outpouring of works on the correspondence of Thomson, Stokes, Faraday, and Maxwell, and there is much more to come. These published volumes offer a unique insight into the scientific imaginations of the figures of the time and are the best guide to how the scientific community functioned and how it saw itself. At the next level are the studies published in journals such as *Isis, Osiris, British Journal for the History of Science, Studies in the History and Philosophy of Science, Historical Studies in the Physical Sciences,* and *Archive for the History of Exact Sciences*, which themselves provide the basis for exploring fully the literature of any issue or idea.[18] Finally, there is an extensive literature of biographies (such as Williams on Faraday, Smith and Wise on Thomson, and Wilson on Rutherford) and treatises on energy conservation, atomism, field theories, aether, and so on. These latter categories constitute the secondary literature, which is invaluable if one does not want to repeat the work of others. I do not shrink from this secondary literature and cite the best of it throughout the book; but readers should remember that secondary sources are not data per se and that knowledge of the nineteenth century grows, evolves, and changes so that what may have seemed clear in 1980 may no longer seem so obvious.

ACKNOWLEDGMENTS

I would like to thank Tulane University for the sabbatical leave that allowed me to finish this book; for a further year in London during which I completed much archival work and rewriting; and for travel support that made it possible to consult the correspondence of Maxwell, Tait, William Thomson (Lord Kelvin), J. J. Thomson, and Rutherford. Special gratitude is due to the staff of the rare books and manuscripts division of Cambridge University Library and especially to its superintendent, Godfrey Waller. The staff of the British Library and the libraries of the Science Museum and Imperial College also provided valuable assistance. For permission to quote from the Whewell-Faraday correspondence in the Wren Library at Trinity College, Cambridge, I thank the Master and Fellows of Trinity College and gratefully acknowledge the assistance of John Smith of the Wren Library. For permission to use the library of the Royal Institution and to quote from Faraday letters in its possession, I thank the Royal Institution and especially Irene McCabe. Permission to use the library at the University of Edinburgh is also gratefully acknowledged. I am thankful for the U.S. Naval Oceanographic Office's support of my work in scattering and propagation, which freed other funds for travel and provided academic-year release time. Conversations with the late Norton Nelkin of the University of New Orleans and Frank Durham of Tulane University are gratefully acknowledged, as are those with Frank James of the Royal Institution. I thank Stephen Brush in particular for his willingness to read the manuscript and for his many valuable suggestions concerning the substance and organization of the book. All errors that remain are my own. The staff of the Howard-Tilton Memorial Library at Tulane, especially Eleanor Elder and Cristina Fowler, provided invaluable assistance. Finally, I offer my gratitude to my meticulous copy editor, Dawn Potter, and to Karen Reeds and Doreen Valentine, my editors at Rutgers, for their hard work and enthusiasm for the project, without which it would not have been completed.

Physics in the Nineteenth Century

CHAPTER 1

Prologue: The Century of Science

*I*t is not particularly remarkable that a century should see in its predecessor the prelude to its own modernism; so we might be excused if we were to view the nineteenth century in this way, finding it interesting primarily for its role in setting the stage for the revolutions of the early twentieth century, ignoring other aspects of the critical heritage of that century and its impact on our own, and, most of all, failing to take it on its own terms. Thus, until recently the nineteenth century has been viewed as merely transitional, a stepping-stone to the twentieth century, and therefore less important. This view of history is rarely, if ever, valid and in the case of physics is especially wide of the mark. If modern physics has been built upon nineteenth-century discoveries, the century is more rather than less important because of that fact.

Physicists and other practicing scientists sometimes have difficulty understanding this point because science, unlike many other intellectual pursuits, proceeds by climbing over the carcasses of its dead—in this case rejected theories and superseded data. Theories are discarded in favor of those with better predictive power, new tools replace old ones, and the past is rarely revisited. As a result, the history of physics is not part of the education of a physicist, and graduate schools explore only the most recent ideas.

The nineteenth century was the century of Maxwell, Helmholtz, Kelvin (William Thomson), Clausius, Faraday, Berzelius, Dalton, Ampère, Joule, Gauss, Laplace, Priestley, and Lavoisier, to select only a few names from the many that enrich the history of nineteenth-century physics and chemistry. But as I shall try to show, one cannot adequately characterize even nineteenth-century science, much less the entire century, by examining the careers of these giants alone. The nature of physics in the nineteenth century, as in any other, is much less the story of a few extraordinary figures and much more that of the dozens of individuals whose work constituted the daily practice of physics. Nevertheless, we must remember that in 1830, or even in 1880, the number of people working at what we would now call physics was far smaller than it is today, with perhaps a few score individuals contributing to the development of the science at any given time in any country. Therefore, when we single out the work of a Faraday or a Maxwell, we are being less selective than one might think. This fact becomes even clearer when we scan the published work in a given field, even though this book notes a number of cases in which important discoveries remained obscure or unknown because the innovator chose not to publish, had his work rejected for publication, or had no opportunity to air it fully.

The nineteenth century was the child of the Industrial Revolution, which as the century opened was already some fifty years old. Upon this revolution was founded nineteenth-century liberalism, both orthodox classical economic theory and Marxist economics, and industrial capitalism. With the head start that it got in England, the Industrial Revolution led to unparalleled British wealth and economic and political ascendancy during the succeeding century. It came later to the European continent, by which time British influence had spread around the world. The chaotic political structure of Europe had prevented effective exploitation of natural resources, and only after the Franco-Prussian war of 1870–71 did Germany industrialize.[1] When the Industrial Revolution did take place in Europe, the ramifications for science and technology in the succeeding half-century were enormous; but coupled with German militarism, it was a material factor in the slaughter of World War I. Thus, driven by industrialization and technology, Europe began to transform from a basically rural society to one that was increasingly urban. Steam power drained the marshes, powered the railroads, and began to span the seas. Machines insinuated themselves into every commercial activity, including agriculture. As we shall see, there was often a strong interaction between natural philosophy or science and technology, to the mutual benefit of each. Beginning in the British textile industry, technology quickly moved to mining, where (among other tasks) pumps driven by steam were employed to clear water from Cornish tin mines. Scientific activity, especially in areas that seemed to support technology and industry, naturally flourished. To offer one example, the theory of heat evolved directly from the need to understand heat engines and the conditions that could produce increased thermodynamic efficiency.

It is hardly arguable that British cultural domination was a direct stimulus to the scientific activity of the nineteenth century. The practical character of the culture's economic basis, industrialization, farflung sea trading, and colonialism, manifested itself in the inclination of British science (which gained ascendancy after about 1845) toward mechanical analogies and concrete models. By the latter part of the nineteenth century, however, British industrial dominance seemed to be on the wane, as dramatized by the Paris Exposition of 1867; and a similar argument might be made about British science, which had been without peer during the half-century after 1840.

In the 1840s railways took over as the dominant form of transportation, the transatlantic cable was first laid in 1858, the Suez Canal opened in 1869, and petroleum was discovered in Pennsylvania in 1859. The automobile was invented at the end of the nineteenth century, and the Wright brothers flew their plane only three years into the twentieth. Progress in medicine began to accelerate with the introduction of anesthesia and the germ theory of disease and, at the end of the nineteenth century, the advent of sulphanilimides. There was increasing urbanization and even reverence for the "metropolis." The explosive expansion of technology in the last half of the century was both the result and the cause of rapid progress in science on many fronts, especially in physics.

Yet despite optimism born of the Industrial Revolution and rapid advances in technology, the closing decades of the century were fully as much epitomized by the uncertainty wrought by the demise of a cohesive and conservative tradition in European intellectual life and the changing political structure of Europe This can best be traced to the new view of humankind and our place in nature, which Darwin and Marx drew, and to the consequent decline in the influence of religion as a cultural force.[2] The influence of Darwin's *Origin of Species* (1859) is evident in the work of writers such as Feuerbach, Strauss, and George Eliot. F.W.H. Myers described a conversation with Eliot at Cambridge in 1873 in which the novelist expressed her own skepticism:

> I remember how, at Cambridge, I walked with her once in the Fellow's Garden of Trinity, on an evening of a rainy May; and she, stirred somewhat beyond her wont, and taking as her text the three words which have been used so often as the inspiring trumpet-calls of men—the words God, Immortality, Duty—pronounced, with terrible earnestness, how inconceivable was the first, how unbelievable the second, and yet how peremptory and absolute the third. Never, perhaps, have sterner accents affirmed the sovereignty of impersonal and unrecompensing law. I listened, and night fell; her grave, majestic countenance turned towards me like a Sibyl's in the gloom; it was as though she withdrew from my grasp, one by one, the two scrolls of promise, and left me the third scroll only, awful with inevitable fates. And when we stood at length and parted, amid that columnar circuit of the forest-trees, beneath the last twilight of starless skies, I seemed to be gazing, like Titus at Jerusalem, on vacant seats and empty halls—on a sanctuary with no Presence to hallow it, and heaven left lonely of a God.[3]

The weakened role of religious belief necessarily led to a search for new foundations, new ways of imagining the world, and for some the hope was to be found in science. The largely nineteenth-century discovery that the creation story of Genesis is epistemologically equivalent to those found in primitive mythology was as profound a psychic shock as the Copernican displacement of Earth and humanity from the center of creation was to sixteenth-century thinkers. A unique and definite act of creation was replaced by indeterminately long natural processes.[4]

It is no accident that the nineteenth century was the golden age of classical physics; for if mercantilism and capitalism have been handmaidens of science almost since the scientific revolution, their synergistic relationship was never more apparent than in that century. The exploitation of colonies by imperial powers required increasingly rapid communication and transportation. The demands of industry for reliable and cheap sources of power; the building of roads, bridges, and rail lines; and the exporting of technology from the laboratory to widespread engineering applications all stimulated science while consuming applied scientific knowledge. In other words, science fed technology, and technology fed science. Engineering challenges moved technology from the laboratory into applications and provided an implicit impetus for more basic scientific studies.

The Classical Paradigm

The nineteenth century is the era of the triumph of classical physics, the century in which each of its classical branches, including (but not limited to) mechanics, electricity and magnetism, optics, heat and thermodynamics, and elasticity and hydrodynamics, not only became clearly defined in substance and methodology but reached some level of completeness.[5] In making this assertion we do not mean to say that they have not continued to develop; nevertheless, in general neither the subject matter nor the mathematical structure of the theories has changed radically. In spite of the dramatic and far-reaching revolutions of the twentieth century and the continuing evolution of these fields, the nineteenth-century delineation of classical physics has remained virtually unchanged. Many, in fact, have stayed thoroughly classical, showing only the faintest impact from quantum theory or relativity. An example is acoustics, which was outlined by Lord Rayleigh (William Strutt) in his *Theory of Sound,* first published in 1877 and revised in 1894. Although acoustics remains a vigorous field of applied physics today, much of it would appear familiar to Rayleigh if he were to read a modern paper. Some of the mathematical techniques are different, of course, as are many of the applications (such as underwater sound). Certainly the data available and the numerical and computational power that can be brought to bear on the problems are enormously different. Yet the physics is essentially the same as it was in 1900.

It might be noted that electromagnetism has undergone a further radical change in the twentieth century and now is thoroughly quantum-mechanical. Yet *quantum electrodynamics,* the quantum field theoretic description of electromagnetism, is so nonclassical that we can almost consider it an entirely new field of postclassical physics, leaving largely intact classical electromagnetic theory. Other fields are, of course, totally nonclassical, including quantum theory and special and general relativity but also atomic and nuclear physics and even, for the most part, condensed matter (solid-state) physics.

More important than the remarkable progress made in the separate fields of classical physics in the nineteenth century is the process by which these individual subjects came to be seen as part of a single science known as physics, for in 1800 they seemed only tenuously related and involved, in practice, very different techniques.[6] This process of definition, or of unification and integration, which in effect created what we know as physics today, had several aspects. By mid-century, moral and natural philosophy had grown apart, and physics and chemistry—or at least the practices that we now combine under those terms—had moved well into the process that gave birth to those separate sciences, with increased specialization and some incipient professionalization. As we shall see, this unification consisted of defining the boundaries of physics and involved unification at a more fundamental level in what was one of the first great triumphs of mathematical or theoretical physics. This unification grew out of the recognition that many fundamental problems in each of the fields we have mentioned could be formulated in similar mathematical terms.[7] This discovery put into the hands of figures such as William Thomson and James Clerk Maxwell a powerful tool that allowed them (and others) to apply results from the theories of heat and fluids to electromagnetism, thus blurring the boundaries between the disparate fields, on the one hand, and, on the other, extending the horizons of physics.

The impulse to unify, while possibly innate, and very often successful, is a two-edged sword. By emphasizing the similarities between or among fields, thereby bringing crucial insights fruitfully to bear on certain problems, it can also obscure essential differences between phenomena, or limits to similarities, if carried too far. If the homologies are mathematical this is less likely to be the case, but if unification is based on physical analogy, more harm than good may be done.

Although this book attempts to trace the development of each of the classical fields of physics during the nineteenth century, some omissions are unavoidable. Most notable, perhaps, is optics, which was a rich and fertile field in the century. Limitations of space make this necessary, a situation that is mitigated by the fact that many issues crucial to the growing science of physical optics, especially the wave theory of light, are dealt with. Continuum mechanics in the form of fluid dynamics, elasticity, and even acoustics was a major preoccupation of the nineteenth century, but I give these subdisciplines short shrift for the same reason. In fact, the evolution of theories of elasticity and hydrodynamics is nearly inseparable from that of electromagnetism and thermodynamics and provided fruitful analogies that have guided research in these somewhat flashier fields. Thus, their omission is all the more regrettable, yet their presence remains if just below the surface, because of those rich interconnections. Although an extended treatment of the aether concept has also been sacrificed, the aether intrudes into my narrative at many points, and many excellent discussions of the subject exist.

Perhaps the field that has changed most in the twentieth century is astronomy (or, as we now call it, astrophysics), which has been thoroughly invaded by both twentieth-century revolutions "against" classical physics, relativity and quantum theory. But astronomy is a special case because it applies all branches of physics, classical and modern, to a broad spectrum of problems arising in the heavens. Its past is unique, for it was largely a descriptive and almost taxonomic science before the eighteenth century—and, indeed, for much of the nineteenth century.[8] That fact, coupled with the existence of excellent histories of astronomy, forces us to relegate it to the background, even to footnotes, until physics and astronomy began to converge in the middle of the nineteenth century and astronomy begins to make the transition to astrophysics. Nevertheless, we must remember that a number of important problems in mathematics and physics have grown directly out of astronomical phenomena.

The process by which each of the branches of classical physics emerged as clearly defined sciences was virtually completed by the 1870s.[9] But understanding the evolution of classical physics from its rather ill-defined and amorphous state in 1800 to its culmination in the last half of the century, or learning how modern physics grew out of classical physics in the 1880s and 1890s, means that we need to do more than read the documents of nineteenth-century physics. We must understand not only the nature of the scientific discoveries through which this maturity was reached but also the social and cultural backdrop against which these changes occurred. Critical to this is the nature of the educational systems, the evolution of scientific institutions, and the growth of the research ethos in universities. The role of schools, universities, and scientific societies and the growing importance of government patronage were crucial as were the concurrent evolution of philosophical and mathematical ideas, which were vital to the pace of scientific discovery and helped determine the way in which scientific discoveries were interpreted.

The crucial roles played by the scientific institutions of the period, especially in Britain, France, and Germany, have been explored by many authors; and the same can be said of the way in which physics was taught in the secondary schools and universities. In ways that are sometimes clear and more often obscure, these institutions and the context in which they operated provided the fertile soil from which the figures of nineteenth-century science sprang. Several questions naturally surface in our attempts to understand their impact. For example, what would have been likely to attract a bright young schoolchild to physics or mathematics in the nineteenth century? What was it about Scottish society or its school system that spawned so many scientific geniuses during this period? What was the role of the Cambridge Tripos or Senate House examinations? Were able students, challenged by the rigor of the Tripos system, drawn to natural philosophy, or did it uniquely train such minds to attack the open problems in physics?

The story of nineteenth-century physics is largely the story of European physics, especially physics in Britain, Germany, and France. Except for a few individual contributions, Italy's decline since the seventeenth century was nearly complete; and barring Benjamin Franklin and his near contemporary Benjamin Thompson, American scientists became prominent only at the very end of the nineteenth century.[10] (Moreover, Benjamin Thompson, Count Rumford of

Bavaria, was an American expatriate who did none of his scientific work in his homeland.)

Philosophical Climate

The philosophical climate of the nineteenth century ranged from the romantic Naturphilosophie of the early part of the century to the positivism of Wittgenstein and others at its end. It encompassed the failure of the French Revolution, the revolutions of 1848, the English romantic movement of the first half of the century, the radicalism and historical materialism of Marx and Engels, the dread and ambiguity of Kierkegaard. The influence of German Naturphilosophie on nineteenth-century science, especially in Britain, continues to be debated, including the extent to which Samuel Taylor Coleridge brought aspects of both German Naturphilosophie (through Schelling) and Kant's transcendental idealism to bear on early nineteenth-century British science, especially through his friendship with Humphrey Davy.[11] Davy, as we shall see, was a strong influence on Michael Faraday; and while Faraday was no philosopher, he introduced one of the most important conceptual revolutions of the century: the field. The importance of Immanuel Kant, the great philosopher of the eighteenth century, is twofold: First, his profound contributions to epistemology clearly influenced discussions of the nature of scientific discovery throughout the nineteenth century. Second, his emphasis on unity and the active powers inherent in matter contrasted with the mechanism and atheism of Laplace yet provided an alternative to the idea of imponderable fluids that invaded matter.

Occasionally, philosophical issues will intrude into the narrative of this book, such as discussions of the tension between realism and idealism, the distinction between observational and theoretical judgments, an examination of mechanistic versus dynamistic views of nature, and the influence of positivist thought on the development of science.[12] The impact of this last concern began to be felt at the end of the nineteenth century, promoted by Comte, the Vienna Circle, and scientist-philosopher-historians such as Pierre Duhem and Ernst Mach. Although the most obvious result was Duhem's and Mach's rejection of atomism, the strongly operationalist character of quantum ontology, imposed largely by Neils Bohr in the twentieth century, reflects philosophical positivism.

Some have argued that an alliance in England between religion and science permitted science to play an exalted role until the time of Darwin, who was clearly one of the most important and influential figures of the century, responsible to a considerable degree for the replacement of Christianity by a scientifically based perspective on life that emphasized material progress, random selection, and struggle for existence. But religion played an important role in the lives of many of the major (and minor) figures in nineteenth-century science, especially in Britain and Germany, even well after the impact of Darwin's *Origin of Species* (1859) had been thoroughly absorbed.

If, in what follows, little more is said about some of these issues, they nonetheless should be kept in the back of one's mind as the external forces that drove, or at least influenced, the scientific activity of the nineteenth century. Ignoring the effects of the Industrial Revolution, developing technology, and social conservatism or radicalism as we attempt to understand the transformations that took place in physics is tantamount to viewing the century with one eye closed. Yet to see the patient progress of the middle period of the century or the scientific radicalism at its end as mere echoes of what was happening in the culture at large is also wrong. Just as the science of the first half of the nineteenth century was built upon the gains of the eighteenth, the revolutions of the early twentieth century could not have happened but for the accomplishments of nineteenth-century classical physics.

The nineteenth century was a period in which institutions were evolving rapidly so that the nature of physics as a profession changed dramatically during the century. It was also a time when philosophical inclination or predisposition was often more influential (upon the science) than it is in our own time, in part because of the way in which a scientist was educated. Indeed, as John Herschel wrote in his *Preliminary Discourse on the Study of Natural Philosophy,* "we have endeavored to explain the spirit of the methods which, since the revival of philosophy, natural science has been indebted for the great and splendid advances it has made."[13]

CHAPTER 2

Nineteenth-Century Science in Context

$\mathcal{T}$he point hardly needs to be made that science does not function in isolation; as much a human endeavor as painting, poetry, and commerce, science reflects the time in which it is carried out. It is not always easy to show that science in a given period was shaped by its cultural context, for its subject matter—the natural world—clearly lies elsewhere. Nevertheless, the spirit of an age ultimately finds expression in all intellectual activity, and science is no exception. In some cases it is possible to correlate a radical social climate with scientific radicalism, creativity, or daring; in other cases the influences are more obscure. With this in mind, we will briefly explore the impact of prevailing moral and natural philosophy and the importance of newly developing scientific institutions on the evolution of nineteenth-century science.

The nineteenth century was a time of searching for new forms of social organization, a time of great concern for the plight of peasants and industrial laborers. If religion began to lose its grip, at least upon the educated classes, and if the first half of the century saw almost unparalleled exploration and the opening up of frontiers, especially in the Americas, Africa, and Asia, so, too, was scientific progress in this period a widening of horizons, the application of new techniques to old problems, and of old techniques to new problems. As we attempt to understand these changes, we must note the major historical and social movements in which that progress was embedded or by which it was conditioned—in particular, the Industrial Revolution, the Scottish Enlightenment, midcentury reform movements, and the French Revolution. Focusing more narrowly, we will examine the rise of scientific institutions in Britain, France, and Germany and the increasing professionalism of science.[1] Prominent among other topics is the role of the educational systems of these countries as well as other aspects of the education of scientists. Finally, we shall examine the philosophical context in which nineteenth-century science flourished, including a study of the influence of Kant's transcendental idealism and its progeny, Schelling's Naturphilosophie and German romanticism, but also a look at other situations in which the interaction between formal philosophy and the philosophy of nature (in the sense of natural philosophy, not specifically Naturphilosophie) was important and strong.

The sociological issues of the role of the professions within the larger cultural context and the extent to which they may or may not operate freely from the restrictions of that context or the state is beyond the scope of this book. It should be apparent, however, that the professions—and the sciences, in particular—are influenced by the relationship of their institutions to the state, their funding arrangements, honors, and so on. Thus, there are two important and inseparable issues: the role of the state as provider of support or patronage, and the influences and constraints of the larger political and cultural context.

The role of the state as patron has a relatively recent origin; arguably, it was not very important before the eighteenth century. The professional societies stand between the professions and the political structure; and although they are products of the profession and their own interests, they are often sanctioned by the state, if not actually supported by it, and may have some political influence or role as state advisor. Societies may lobby the state for support of their profession, provide expert counsel, and so on. In fact, they are often anything but independent of politics and culture. Furthermore, we should be careful, as many other people have pointed out, not to equate institutionalism and professionalization with excellence and progress.[2]

The two most important scientific societies through the 1830s were the Royal Society of London, founded in 1660, and the Paris Academy of Sciences (Académie Royale des Sciences), founded in 1666.[3] The Royal Society of Edinburgh was founded more than a century later in 1783 (its predecessor was the Philosophical Society), the Royal Institution of Great Britain in 1799, and the British Association for the Advancement of Science in 1831. In Germany, the Berlin Physical Society was founded in 1843 and became the German Physical Society (Deutsche Physkalische Gesellschaft) in 1899. Founded in 1700, the Berlin Academy became the Berlin Academy of Sciences in 1804 and later the Prussian Academy of Sciences (Konigliche Preussische Akademie der Wissenschaften). In Paris, the venerable Academy of Sciences was reestablished in 1816 after having been suppressed from 1795 to 1815 in the wake of the French Revolution. The First Class of the Institut National was in effect its continuation in Republican form, and the Société Philomatique also served an expanded function.[4] The journals published by these and other societies, along with meetings of the societies themselves, were a vital means of exchanging scientific information, although some of the most important journals, such as the *Philosophical Magazine* and *Nature,* were not published by scientific societies.

Institutionalization and Professionalism in French Science

The process of institutionalization has been best studied in France.[5] The first structures that accomplished this were the schools and universities: the Ecole Militaire (where Laplace became professor of mathematics in 1769), the Ecole Normale, the Ecole Polytechnique (the leading institution for higher education in French science), and the Collège de France. Also important were the official institutions: the Academy of Sciences, the Institut de France (founded in 1795), the Philomatic Society, and the Paris Faculty of Science (founded in 1808). Finally, the informal societies, such as Société d'Arcueil, were influential, as were the many salons.[6] These institutions presided over the transition of science as an elite intellectual endeavor to its status as a profession and a well-defined career with appropriate certification. As the state came to play a larger role in science, some of its freedom and universality was inevitably sacrificed. Robert Fox has argued that this relationship led to parochialism, centralization of intellectual life, and real or potential control of the learned community by the French state.[7] He also suggests that the process of selection to the profession encouraged "glibness and mastery of received truths." Not surprisingly, there was eventually a reaction to the stultification of French science after midcentury.

French science had been preeminent during the half-century before the death of Laplace in 1827 and the revolution of 1830; and this period saw the "incorporation of science into polity," initiating what Gillispie calls a "second scientific revolution," one that was organizational rather than intellectual.[8] Before 1870, when the support of science depended almost solely on the government, the professional scientist dealt with increasing bureaucracy and the necessity of pandering to it; after the Franco-Prussian war, a new alliance between academic science and industry began, which meant catering to a different master.[9] This shift also led, as some have claimed, to an increased interest in gadgetry and "pedestrian empiricism" at the expense of theoretical insights.[10] For example, in 1876 Léon Brillouin's interest in kinetic theory was ridiculed as impractical by the president of the Société Française de Physique. In general, while French science grew massively under the patronage of state, commerce, and industry, it suffered from bureaucratization and isolation, especially in relation to the more diffuse university structures of Germany and Britain.[11]

Attempts at reform in French science from the 1880s onward bore little fruit; scientists had exchanged the whims of a Laplace for the indifference of the bureaucracy.[12] Philosopher and scientist Pierre Duhem, a Republican and a Catholic, was one of those who suffered from the prejudices of the Second Empire. But an important development in the 1860s was a greater emphasis on research in French universities, following the German model. The change came in response to German advances and later to France's 1870 defeat in the Franco-Prussian war, which gave impetus to demands for educational reform.[13] One consequence was the founding of the French Society of Physics (Société Française de Physique) in 1873.

Scientific Culture in Britain

In Britain, where science never became as institutionalized or professionalized as it had in France, the educational matrix was quite different. Although there was a time-honored tradition of wealthy gentlemen amateurs who contributed to natural philosophy, a career in science was another matter; its suitability for a gentleman was in doubt.[14] Well into the nineteenth century the Royal Society was largely comprised of amateurs, a fact that lay behind the founding of the British Association for the Advancement of Science (BAAS) in 1831 after a failed reform movement in the Royal Society and the defeat of John Herschel (1792–1871) as candidate for its presidency.[15] Charles Babbage (1791–1871), best known for the seminal role he played in the evolution of the automatic computer, sharpened the controversy through his *Reflections on the Decline of Science in England* (1830), which deplored the country's neglect of science and, in particular, the Royal Society's role in the decline. The society had never offered much tangible support for scientific research, having received from the Crown nothing but a charter. This made the situation quite unlike that in France, where the Academy of Sciences received state support. On the contrary, science in Britain was financed by private patronage, which was quite substantial from industrial sources in the last half of the nineteenth century.

Early in the century, many scientists were Anglican clergymen with ample time to pursue their scientific interests, although some believed that experimental philosophy bred skeptics.[16] Science, however, could not be confined indefinitely to the parsonage, the great universities, or the Royal Society and its patrons. This issue is explored in Larry Stewart's *The Rise of Public Science,* which describes the diffusion of science into the coffee houses of London and the country towns.[17]

An important difference between the situations in France and Britain was that Britain had no revolution: Reform came gradually, and a university education was still available only to the privileged. As a result, there was no institutionalized method for training specialists.[18] Gillispie has called the French and British institutional models "the official and the voluntary," and Donald Cardwell notes that industry began to see the benefits of scientific specialization only after the university system had begun to produce them.[19]

The universities of Oxford, Edinburgh, Glasgow, and especially Cambridge were at the center of British science, particularly after the 1812 reform movement led by John Herschel, Charles Babbage, and George Peacock, which revived English science by introducing French physics and mathematics into Cambridge.[20] Also involved were George Airy, a professor of astronomy at Cambridge, where he was instrumental in reviving experimental physics and subsequently appointed Astronomer Royal, and William Whewell, eventually Master of Trinity.[21] Airy was briefly Lucasian professor of mathematics (1826–28) and was followed in the position by Babbage, who was by then totally preoccupied with his computing machine.

The powerful and influential Cambridge network included Herschel, Peacock, Babbage, Whewell, Sedgewick, Forbes, William Hamilton, and William Rowan Hamilton. Professors James Forbes and William Hamilton taught both Peter Guthrie Tait and James Clerk Maxwell at Edinburgh.[22] On the advice of Henslow, Peacock (then Dean of Ely) recommended Charles Darwin as naturalist on the *Beagle.* Yet Oxford and Cambridge in the 1860s were not primarily research universities in the modern (or even the contemporary German) sense but remained "in many respects a club for young men of the nobility and gentry, or at least of wealth."[23]

The flowering of Scottish intellect known as the Scottish Enlightenment, which began about 1707 after the country's union with England, is an extraordinarily interesting episode in the history of science (and the history of ideas) but does not give much support to the idea that the great scientific progress of the eighteenth and nineteenth centuries was substantially due to the institutionalization of science. It is sometimes assumed that scientific output increased (thus leading to important and sustained scientific discovery) as various scientific disciplines became professions with regularized institutional patronage (from scientific societies or academic institutions) and as the number of active scientists increased, meaning more support for laboratories and the added benefit of interaction with colleagues. This assumption, however, seems invalid as applied to British science in the nineteenth century. Although British science was at its zenith in the 1840s through the 1880s, the fact is that as late as the early second half of the century it was only marginally institutionalized.[24] Among nineteenth-century British scientific laboratories, only three come to mind as important: Humphrey Davy's at the Royal Institution (where Faraday, Tyndall, and the Braggs spent much of their careers), founded at the turn of the century; the Cavendish Laboratory, established under Maxwell's direction in 1871; and the Clarendon Laboratory at Oxford, established in 1872.[25] After the Cavendish was founded (with Maxwell as the first Cavendish professor) and especially under the direction of J. J. Thomson from 1884 to 1919, it became the centerpiece of British experimental science; indeed, its prestige continues today. When Niels Bohr finished his doctorate in 1911, he had little difficulty deciding in favor of Cambridge and the Cavendish Laboratory for his year abroad rather than the great Lorentz at Leiden.[26] In Bohr's time the laboratory's main attraction was Thomson; but during his brief stay there, Bohr also attended the lectures of James Jeans and Joseph Larmour. Lord Rayleigh (William Strutt), Thomson's predecessor, who had been a pupil of Routh and Stokes, not only assumed the Cavendish professorship on Maxwell's death but also was a major benefactor of the laboratory.

Informal quasi-scientific gatherings such as the Newcastle and Northumberland circles in seventeenth-century England fostered discussion of important scientific issues of the time and provided a limited amount of patronage. At the end of the eighteenth century, the Lunar Society of Birmingham (1765–95), whose membership included Matthew Boulton, James Watt, Erasmus Darwin,

Josiah Wedgwood, James Keir, and Joseph Priestley, and which had strong ties to Benjamin Franklin and other luminaries, played a major role in the development of the steam engine and the birth of British chemistry.[27] Another provincial society, the Manchester Literary and Philosophical Society (founded in 1781), was important to both John Dalton, who dominated it for a number of years, and his protégé William Prescott Joule. Intellectual and scientific societies such as the Select Society of Edinburgh, the Philosophical Society of Edinburgh, and the Royal Society of Edinburgh brought scholars together and disseminated the latest ideas.

But let us return to Scotland for a moment, which provides one of the most dramatic examples of the intellectual productivity of a single nation, in this case during the eighteenth and nineteenth centuries, especially after 1750.[28] Our immediate interest is philosophy and science (that is, natural philosophy), but Adam Smith was a product of the same culture. To some, the nineteenth century seems to be the century of Scottish science, highlighted by the great contributions of William Thomson and James Clerk Maxwell (Thomson was born in Belfast but moved to Glasgow at the age of eight) and the eminence of the universities of Edinburgh and Glasgow. Thomson's father, Dr. James Thomson, was a mathematician at the University of Glasgow from 1832 until his death in 1849; and William was professor of natural philosophy there from 1846 until 1899. In Maxwell's case, although his family were landed Scottish gentry, much of his achievement came through the channel of the Cambridge Tripos system, as did Thomson's. Indeed, Maxwell, educated at both Edinburgh (Edinburgh Academy and the University of Edinburgh) and Cambridge, combined the traditional British predilection for concrete models, the Scottish regard for philosophical depth rather than manipulation and analysis, and the wrangler's mathematical sophistication and love of rigor.[29] He seems to have been unhindered by the prejudice against speculative thought in the sciences, which was an element of eighteenth-century Scottish thinking that Adam Smith argued against.[30]

Maxwell profited from the broad, unhurried, perhaps less challenging education he received at Edinburgh (where he was influenced by James Forbes and William Hamilton) but also from the intense, specialized, uncompromising Cambridge route to the Mathematical Tripos, "the glory of Victorian Cambridge."[31] Other notable products of the Tripos system were Arthur Cayley, George Gabriel Stokes, William Thomson, Peter Guthrie Tait, and Edward Routh; like Maxwell, nearly all were profoundly influenced by the superlative tutor William Hopkins. Furthermore, Stokes was Lucasian professor at Cambridge when Maxwell was there, and William Whewell was then Master of Trinity College.[32] Maxwell's interaction with them is known to have been important. It is tempting, then, to argue that his speculative and experimental side came from Edinburgh, his analytical faculties from Cambridge.[33]

If this shows that Maxwell's greatness (and Thomson's) was shaped by the quite different educational systems of Scotland and England, we shall also see that he was greatly influenced by the prevailing natural and moral philosophy that he absorbed in both places. The rest, however, is beyond our ken. Why did Maxwell or Clausius or Helmholtz or even Faraday (who was self-taught), rather than someone else, develop into powerful and original thinkers? We cannot say; for even if we understand the impact of local culture, the educational system, religious heritage, and influential teachers and colleagues, even if we have a glimmer of understanding into how a family shaped a life, we can never do more than guess about the role of the raw material—the genetic heritage—that may have been the sine qua non of a scientific life.

It has been suggested that Maxwell's intellectual activity was fostered by Scottish landed society's identification with intellectual culture, perhaps a substitute for the political activity that was foreclosed by the union of Scotland and England.[34] This provides a measure of explanation for more than a century of Scottish intellectual vigor. But there clearly was, in addition, an appreciation of the social role of science and an attention to the uses of knowledge.[35]

If we turn to other parts of Britain and revisit the question of institutionalization of English science, we find that the new century began with the establishment (by Benjamin Thompson, later Count Rumford) of the Royal Institution in 1800.[36] Humphrey Davy later became its most illustrious lecturer, and Thomas Young lectured there from 1801 to 1803. Davy grew up in Cornwall and was not university educated, but he was strongly influenced by Cornish practical engineering, the Cornish copper and tin mining industry, and the success of James Watts's engines in pumping the mines.[37] Although he resigned his chemistry chair at the institution in 1812, Davy continued to serve as director of the laboratories.

In the audience as Davy gave his farewell lecture to the Royal Institution (1812) was twenty-year-old Michael Faraday, soon to become his assistant and eventually a world-renowned scientist. His place at the institution was the only professional position that Faraday ever held, and he became director of the laboratory in 1825 after Davy's retirement. Although Faraday had, by any reasonable definition of the term, become a professional scientist by 1820, his route to that state was extraordinary and untypical. By 1812, when he was Davy's apprentice, *natural philosopher* was already a professional title won by specialized university study and a teaching position at a university or center for research and lecturing such as the Royal Institution or elsewhere. This may or may not have meant earning a living at science, as Davy initially did as an assistant lecturer in chemistry at the Royal Institution. Faraday would have almost surely have foundered in his attempts to become a scientist without the existence of the institution. In addition to Davy, Young, and Faraday, a long list of notable chemists and physicists led or worked at the Royal Institution, including Lord Rayleigh and the Braggs. Founded with the purpose of exposing the general public to ideas (scientific ideas, in particular), the Institution carries out a similar function today. But Faraday's contacts with other scientists, especially William Thomson, were also facilitated by the existence of the British Association for the Advancement of Science.[38]

Countering this positive view of the impact of professionalism in science and the slow development of scientific institutions in England is Martin Weiner's claim in *English Culture and the Decline of the Industrial Spirit* that the rise of middle-class professionalism in the mid-nineteenth century led to the adoption of aristocratic values and a rejection of the practical arts, engineering, and the natural sciences. Accompanying this change was rapid growth in the public schools, which emphasized classical learning and taught little natural science. Three new English universities were founded in the nineteenth century to meet the needs of both the middle class and religious nonconformists, who were not admitted to Oxford and Cambridge until after 1850: Durham, London, and Victoria.[39] Toward the end of the century, other universities were established in the industrial cities of Manchester, Birmingham, Liverpool, Leeds, Sheffield, and Bristol. Nevertheless, the support of science in Britain remained a laissez-faire proposition. Private patronage continued to be important; and although scientific societies had proliferated, their ability to finance research was limited.[40] Eventually, the founding of the Physical Society of London in 1874 opened a new era in the professionalizing of physics in Britain.

Scientific Institutions in Germany

The chapters that follow do not mention German physicists much before Weber, Clausius, and Helmholtz, whose contributions belong mostly to the last half of the nineteenth century.[41] The reasons are at least twofold, not the least of which is the fact, already noted, that the Industrial Revolution came late to Germany. But German science in the late eighteenth century was strongly influenced by the romantic movement, which took the form of a scientific philosophy grounded in works such as Kant's *Critique of Pure Reason* (1781) and his *Metaphysics of Natural Science* (1786) but soon developed into a form that would have left Kant skeptical. (See the discussion of Naturphilosophie later in this chapter.) Typical were Schelling's *Ideas for a Philosophy of Nature,* published in 1797, and Hegel's *Philosophy of Nature* of 1817.[42]

Two of the most important innovations of nineteenth-century science, the teaching laboratory (by Justus von Liebig) and the association of professional scientists (Association for the Advancement of Science, by Oken), were German creations.[43] The first university physics laboratory in Germany was Wilhelm Weber's at Göttingen, established in 1833–34.[44] In the 1840s, other laboratories were established at Leipzig (also by Weber), Berlin, Königsberg, and Heidelberg.[45] At about the same time, research institutes with staff members and seminar programs became widespread; among the first were the 1834 mathematics-physics seminar at Königsberg under Franz Neumann and Gustav Dirichlet's in Berlin.[46] A growing research ethos was transforming the professoriate so that by 1848 research was an integral part of the role of the university instructor.[47] Although the German institutes slowly developed from the typical German cabinet after 1830, when Carl Friedrich Gauss brought Weber to Göttingen, it was not until after

1870 that most physics institutes received separate buildings.[48] Milestones in this establishment of research professorships in Germany include Gustav Kirchhoff's arrival at Heidelberg in 1854, Rudolph Clausius's joining Würzburg in 1867, and Wilhelm Conrad Röntgen's assumption of the professorship at Giessen in 1878. Neumann's institute at Königsberg was finally built in 1884, after fifty years of pleading, when he was almost ninety years old. The subdiscipline, or specialty, of theoretical physics began to develop slowly in Germany in the 1840s and 1850s, with Clausius, Helmholtz, and Kirchhoff dominating other scientists; nevertheless, the creation of extraordinary professorships in theoretical physics became widespread only after 1870, and it was the end of the century before chairs of theoretical physics were seriously contemplated.[49] Generally, these positions were not meant to compete with ordinary professorships but to supplement them, and usually the extraordinary professor was expected to rise to an ordinary professorship in experimental physics.[50]

By the 1870s German physics journals were citing German authors with great frequency, especially Helmholtz. Kirchhoff joined Helmholtz at Berlin University in 1875 as ordinary professor of theoretical physics, and Planck matriculated at Berlin in the same year. He received his doctorate at Munich in 1879 and stayed on as Privatdozent for theoretical physics.[51] In 1880 Hertz received his doctorate under Helmholtz while working on a problem concerning the inertia of electricity. After assisting Helmholtz for three years, Hertz went to Kiel in 1884 and to Karlsrühe the next year. When he left Kiel, Planck replaced him as extraordinary professor of theoretical physics and then took the same position in Berlin in 1889, the year that Hertz moved to Bonn to assume the chair of experimental physics. In other words, from the 1830s until the Congress of Vienna and the unification of the German states at the end of the Franco-Prussian war (1871), both experimental and theoretical physics flourished in German universities. State support of science grew rapidly after the Franco-Prussian war and unification, paving the way for other European countries, which began to provide support only after the turn of the twentieth century.[52] Technology played an important role in late nineteenth-century discoveries in experimental physics, and German industry developed rapidly.

The Education of a Physicist in England and Germany

By 1870 the ordinary and extraordinary physics professorships in Prussian and Bavarian universities were filled by productive and promising physicists, including Weber, Helmholtz, Kirchhoff, Clausius, and Boltzmann (in Austria). We may contrast this situation with the aristocratic British educational system dominated by Oxford and Cambridge but rivaled by Glasgow and Edinburgh. Provincial universities such as Manchester and Leeds had begun to play important roles, especially in applied science, but usually had no chairs of natural philosophy.

As a result of these differing systems, it relatively easy to trace the influences of particular teachers and researchers on important German physicists in the late nineteenth and early twentieth centuries; in the case of Britain, it is nearly impossible. We can note the influence of William Hopkins's tutelage, whose stellar students we have already mentioned, and of mathematician Edward Routh, especially on Lord Rayleigh and J. J. Thomson. George Gabriel Stokes had a corresponding influence on Maxwell. But neither Maxwell nor Thomson, to mention only two examples, had any important students. In Germany and Austria, however, we can trace complete lineages from Stefan to Exner to Hasenöhrl to Schrödinger, or Stefan to Boltzmann to Hasenöhrl to Ehrenfest. Boltzmann worked with Bunsen, Kirchhoff, Helmholtz, and Königsburger; Meitner, Arrhenius, and Nernst were other students of Boltzmann. Wein studied under Helmholtz and Qunicke, and Planck attended lectures by Kirchhoff and Helmholtz (although he claimed to have been bored by them). Today it is the German system that has become the norm, with a research ethos and clearly defined system of mentorship in which a student has a definite research director. Certainly, the British system (that is, the Cambridge system) produced some remarkable scientists, especially in the 1830s through 1850s (including, of course, giants such as Charles Darwin), but that productivity was not sustained. One may argue, in fact, that the brilliance characterized by Thomson, Maxwell, and Stokes and later Rayleigh, Jeans, and Eddington was not very deep. In large part, science was tapping a resource consisting only of the upper classes. Figures such as Faraday and Joule are obvious exceptions; but they were essentially part of an entirely different tradition, the practical approach that drove the Industrial Revolution.[53] Neither had any protégés of note.

Scientific Communication through Journals

The founding of scholarly journals made possible widespread and increasingly rapid communication and, with the societies that published them, initiated the process that created scholarly communities. The Royal Society of London began publishing the first scientific journal, *Philosophical Transactions,* in 1665.[54] The *Memoirs of the Manchester Literary and Philosophical Society* dates from 1785, the *Transactions of the Royal Society of Edinburgh* from 1788. The *Philosophical Magazine*, whose title varied somewhat over the years, was founded in 1798 and before midcentury became the most important British journal in the physical sciences. The weekly magazine *Nature* began publication in 1869. When the Physical Society of London was founded in 1874, it immediately began publishing its *Proceedings.*

The Paris Academy began issuing *Histoire et Mémoires* in 1699, while *Comptes Rendus,* which began publication in 1835, soon became the largest of all scientific periodicals.[55] The *Journal de Physique Théorique et Appliquée* was founded in 1872, a year before the founding of the Société Française de Physique. In Germany, *Annalen der Physik* had been founded in 1790 as *Journal der Physik,* assuming its present title (which had changed often in the interim) in 1798. L. W. Gilbert edited the *Annalen* for twenty-five years, beginning in 1799; and when J. C. Poggendorff died in 1877, he had been editor of *Annalen der Physik und Chemie,* commonly known as *Poggendorff's Annalen,* for nearly half a century. The Berlin Physical Society took over the *Annalen* at Poggendorff's death, and in 1900 it again became known as *Annalen der Physik.*[56]

Among the more specialized periodicals was *Fortschritte der Physik,* which the Berlin Physical Society began publishing as a review journal in 1845. The journal of the Prussian Academy of Sciences primarily published articles on experimental physics. Other important journals were *Zeitschrift fur Mathematik und Physik* (founded in 1826), *Physikalische Zeitschrift,* and in France the *Annales de Chimie et de Physique* and *Journal de Physique.* The growing importance of German science was highlighted by the fact that the English journal of translations, *Taylor's Scientific Memoirs,* published eleven papers by German authors in 1831 and devoted its last volume, in 1853, entirely to German authors.[57] The *Sitzungsberichte* of the Vienna Academy became the vehicle through which the papers of Boltzmann, Loschmidt, and other Austrian scientists achieved notice. In England, the *Cambridge Mathematical Journal* was founded in 1837 with Gregory as its first editor; and when William Thomson became editor in 1846, it became the *Cambridge and Dublin Mathematical Journal.* The Italian journal *Il Nuovo Cimento* was founded in the decade 1844–55. The first American scientific periodical, the *Transactions of the American Philosophical Society,* began circulation in 1769. In astronomy, the first journals began to appear in the 1820s, notably the *Astronomische Nachrichten* and the *Memoirs* and *Monthly Notices* of the Astronomical Society of London.

Nineteenth-Century Scientific Philosophy

Philosophical questions have played a more important role in shaping the course of scientific discovery during some periods and in some fields than they have in others. Discerning the aims of physical theory has been an important goal since the Greeks, with realist rather than positivist or intrumentalist views dominating at one time or another; Duhem, Popper, and others notwithstanding, the question remains undecided. The truth is that scientists in their productive youth are often not consciously guided by any philosophical inclination, and the philosophical ruminations of their later years may not accurately inform any analysis of their earlier work. Many great scientists have vigorously denied the influence of philosophical thought on their work—have even denied its usefulness anywhere in science. Yet scientists are often not aware of how the philosophical views they hold affect their scientific thinking.

Although philosophical biases rarely guide scientific investigation in an overt and conscious way, it would be a mistake to argue that science is not at all influenced by such commitments. Although distinctions later drawn by philosophers of science were not always apparent to the scientists themselves, this in no way lessens their potential validity or value. It does suggest, however, that such distinctions operate on a level that is more subtle than some observers realize. As others have pointed out, scientific methodology is perhaps most strongly characterized by its ad hoc character. What contemporaries or near contemporaries view as simply a style of physics, a predilection for this or that technique, may later appear as an important philosophical commitment or a change in philosophical underpinnings.

As an example of some of these issues, let us consider the conflict between dynamism and mechanism in the eighteenth and nineteenth centuries. To some extent, this division, which is drawn by Trevor Levere, John Hendry, and others, is oversimplified and even artificial.[58] It appeals to those who like to see the world (politically or intellectually) in polar terms. Yet one need not take such distinctions too seriously. If we resist the temptation to label each individual a dynamist or a mechanist but recognize those terms as two poles on a continuum and acknowledge that there are other dimensions to this graphical representation of philosophical inclination, then more good than harm may result from it.[59]

For the sake of clarifying some of the issues in nineteenth-century science, especially in England, we will accept for the moment the division of scientific activity in the early nineteenth century into these two philosophical predispositions: the dynamical and the mechanistic. The *mechanistic* tradition emphasized mechanical explanation, models, and hypotheses—that is, a world behind or beneath experience, including, for example, real atoms. The *dynamical,* on the other hand, eschewed hypotheses and assumed a positivist stance toward the microscopic structure of matter.[60] In denying the mechanical, it was susceptible to the assumption of active powers inherent in matter.

Often, scientists were neither fish nor fowl; or if they were mechanists, they may have entertained thoroughly dynamical ideas. Nevertheless, we can loosely associate Galileo, Dalton, Bacon, Locke, Lavoisier, Laplace, Biot, Cauchy, and Poisson with the mechanistic approach; Kepler, Leibniz, Kant, Davy, Lagrange, Fresnel, Fourier, and Ampère with the dynamistic.

Laplace's mechanistic point of view was typical: "I have wanted to establish that the phenomena of nature reduce in the final analysis to action-at-a-distance from molecule to molecule, and that the consideration of these actions ought to serve as the basis of the mathematical theory of these phenomena."[61] Fourier, however, represents the dynamical: "Primary causes are unknown to us. . . . The knowledge of the mathematical laws . . . is independent of all hypotheses. . . . Mechanical theories . . . do not apply to the effects of heat."[62]

The late nineteenth-century philosopher-scientist and positivist Pierre Duhem found Fresnel a good example of his own view of scientific epistemology: "Fresnel did not assign, more than Ampère or Fourier, any metaphysical explanation as the aim of the theory. He saw in theory a powerful means of discovery because it is a summary and classified representation of experimental knowledge."[63] The positivism of the dynamical philosophy had one great advantage: Before atomic theory became firmly established, and when physics could study *only* macroscopic phenomena, mechanical models and speculative hypotheses about underlying structure could be counterproductive. A theory might fail because of such a model, while a macroscopic model only had to describe or reproduce the phenomena. Duhem's rhetoric becomes strident when he discusses the English predilection for models: "These abstract notions of material points, force, lines of force, and equipotential surface do not satisfy the [English] need to imagine concrete, material, visible, and tangible things. . . . we thought we were entering the tranquil and neatly ordered abode of reason, but we find ourselves in a factory."[64]

In his 1884 Baltimore Lectures, William Thomson said: "It seems to me that the test of 'Do we or do we not understand a particular subject in physics?' is 'Can we make a mechanical model of it?'. . . I never satisfy myself until I can make a mechanical model of a thing."[65] Duhem quotes Henri Poincaré as saying: "The first time a French reader opens Maxwell's book [*Treatise on Electricity*], a feeling of discomfort, and even of distrust, is at first mingled with his admiration."[66] Of course, to Duhem, "strong minds . . . do not need to embody an idea in a concrete image [but they] cannot reasonably deny to ample but weak minds . . . the right to sketch and paint the objects of physical theories in their visual imaginations."[67] Although he goes on to claim the mantle of intellectual liberalism when he says that he does not compel the English to think in the French manner, Helmholtz's statement rings truer: "As for me, I must confess that I remain attached to [the] representation of facts and their laws by the system of differential equations of physics . . . and I place more confidence in it than in the other. But I cannot raise any objection in principle against a method pursued by such great physicists."[68]

It is worth noting that Scottish commonsense philosophy also sought to avoid hypothetical entities, establishing a tradition that influenced thinkers from Robison and Playfair to Rankine and Maxwell. Yet Maxwell was one of those British physicists whose methodology Helmholtz and Poincaré suspected.

Robert Fox claims that the Laplacian mechanistic tradition was a blind alley and that Laplacian physics had collapsed by the mid-1820s.[69] He goes on to argue that the scientists responsible for the revolt against Laplacian physics (1815–25) were precisely those whom we connected with the dynamistic, Lagrangian view: Fourier and Fresnel along with Arago, Petit, and Dulong.[70] Some of the factors that led to this turn of events were the ascendancy of the undulatory theory of light (largely due to Young and Fresnel), the weakening of the caloric theory of heat, the brief introduction of an undulatory theory of heat, and the growing list of adherents to Daltonian atomism. In Britain, George Green, whose work was based on Poisson's, nonetheless adopted Fourier's indifference to mechanical assumptions; the success of Fourier's analytical theory of heat, combined with Young and Fresnel's earlier wave theory of light, led to the abandonment of atomistic reduction for a decade after 1825.[71] On the philosophical side, Auguste Comte was an especially harsh critic of Laplacian imponderable fluids. To some extent, French dynamistic natural philosophy died with its adherents; Ampère was the last to depart, in 1836, just as British science (especially in electromagnetism but also in heat and energy conservation) was beginning its supremacy.[72]

In Germany, where Kant's dynamical system of physics was revered and where romantic Naturphilosophie evolved, dynamical philosophy intruded into textbooks. Professor G. W. Muncke of Heidelberg claimed in his *System* that "soon after its founding by the immortal Kant, the system of dynamical physics [had] . . . acquired such an authority that the atomistic system was in disrepute."[73] Muncke, however, argued that the atomistic view that matter is made up of particles rather than forces or powers could indeed be applied successfully. Professor G. F. Parrot's antipathy for the dynamical approach was even stronger: "everything reasonable that can be said about a dynamical system, Boscovich, Priestly, [and] Kant have said, and it can serve to compose for oneself at best a meaningful fiction beyond experience. But to want to deduce the laws of nature from this fiction, to tell nature a priori how it should behave—that would be to make a satire of the human mind."[74]

In a loose sense, the distinction between dynamists and mechanists was one between positivists and realists, even though the ideas are not equivalent.[75] A positivist essentially sees the aim of physical theory as economically summarizing empirical results: as the Greeks saw it, "saving the appearances." No mechanical hypotheses are introduced that are not justified by what is observable. The realist, of course, sees the entities introduced to explain the experimental results as objectively real. The mechanist, in trying to explain the properties of matter on the basis of the nature of its smallest parts, often has recourse to entities that are not accessible to observation. In some cases these entities may ultimately be observed and become part of the empirical world; in others they may disappear from the literature or survive only as heuristic elements. In some cases, of course, the entities are not intended to be real and serve only as analogy, as an aid in reasoning. This is sometimes the case in Maxwell's use of models, which included elements that he never claimed to exist fully. Yet there is no doubt of Maxwell's commitment to the reality of the molecular vortices on which much of his theory of electromagnetism was based. Although we shall explore this issue further in chapter 3, it is clear that Maxwell seems at times to be a mechanist, at others to be a dynamist. One may take this as evidence of the weakness of the categorization or as a sign of the complexity and subtlety of Maxwell's thought.

The analytical tools of the great French mathematical physicists Lagrange and Laplace were similar to each other, yet their views of mechanics as an empirically derived science were quite different.[76] While Laplace believed that the laws of mechanics sprang from observation, Lagrange regarded mechanics as an a priori axiomatic science.[77] Somewhat the same pairing can be made of Fourier and Poisson, with Fourier's analytical theory of heat, as an outgrowth of Lagrange's purely analytical techniques, being seen as dynamistic and the speculations of Poisson, a follower of Laplace and a member of the Society of Arcueil, seen as clearly mechanistic in the way in which they depend on detailed hypotheses about the small-scale structure of matter.

At the beginning of the nineteenth century, this issue divided England and the continent—in particular, England and France. Before 1800 Britain was not much influenced by the French mathematical physicists or the analytical methods of Laplace and Lagrange (or, for that matter, of their predecessors, d'Alembert and Maupertuis). Yet even as British mathematicians began to absorb and elaborate upon these techniques, the goals of Whewell and Maxwell were quite different. Maxwell wrote: "The aim of Lagrange was to bring dynamics under the power of the calculus. . . . Our aim, on the other hand, is to cultivate our dynamical ideas. We therefore avail ourselves of the labours of the mathematicians, and retranslate their result from the language of the calculus into the language of dynamics, so that our words may call up mental images, not of some algebraic process, but of some property of moving bodies."[78]

For more than a century, the geometric methods of Newton had held sway, and the results were far from impressive. The great resurgence of British natural philosophy at the hands of Thomson, Tait, Stokes, Green, and Maxwell was based on a thorough absorption of continental tools of analytical geometry and partial differential equations, which we now describe as analysis. At this point an almost unbridgeable gap between physicists and mathematicians opened, wider in Britain than in France or Germany. Although British scientists adopted continental analytical techniques, they continued to have little interest in mathematics for its own sake—this when Cauchy was revolutionizing mathematics in France. Thus, British physicists continued to think in geometrical terms and to require geometrical legitimization of the calculus.[79] Perhaps one should not talk, as Norton Wise does, of the "failure, even . . . unwillingness, to replace physical reasoning with abstract algebra"; for no method should be inherently preferred over another in science, and Scottish and Cambridge physicists were clearly successful during more than a half-century of ascendancy in British physics. It is nonetheless true that this reluctance gave mathematical physics in Britain a kind of conservatism that did not serve it well at the end of the nineteenth century. The reforms of Babbage, Peacock, and Herschel in the 1820s and 1830s, of which Thomson and Maxwell were beneficiaries, had by then run their course.[80]

John Hendry claims that no philosopher since Aquinas (Bacon excepted) has influenced scientific thought as much as Kant; certainly, he had an enormous impact on the physical sciences in the nineteenth century.[81] Two of Maxwell's teachers, William Hamilton and William Whewell, were strongly influenced by Kant's ideas, which are especially evident in Whewell's *Metaphysical Foundations of Natural Science.*[82] Hamilton was an advocate of the method of analogy in science, used so fruitfully by Faraday, Thomson, and Maxwell.[83]

Whewell provided a philosophical framework, largely Kantian, for British science in the middle of the century. A respected natural philosopher, he was Master of Trinity College, Cambridge, from 1841 until his death in 1866,

PLATE 1. William Whewell (1794–1866). *Reprinted by permission of the Master and Fellows of Trinity College, Cambridge University.*

conversant with contemporary scientific developments and aware of their historical origins, and correspondent with a wide range of scientific figures. Although he was neither a serious practicing physicist nor an original philosophical thinker, his synthesis in *Philosophy of the Inductive Sciences* (first published in 1840) had an enormous influence on nineteenth-century science.[84] Using what Yeo has called metascience, Whewell's writings sought to preserve a moral dimension in science and bridge the widening gulf between philosophy and science. He was as close to an associate as Faraday ever had and one of the few scientific figures with whom he had a regular correspondence. Both viewed only facts as certain, although Whewell alone thought of established theories as facts. Whewell certainly believed that hypotheses had heuristic value and were essential to scientific progress, even though he remained skeptical of their truth value.[85] Yet Faraday was far more willing to introduce speculative, hypothetical elements into his theorizing despite the dynamistic traditions that both he and Whewell inherited from Kant.[86] Whewell's influence extended to Green, Stokes, Kelvin, and Maxwell. (Another example of the influence of Kant, albeit indirect, was Samuel Taylor Coleridge, an ardent Kantian. Through him Humphrey Davy received Kant's ideas of unity and active powers inherent in nature, which in turn he passed on to Michael Faraday.)

The role of philosophy in guiding scientific inquiry (as opposed to interpreting and understanding it) is often obscure. Yet it is clear that in the nineteenth century there was an important interaction between science and philosophy. One may debate whether that interaction was, in the main, felicitous, but Elkana may be correct in arguing that it prepared the ground for acceptance of the energy principle.[87]

Mechanistic Philosophy and Imponderable Fluids

The theory that certain properties of matter are due to the existence of imponderable fluids was an important element in the mechanistic Laplacian view of matter. For three-quarters of a century (1757 to about 1825) the theory of heat provided the best example of the utility of the concept of a material fluid—in this case, the caloric—which passed from hot to cold bodies; electricity provided another opportunity to apply this idea in the form of electrical fluids. The future of these two implementations of imponderable fluids were to be very different, but it is worth noting that as the tide shifted away from the caloric to the dynamical theory of heat in the second quarter of the nineteenth century, there was also an inclination to view electricity as some sort of vibratory phenomenon.

The aether as the medium of the propagation of light (luminiferous aether), and later of the electromagnetic field, is part of this tradition of an all-pervasive substance, although in this case the medium was necessary only to *transmit* force; it was not passed from one body to another as the caloric was. How Faraday's field, an influence existing in space independent of its source and transmitting electromagnetic force, is related to the mechanical device of the imponderable fluids is controversial. For Faraday not only rejected electrical fluids but the aether as well, and he was certainly an inheritor of the dynamic tradition of powers inherent in matter.

In the course of this book, I will discuss imponderable fluids only in the context of a particular physical theory—for example, the theory of heat. For further details see Robert Fox's *The Caloric Theory of Gases from Lavoisier to Regnault* and Edmund Whittaker's *A History of the Theories of Aether and Electricity.*[88]

Scientific Philosophy in Germany

We have noted that Kant had a significant influence on British natural philosophy and science during the first half of the century. A great deal of his philosophy is indeed *natural,* and his *Metaphysics of Natural Philosophy* provided a foundation for science that appealed to professionals and amateurs alike. More important, it provided an alternative to the mechanical, atomistic tradition of Locke, Dalton, Lavoisier, and Laplace. According to Kant, the essence of matter could be found in active powers or forces inherent in matter—an idea, along with a positivist attitude toward mechanical hypotheses, that garnered much support in the early nineteenth century when British science began its resurgence. Moreover, it survived the importation of French mathematics and physics into Britain early in the second decade of the century, in part because it absorbed the influence of both Laplace and Lagrange, who had fundamental differences on this point.

But what was the situation in Germany? Kant and Goethe dominate any discussion of natural philosophy in late eighteenth- and early nineteenth-century Germany. (Kant died in 1804, Goethe in 1832.) The Naturphilosophie of Schelling and his contemporaries had an enormous impact on the romantic period (affecting science as Hegel never did) and, in particular, influenced the scientists Oersted and Ritter. Lorenz Oken (1779–1851), whom Williams and Steffens call the "arch-priest" of Naturphilosophie, indulged in largely fantastic speculations that included thoughts on polarity, aether, and heat.[89] Helmholtz, however, was greatly influenced by Goethe, even though he freely criticized the latter's attacks on Newton's optics.[90]

A philosophical war waged between the adherents of the materialist philosophy, typified by Helmholtz, who deplored "the philosophical vaporing and consequent hysteria of the nature-systems' of Hegel and Schelling" that had "brought philosophy into disrepute," and the admirers of Kant's transcendental idealism, from which (with strong, direct influence from Leibniz's idea that nature is dynamic and unified) emerged the Naturphilosophie of Schelling and the views of Hegel.[91] Schelling's *Ideas for a Philosophy of Nature,* first published in 1797, drew on Kant's view that matter is not dead but a balance between opposing forces, that its impenetrability is due to a repulsion that must ultimately be balanced by an attractive force to yield equilibrium.[92] Specifically, "matter and bodies . . . are themselves nothing but products of opposing forces, or rather, are themselves nothing else but these forces."[93] Believing that force rather than matter was primary, Naturphilosophie rejected eighteenth-century rationalistic materialism.[94] Of the mechanistic description of electricity, Schelling said, "the theory of electricity has become almost more an enumeration of the machines and instruments, which have been invented on its behalf, than an explanation of its phenomena."[95]

The influence of Naturphilosophie on the romantic movement is well known. (I have already mentioned the influence of Kant and Schelling on Coleridge and thus Humphrey Davy.) As others have pointed out, it is a testimony to the tenacity and pervasiveness of Naturphilosophie that Helmholtz still railed against it long after the philosophy had conceded the field to the mechanistic, atomistic science that dominated the late nineteenth century.[96] According to David Knight, Naturphilosophie "was clearly a step towards the unity, simplicity and harmony of nature which was predicted in the Platonic tradition; in the world of the Romantic scientist, there is nothing cold and inanimate. We seem to encounter brute matter, but really we meet living forces. It appeared that chemists and physiologists had stormed the Bastille of materialism in which Rationalists had allowed their minds to be imprisoned. . . . materialism was viewed by Romantics in Germany and England as a particularly French kind of intellectual pox."[97]

But the *Annalen der Physik*'s obituary for its editor, Gilbert, in 1824 harshly denounced the romanticism of Naturphilosophie: "This mixing of fiction and truth, of poetry and science, this playing with empty, half-true analogies, this guessing and suggesting instead of knowing and understanding has ruined our good name for us Germans abroad, has led us away from thorough science, and has brought us to believing that we know everything, while we have fallen behind in real knowledge."[98] Thus, while post-Kantian idealism may have helped create theoretical physics, it also led to unbridled metaphysical speculation that in no way aided scientific progress.[99] The reactions against it were antitheoretical and positivistic or explicitly materialistic. Even the neo-Kantians harshly attacked the remnants of Naturphilosophie.

Set against this intense interest in Naturphilosophie's affect on the nineteenth century are the arguments of Barry Gower, who disputes its influence on the actual course of science. He notes that Mayer, Kastner, Schmidt, and Fischer gave essentially mechanical explanations of phenomena in their textbooks and that the title of Fischer's book was *Textbook of Mechanical Natural Philosophy*. The issue will remain controversial as will the extent of the philosophy's impact on figures such as Oersted and Davy and its role in directing physics toward the energy principle.[100] But Kant's influence, we should remember, was by no means limited to the early nineteenth century, nor was it felt only through the popularity of Naturphilosophie. Members of the Vienna Circle, such as Carnap and Reichenbach, were Kantians before their conversion to Machian and logical positivism.

Unity and the Role of Analogy in Nineteenth-Century Physics

To examine in detail the role of analogy in scientific discovery would take us far afield into questions concerning the nature of creativity (scientific and otherwise) and the mind and the source of the process of argument by analogy, especially physical analogy. It is nonetheless important to understand the role of analogy in nineteenth-century physics because it was used deliberately and self-consciously by some of the most important scientific figures of the time, especially Faraday, Thomson, and Maxwell. Indeed, Maxwell not only used a method of physical analogy with great success but also speculated extensively about it, especially the question of whether analogies lie in the natural world or the human mind. (In his case the question of the role of analogy and its concrete realization in the form of mathematical models is complex, and it will be discussed further in chapter 3.)[101] While Maxwell employed mechanical models to whose reality he was committed in differing degrees at different times, the case of his mentor, William Thomson (to whom Maxwell was indebted for concrete models such the molecular vortices and also, to some extent, his way of doing physics), is much clearer.

We have already noted that an important legacy of early nineteenth-century science was the search for unity in natural philosophy, a search exemplified by the careers of Faraday and Thomson (to name only two). This urge had several sources: the shared view of the followers of Kant and Scottish commonsense philosophers such as John Playfair, who saw unity emerging from the fact that all physical laws have their origin in the human mind; the discovery by Thomson (and his predecessors) that the mathematical descriptions of seemingly disparate phenomena are quite similar; and the use of analogy by those, like Faraday, whose search for unity had no mathematical basis. We can identify two kinds of analogical reasoning: *mathematical* and *physical.* Faraday's powerful use of analogy, which derived in part from Kant, was of the latter type and in part the result of a predilection toward unification whose origin is necessarily obscure but that some have argued may lie in the structure or function of the brain rather than being a product of education or training. Hendry sees two dimensions to this use of analogy: that between or among the various sciences and that between nature and the mind.[102] It was as remarkable to Coleridge as it has since been to many others "that the material world is found to obey the same laws as had been deduced independently from reason." Often that idea is restricted to the unreasonable success of mathematics, beautiful tautology that it is, in describing nature.[103] The solution (some would say partial solution) to this dilemma was one of Kant's greatest triumphs. That the laws of nature as we understand them are not independent of the human mind was not lost on Maxwell, who commented that "the whole framework of science . . . seems sometimes a dissected model of nature, and sometimes a natural growth on the inner surface of the mind."[104]

The use of analogy as a methodological tool is related to a commitment to the unification of similar or even superficially disparate fields of physics, although the latter commitment implies some kind of homology of causes while the former may be little more than metaphor. In seeking to unify seemingly disparate areas of physics, one hopes to bring them under a single umbrella to demonstrate that they are in a larger sense the same or similar. Good examples are Maxwell's unification of electricity and magnetism, Einstein's attempts to unify electromagnetism and gravity, and the unification during the 1960s and 1970s known as "grand unification" in which all the forces of nature except for gravity were in some sense unified. The use of analogy, on the other hand, while following in the same spirit, attempts to argue from the structure of one theory or set of phenomena to another in which the subject matter is apparently quite different, as when Thomson drew an analogy between heat conduction and electrostatics.

Ideas of unity have guided the research of otherwise dissimilar scientists such as Faraday and Einstein and were a major preoccupation of the dynamic natural philosophers and scientists. In a paper announcing the discovery that heat could be reflected and polarized like light, James Forbes, who taught Maxwell natural philosophy at Edinburgh, concluded: "The importance of analogies in science has not, perhaps, been sufficiently insisted upon by writers on the method of philosophizing. A clear perception of *connexion* has been by far the most fertile source of discovery."[105] For Faraday, the use of analogy, although devoid of the mathematical characteristics it later assumed in the hands of Thomson and Maxwell, was an indispensable guide to experimentation.

Maxwell's ideas on the use of analogy are best expressed in the essay "Are There Real Analogies in Nature?" which he prepared in February 1856 for the Apostle's Club in Cambridge.[106] In it he wrote: "Now, if in examining the admitted truths in science and philosophy, we find certain general principles appearing throughout a vast range of subjects, and sometimes reappearing in some quite distinct part of human knowledge; and if, on turning to the constitution of the intellect itself, we think we can discern there the reason of this uniformity, in the form of a fundamental law of the right action of the intellect, are we to conclude that these various departments of nature in which analogous laws exist, have a real interdependence; or that their relation is only apparent and owing to the necessary conditions of human thought?"[107] He concluded by saying that "the only laws of matter are those which our minds must fabricate, and the only laws of mind are fabricated for it by matter."

The use of analogy led to the use of mechanical models, which could be seen as objective representations of nature or simply as heuristic devices. In Hendry's paraphrase of William Rowan Hamilton, "the models might well act as fit media between nature and the human mind."[108] Although both Thomson and Maxwell made fruitful use of mechanical models, they were divided on the issue of the introduction of concepts not subject to empirical study.

Theory and Experiment

Although the question of the relative importance of theory and experiment—that is, the epistemological status of one as opposed to the other—need not be explored here at length, we want to make the point that the view held at the close of the nineteenth century was very different from the view that is popular today. Much of our change in understanding has been salutary, emerging from detailed studies of scientific discovery showing that (in physics, at least) undirected experimentation is very rare and not often productive. Experimental physics, whether its purpose is the testing of a theory, explorations at the margin of a field, or using new techniques or better instrumentation that offer the possibility of the discovery of new phenomena, must be guided by theory. The problem is that this awareness of the role of theory in channeling experimental studies encourages the modern predilection to raise theory and theorizing to the status of a higher form of scientific endeavor and to imbue theoretical physics with an aura of depth and profundity it cannot really claim—the "theoretical highground" of Jed Buchwald.[109] It is not just a case of acknowledging (with apologies to Kant) that "theory without experiment is empty, and experimental science without theory is blind," an epigram that needs no defense. Rather, we need to emphasize now and again those episodes in the history of science in which discoveries are largely the product of a qualitatively improved ability to measure certain quantities because of better instrumentation or a technological revolution. In the nineteenth century, those episodes became frequent only after the 1870s. This is an important lesson, but one that must be kept in some perspective, as perhaps Derek de Solla Price did not in arguing for the autonomy of experimental investigation in his 1983 AAAS Sarton Lecture "Sealing Wax and String." Without seriously addressing the philosophical issues surrounding the question of how theory shapes experiment and vice versa, I nonetheless want to emphasize how important a role experiment played in forcing theoretical innovation rather than merely verifying or falsifying.[110] This was especially true in the late nineteenth century, but it applies as well, in a different sense, to the interaction between steam technology and the theory of heat almost a century earlier.

On the purely theoretical side, the dichotomy between discrete particles and continuous fields is another theme that runs throughout most of the century, with both corpuscular and continuum models being offered in attempts to solve the problems of heat and wave propagation in various media. The widespread application of partial differential equations is one of the clearest characteristics of the mathematical physics of the period. This controversy over the underlying discrete or continuous nature of the microscopic world came to a head at the very end of the century, when Planck and Einstein introduced the quantum hypothesis—in each case, driven by experiment.[111] The statistical mechanics of Boltzmann, Gibbs, and Einstein also played an especially important role, offering on the one hand a convincing atomistic foundation for thermodynamics and on the other the techniques that would be crucial in developing a quantum description of radiation.

*B*y now we are familiar with the idea that scientific progress is guided by two imperatives that we might loosely label *internal* and *external*. There is no doubt that at any given time the course of scientific development is strongly shaped by the status of the science, a common recognition of the critical problems and the avenues of attack most likely to be fruitful, new discoveries, and new tools and techniques. These factors integral to a discipline are largely independent of the outside world. Nevertheless, as I hope this chapter has made clear, no scholar is really free from the culture in which he or she resides. Scientists are shaped by family, educational systems, intellectual and religious movements, social ferment, and institutions that support research. Of course, no two individuals respond in identical ways to these influences; and it is this fact that makes the study of contemporaries such as William Thomson and James Clerk Maxwell, who had similar backgrounds and education but quite different styles of doing physics, so fascinating.

The subsequent chapters may not always be explicit about the external influences at work in the development of physics: that is, the impact of the Industrial Revolution and its attendant technology in early nineteenth-century England, the dominance of the British Empire during much of the century, the rise of German nationalism, and so on. But the careful reader will keep the issues, raised in this and the previous chapter, in mind as we proceed to follow the course of scientific discovery in this last great century of classical physics.[112]

CHAPTER 3

Electromagnetism

*E*lectromagnetic theory and the science of heat or thermodynamics (see chapter 4) in many ways epitomize nineteenth-century physics. Each grew out of humble origins in the last half of the eighteenth century and became a mature field with well-developed theoretical foundations by the last quarter of the nineteenth century. Although the story of their evolution is inseparable from several issues addressed in later chapters (notably, atomism and the energy principle), they stand apart as the crowning achievements of physics in the century and form a background for the careers of many eminent physicists, including Laplace, Poisson, Fourier, Fresnel, Ampère, Faraday, Berzelius, Thomson, Joule, Maxwell, Clausius, and Boltzmann.

Electromagnetism did not exist as an area of investigation in the eighteenth century because electricity and magnetism were not generally recognized as two manifestations of one force. By the middle of the seventeenth century, understanding of electricity had not advanced greatly since the earliest speculations of Thales and Theophrastus about static electricity; but that stagnation began to shift as philosophers such as Robert Boyle (1627–91) occasionally turned their attention to electrical problems. In the absence of any source of sustained current, studies of magnetism were confined to permanent magnets; but by the middle of the eighteenth century Benjamin Franklin (1706–90) and others had shown that lightning or a discharging Leyden jar could magnetize or demagnetize a magnetic needle, suggesting a connection between the two phenomena.[1] In the end, it took the experiments of Oersted, Faraday, and Ampère to demonstrate the connection conclusively. Ampère was thoroughly convinced that magnetism was electricity in motion, that electricity and magnetism were in some sense the same thing, and that magnetism was due to circulating molecular currents. Faraday's discovery of magnetic induction resulted from his belief that, if an electric current could cause a magnetic field, then by symmetry a magnetic field should cause a current to flow in a conductor. Nonetheless, it was not until the early 1860s that Maxwell provided the theoretical underpinning for what had become evident experimentally: that electricity and magnetism were in some sense the same phenomenon.

The question of the relation of electricity to light—or, alternatively, of the electrical fluid to the luminiferous aether, the medium in which light was supposed to propagate—was rich in significance, and Franklin in particular spent much time speculating on it. It was common to hear the argument that the electrical fluid or fluids and light (at least sunlight) were in some sense identical. Light had been systematically studied by Hooke, Newton, and Huygens in the seventeenth century; and if Newton's corpuscular view prevailed because of its explanatory power and Newton's own authority, its acceptance was far from universal. Franklin, for example, himself a great follower of Newton, advocated an undulatory theory of light. In the nineteenth century, Thomas Young in England and Augustin Jean Fresnel in France would demonstrate the power of the wave theory.

Electricity and Magnetism in the Eighteenth Century

Much of the groundwork for nineteenth-century electromagnetic theory was carried out in the eighteenth century. Indeed, Robert Boyle's "Experiments and Notes about the Mechanical Origine or Production of Electricity" (1675), set firmly in the mechanical and corpuscular philosophies, provided a kind of unified theory of heat and electricity. Like Newton (particularly through the *Opticks*), Boyle exerted a strong influence on Franklin.

The early studies of Jean Theophilus Desaguliers (1683–1744) on electrical induction (1729), along with those of his predecessor Stephen Gray (1666–1736), mark the beginning of a systematic science of electrostatics.[2] Reverend Stephen Hales, whose work Franklin knew, showed that in the case of a conductor the electrical fluid must reside on the surface of a body, thus establishing that, in spite of similarities between heat and electrification, there were also important differences.[3] Charles-François du Fay (1698–1739) offered a theory of the imponderable electrical fluid and first recognized (or hypothesized) the existence of two kinds of electrification (*vitreous* and *resinous*). Both Franklin and his British contemporary William Watson (1715–87) initially held that there was only one electrical fluid or aether. In Franklin's view, a deficiency or excess of the electrical fluid or "electrical fire" produced the observed phenomena of (Franklin's terms) positive and negative electrification. The problem was central to debates about electricity in the late eighteenth century.

Experiments in electricity were greatly facilitated by the Leyden jar, the first device for storing electrical charge, invented by Pieter van Musschenbroek and others. Franklin studied the nature of induced charge, experimented with the Leyden jar, and concluded that "electrical fire," or charge, consisted of positive and negative kinds. He was committed to the idea that electrical matter consisted of "extremely subtle" particles. Perhaps the most important element in his theory was his commitment to charge conservation. Thus, electrical phenomena resulted from movement of the electric substance, or its collection (as by friction when a rod is rubbed), rather than its creation (see Franklin's *New*

Experiments and Observations on Electricity, published in 1750). The inscription on Houdon's bust of Franklin, attributed to Turgot, "He snatched lightning from the heavens and scepter from the tyrant," honored his great versatility.

Ulrich Theodor Aepinus (1724–1802) guessed that electrical action might involve an inverse-square force; and in 1767, Joseph Priestly (1733–1804), physician, dissenting cleric, and discover of oxygen, concluded—from the absence of electrical force inside a conductor in Franklin's experiments and by analogy with gravitation in which the force on a mass inside a spherical shell is zero—that the force between two charges must be inversely proportional to the square of the distance.[4] Although John Robison (1739–1805), professor of natural philosophy at Edinburgh, found experimentally that the force law was approximately $1/r^2$, his researches were not published until 1822.[5] In 1771 Henry Cavendish (1731–1850) determined that the exponent of r was less than 3 and eventually concluded that it was in fact 2 (within 1 part in 50) by showing experimentally that the charge on the inner surface of a conducting sphere was nearly zero. Again, however, this was revealed only in 1879 when Maxwell, as the first Cavendish professor at Cambridge, edited the Cavendish papers.[6] So in spite of three direct determinations of the inverse-square law between 1767 and 1771, it was a novel result when Charles Augustin Coulomb (1736–1806) again deduced the law from experiment in 1785.[7] He verified the $1/r^2$ dependence using the torsion balance, which he and John Michell (1724–93), the first to determine the force law between magnetic poles, had invented independently.[8] Tobias Mayer (1723–62) of Göttingen and the Swiss-German J. H. Lambert (1728–77) were among others in this period who thought that the electrical-force law was an inverse square.

Coulomb—following Robert Symmer (c. 1707–63), pupil of Aepinus and translator of Franklin's *New Experiments* into German; Johann Carl Wilke (1732–96); and Anton Brugmans (1732–89)—advocated a two-fluid model of electricity in which inverse-square attraction or repulsion between portions of different or the same kinds of fluid accounted for electrical phenomena. Aepinus attributed magnetism to an imponderable magnetic fluid in 1759, and a two-fluid theory was advanced by Anton Brugmans and Johan Wilcke, essentially by symmetry with electricity.[9] The fact that breaking a magnet into pieces produced additional magnets led Coulomb to conclude that the magnetic fluids did not escape because they were locked within the molecules of the magnetic bodies.

Opening the Nineteenth Century

Until Alessandro Volta (1745–1827), professor of natural philosophy at the University of Pavia, began the work that led to the voltaic pile (1800), experimentation with sustained currents was impossible.[10] All that was available as a source of stored electrical energy was the Leyden jar, which permitted only brief, transient currents when it was discharged. The voltaic pile, or battery, ushered in the modern era of electrical investigation. It also led to the creation of

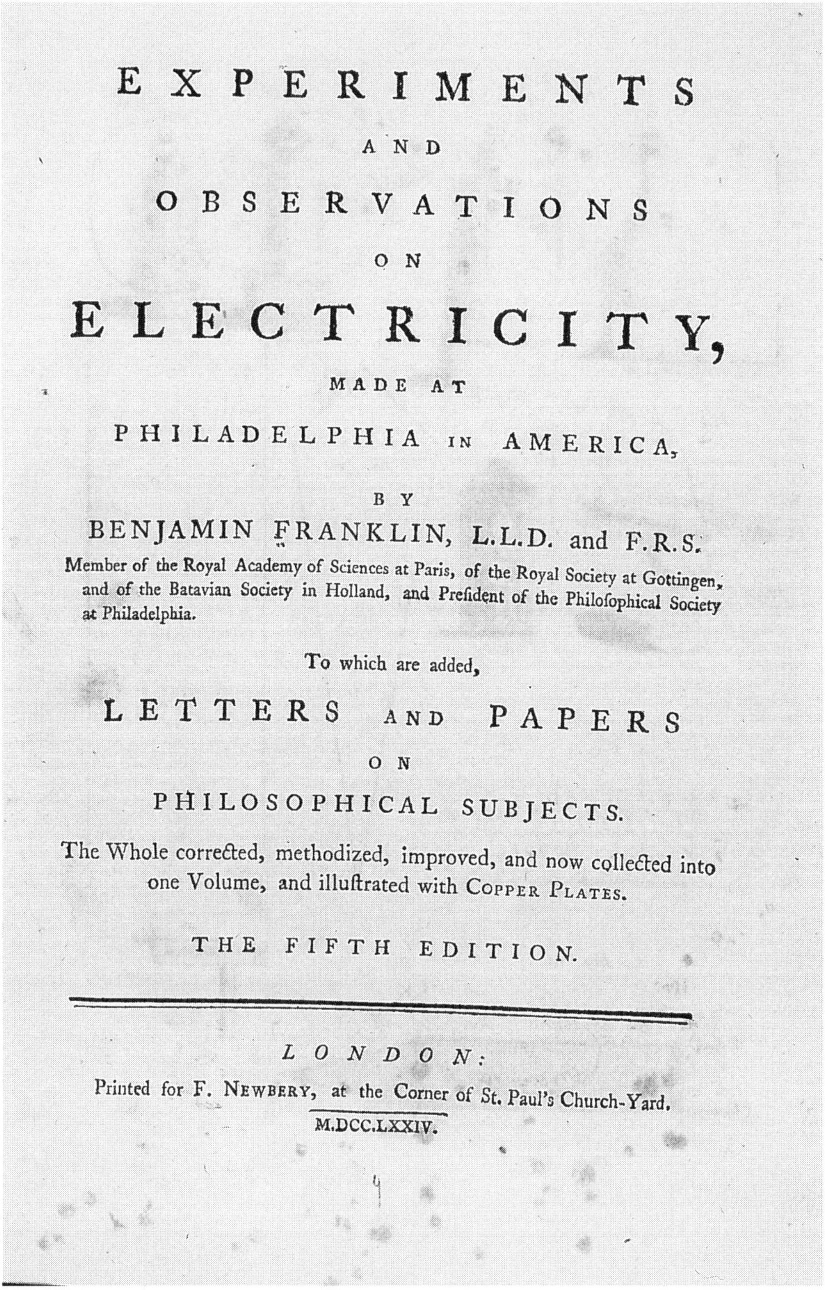

EXPERIMENTS

AND

OBSERVATIONS

ON

ELECTRICITY,

MADE AT

PHILADELPHIA in AMERICA,

BY

BENJAMIN FRANKLIN, L.L.D. and F.R.S.

Member of the Royal Academy of Sciences at Paris, of the Royal Society at Gottingen, and of the Batavian Society in Holland, and Prefident of the Philofophical Society at Philadelphia.

To which are added,

LETTERS AND PAPERS

ON

PHILOSOPHICAL SUBJECTS.

The Whole corrected, methodized, improved, and now collected into one Volume, and illuftrated with COPPER PLATES.

THE FIFTH EDITION.

LONDON:

Printed for F. NEWBERY, at the Corner of St. Paul's Church-Yard.

M.DCC.LXXIV.

PLATE 2. Title page of Benjamin Franklin's *Experiments and Observations on Electricity* (1774 ed.). *Courtesy of the Library of Congress.*

PLATE 3. Charles Augustin Coulomb (1736–1806). *Courtesy of the Burndy Library, Dibner Institute for the History of Science and Technology.*

electrochemistry, when Nicholson and Carlisle, after being informed of Volta's discovery, found that a flammable gas was evolved when the electrodes from the voltaic pile were placed in water.[11] This work was a direct stimulus to Humphrey Davy's (1778–1829) initial experiments in electrochemistry and ultimately led to the decisive work of Faraday, Berzelius, and others.[12] Volta's explanation of the operation of the voltaic pile was based on the assumption of a single electric fluid, which, because it was imponderable, could circulate indefinitely.[13] His contact theory of the pile was opposed by Davy, William Hyde Wollaston (1766–1828), and William Nicholson (1753–1815), who advocated a theory of chemical action at the metal surfaces and in the liquid separating them.[14]

Still, experimentation with stationary currents did not in itself yield much that was new or fundamental. (The most notable exceptions were in electrochemistry.) Sustained currents began to yield important results only after Oersted's discovery of electromagnetism in 1820, the "invention" of the electromagnet by Johann Schweigger in Germany and William Sturgeon in England, and the experiments of Faraday, Ampère, and their contemporaries.[15]

Between the 1730s and the 1820s, then, the only fundamental experiments possible were in electrostatics. It is surprising that given Coulomb's 1785 establishment of the inverse-square nature of electric force, no one before Poisson in 1811–13 applied the Legendre-Laplace gravitational potential theory of 1784 to electrostatics. In 1811 Poisson wrote:

> The theory of electricity which is most generally accepted is that which attributes the phenomena to two different fluids, which are contained in all material bodies. It is supposed that the molecules of the same fluid repel each other and attract the molecules of the other fluid; these forces of attraction and repulsion obey the law of the inverse square of the distance; and at the same distance the attractive power is equal to the repellant power; whence it follows that, when all the parts of a body contain equal quantities of the two fluids, the latter do not exert any influence on the fluids contained in the neighboring bodies, and consequently no electrical effects are discernible.[16]

Laplace had shown in 1784 that the scalar function V (of the form $1/r$) satisfied what we now call Laplace's equation, $\nabla^2 V = 0$.[17] Poisson generalized this to $\nabla^2 V = -4\pi\rho$ when matter is present (ρ being the mass density) and then applied it to the $1/r^2$ force in electrostatics.[18] He also showed that conductors are equipotential surfaces. In 1824, he developed a mathematical description of magnetism in terms of a magnetic scalar potential based on his theory of two imponderable magnetic fluids.[19]

By 1828 George Green (1793–1841) had generalized many of Poisson's results and introduced the name *potential* for the function V of Laplace, Lagrange, and Poisson.[20] Green's work, privately published in Nottingham by subscription (there were fifty-two subscribers), virtually escaped notice, not only on the continent but even in Britain, until William Thomson introduced it to the French via Liouville's *Journal* in 1845. Green was largely self-taught, entering Gonville and Caius College, Cambridge, in 1833 at the age of forty. In 1839 he became a Fellow of Caius but died two years later at age forty-seven.[21] What is now known as Green's Theorem, a mainstay for discussions of solutions to partial differential equations of physics, was introduced in this privately published work. Most of his other work was published in *Transactions of the Cambridge Philosophical Society* between 1833 and 1839.

By as late as 1820, then, experimental studies had yielded many interesting results; but except for a moderately well developed theory of electrostatics, little else existed. The next ten years, however, saw remarkable discoveries by Oersted and especially Ampère and Faraday; and well before 1850 many issues had been clarified, including the generation of magnetic fields by currents, the creation of currents by changing magnetic fields (magnetic induction), and the effect of magnetism on light and hence the connection between electromagnetism and light. There were related discoveries during that fertile quarter-century leading up to 1850, especially in the theory of heat (1824–47) and the understanding of energy conservation in the late 1840s, which later influenced the development of the theory of electromagnetism.

The history of electromagnetism has two easily distinguished elements. One involves the experimental tradition epitomized by Franklin, Cavendish, Coulomb, Volta, and, above all, Faraday. The other is characterized by the developments in potential theory and the theory and application of partial differential equations—primarily the work of Lagrange, Laplace, Poisson, Green, and Gauss. The growing tendency, as exemplified by the French mathematical physicists, was toward increasing mathematical complexity of the theories of heat and electricity, which widened the gulf between those who were primarily experimentalists and those intrigued by mathematical physics. There were exceptions even in the last half of the nineteenth century, notably Maxwell, Helmholtz, and Kirchhoff. But by the end of the century, it was rare for a scientist to combine mathematical prowess and originality with an interest in experimentation.

As the eighteenth century drew to a close, Newton's influence was hardly less strong than it had been at the beginning. But what Newton meant in Britain and what he meant in France were different. The distinct traditions of the *Opticks* and the *Principia* exhibited themselves differently in the two nations. More generally, Franklin, Priestley, Coulomb, Cavendish, and Volta carried forward the traditions of Newtonian experimental philosophy, while Poisson, Biot, and others (following in the footsteps of Maupertuis, Laplace, and Lagrange) emphasized and extended the mathematical and theoretical side of the Newtonian method. We shall follow those two threads through the nineteenth century in the experiments of Oersted, Ampère, and Faraday; the development of the theory of electricity; and the progress of electrostatics, magnetism, and electromagnetism from Green through Thomson and Maxwell to Helmholtz.

The Wave Theory of Light

The controversy over the nature of light belongs largely to the seventeenth and eighteenth centuries and was essentially resolved by the first quarter of the nineteenth century.[22] Although Grimaldi, Hooke, and Huygens, and later John Bernoulli and Euler, advocated a wave theory of light, Newton's corpuscular view generally held sway. Yet even Newton's followers were divided on the question: Euler was sure that "light is in the aether the same thing as sound in air," while Laplace adhered to the corpuscular theory as late 1817.[23] When the issue was decided by about 1820, attention turned to the nature of the luminiferous aether that supported light waves, then to the study of diffraction and polarization phenomena, and eventually to the demonstration that light and electromagnetic waves are the same.[24]

Many of the basic phenomena of light and its interaction with matter were discovered in the seventeenth century. In 1621 W. Snell determined the law of refraction that carries his name; Grimaldi first and later Hooke discovered the phenomena of diffraction, Hooke and Boyle independently observed interference, and Huygens discovered the polarization of light.[25] Birefringence was first observed in 1670 by Erasmus Bartholin, and Olaf Römer demonstrated the finite speed of light by 1676. These properties of light were interpreted in the terms of either the corpuscular or wave theories; but after a century of Newton's dominance, the former was so firmly entrenched that, when Thomas Young (1773–1829) advanced his wave theory in 1801, the hostile reaction drove him out of optics altogether and into antiquities.[26] The discovery of polarization of light by reflection in 1808–10 by Etienne Malus (1775–1812) was initially taken as decisive evidence against the wave theory.[27] Its effect on Young, however, was to rekindle his interest in the question. When François Arago showed in 1816 that two beams of light polarized in perpendicular directions cannot interfere, Young proposed that light consists of transverse rather than longitudinal vibrations, providing a way of explaining the observed polarization but at the cost of making the aether an elastic solid rather than a fluid. Fresnel (1788–1827) developed Young's ideas, but he faced hostile opposition from Laplace and Biot, who, it has been said, drove him to an early grave.[28] In the French Academy of Sciences competition of 1818, Fresnel offered the first mathematical theory of diffraction, which opposed the corpuscular "emission theory" of Laplace, Biot, and others. Poisson soon showed that the theory predicted that a bright spot should occur in the center of the shadow of a disk illuminated by a light source, a fact that was considered a blow to the theory until the prediction was verified by Arago.[29] Finally, nearly forty years later, Fizeau and Foucault measured the speed of light in air and water and showed that light traveled more slowly in an optically dense medium, contrary to Newton's predictions. By that time, however, it was generally agreed that light was an undulatory phenomenon.

Among advocates of the wave theory there was disagreement about whether the aethers of heat, light, and electricity were the same. The nature of the connection between electromagnetism and light, if any, was not addressed experimentally until Faraday's discovery of the magneto-optic effect in 1845, and the question was only "settled" when Maxwell wrote "A Dynamical Theory of the Electromagnetic Field" in 1864.[30]

Hans Christian Oersted

The experiment in which electromagnetism is often considered to have been discovered was carried out by Hans Christian Oersted (1777–1851) in the winter of 1819. An ardent Kantian and Naturphilosoph, Oersted found that a current could deflect a compass needle.[31] His interest in the problem dated back to at least 1807, and in 1813 he wrote that "it is necessary to test whether electricity in its most latent state has any action upon a magnet as such."[32] In the course of a public lecture-demonstration on magnetism, Oersted observed a small effect when a current-carrying wire was laid over a compass, although he noted that it "made no strong impression on the audience."[33] In his brief account of subsequent experiments, "Experimenta circa effectum confluctus electrici in acum magneticam," written on 21 July 1820 and published that same month, he noted the dependence of the effect on distance and on the relative position of the wire and the needle: no effect if the wire is in the same plane as that in which the needle rotates or if the wire is perpendicular to the needle (see figure 3–1).[34] He also showed that interposing various materials between the wire and the needle had little or no effect (unless they were magnetic) and concluded that "the effect cannot be ascribed to attraction." It was clear that what he called the "electric conflict" was not confined to the conductor and had a circular character. Oersted used the impenetrability of magnetic bodies to the electric conflict to explain the effect on the magnetized needle: "All non-magnetic bodies appear penetrable by the electric conflict, while magnetic bodies, or rather their magnetic particles, resist the passage of this conflict. Hence they can be moved by the impetus of the contending powers."[35] Finally, recalling that he had previously written that "heat and light consist of the conflict of the electricities," he speculated upon a connection between the circular effect noticed in this experiment and the polarization of light.

These studies were guided by Oersted's and his colleague Wilhelm Ritter's belief that "the same forces manifest themselves in magnetism as in electricity" as well as by the premise of Naturphilosophie that polar or opposing forces were at the heart of matter and material phenomena.[36] In Brewster's *Edinburgh Encyclopaedia* Oersted wrote (in the third person) that "he did not consider the transmission of electricity through a conductor as a uniform stream, but as a succession of interruptions and re-establishments of equilibrium, in such a manner that the electrical powers in the current were not in quiet equilibrium but in a state of continual conflict."[37] Oersted believed that, just as a current influenced a magnet, a magnetic pole should exert a force on a current, a conjecture that Davy verified in 1821.

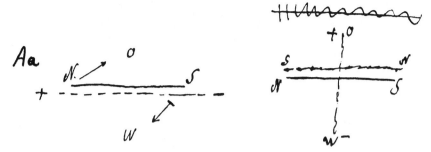

FIGURE 3–1. Oersted's notes on the experiment in which he "discovered electromagnetism." The magnetized needle aligns itself along the magnetic field lines—essentially, perpendicular to the current-carrying wire. *From K. Meyer, ed.,* Oersted: Scientific Papers *(Copenhagen: Host, 1920), lxxiv.*

Although Oersted's discovery soon became a sensation, the initial French reaction was hostile, as Dulong's dismissal exemplifies: "just another German dream." But François Arago (1786–1853), who had been present when August de la Rive (1801–73) reproduced Oersted's experiment in Geneva, was so impressed that he repeated the experiments before the French Academy on 11 September 1820.[38] As sensational as the discovery itself was the confirmation that the magnetic force was circular. Oersted was not alone in expecting some sort of interaction between electricity and magnetism; but the authority of Coulomb, who categorically stated that such an interaction was impossible, was so great that Oersted's simple experiment had not been attempted during the decades after the invention of the voltaic pile. Ampère was very clear: "I believe . . . that I can assign a cause for this; it lies in Coulomb's hypothesis on the nature of magnetic action; this hypothesis was believed as though it were a fact. . . . This prohibition was such that when M. Arago spoke of these new phenomena at the Institute, they were rejected. . . . Everyone decided that they were impossible."[39]

Hardly a month after De la Rive's demonstration, Jean-Baptiste Biot (1774–1862) and Felix Savart (1791–1841) experimentally determined the force law that described the force exerted on a magnetic pole due to a current. This they accomplished by measuring the rate of oscillation of a small magnet placed in various relationships to a current-carrying wire. The two men concluded that the force was inverse square ($1/r^2$) and perpendicular to the line joining the magnetic pole and the wire and to the axis of the wire.[40] The fact that this force exhibits an inverse-square dependence on the distance had been established earlier by Coulomb, Michell, and Laplace.[41]

André-Marie Ampère

Like so many other French scientists of this period, Ampère's life (1775–1836) was profoundly affected by the French Revolution: His father was

PLATE 4. Hans Christian Oersted (1777–1851). *Portrait by D. Hvidt. Courtesy of the Royal Library, Copenhagen.*

PLATE 5. André-Marie Ampère (1775–1836). *Courtesy of the Burndy Library, Dibner Institute for the History of Science and Technology.*

guillotined by the Republicans in 1793. Ampère's education took place at home in his father's considerable library. Among his strongest influences was Diderot's *Encyclopédie.* In 1802 he began his professional career as professor of physics and mathematics in Bourg-en-Bresse; after his wife died, he moved to the Ecole Polytechnique in Paris, where he lived until his death. Ampère's career included membership in the Institut Imperial, teaching at the University of Paris, and assuming the chair of experimental physics at the Collège de France in 1824. Between 1800 and 1814 his studies were mostly in mathematics and chemistry, but from 1820 to 1827 he devoted his efforts to electrodynamics.

Of all the figures who continued Oersted's inquiries (such as Biot, Arago, Davy, Faraday, Seebeck, and Prechtl), Ampère experienced the most immediate and spectacular success.[42] After his years as a teacher at Bourg-en-Bresse, he had shown little interest in the phenomena of electricity and magnetism. But by the 18 September 1820 meeting of the Academy, Ampère was ready to report on a revolutionary set of experiments. In the words of Englishman John Tyndall, who succeeded Faraday at the Royal Institution, "in this single week he developed the laws of what are called electrodynamics."[43] Sketching Ampère's life and work, Pearce Williams observes rightly that "had he died before September [1820] he would be a minor figure in the history of science. It was the discovery of electromagnetism by Hans Christian Oersted in the winter of 1819 which opened up a whole new world to Ampère and gave him the opportunity to show the full power of his method of discovery."[44] Pierre Duhem wrote that, through Ampère, "Newtonian physics conquered a new empire by submitting electrodynamic and electromagnetic forces to the rules of this Physics."[45]

At the 18 September meeting Ampère reported his discovery that a compass needle will set at 90° to a current-carrying wire (a result implicit in Oersted's experiment) and the more important discovery of the interaction of two parallel currents. He formally announced these discoveries in his first memoir on electricity, published the same year, which included speculations about the currents that produce the Earth's magnetic field and his conclusion that the electrostatic theory of the electric pile was wrong.[46]

Ampère's program to determine the exact force law describing the interaction of two currents was careful and painstaking. The problem was very difficult, as we now know, because of the vectorial nature of the force (the law of Biot and Savart) coupled with the vectorial character of the law for the magnetic field due to a current element.[47] Today we write Ampère's law for the $\mathbf{dF}$ force exerted on a current element $di_1 = i_1 dl_1$ due to another current element $di_2 = i_2 dl_2$

$$\mathbf{dF} = i_1 \, i_2 \, dl_1 \times (dl_2 \times \hat{\mathbf{r}})/r^2,$$

where the two currents are i_1 and i_2 and the infinitesimal current elements are dl_1 and dl_2. The expression involves two angles and the distance $\mathbf{r}$ between the current elements. Ampère expressed his result in terms of three angles: the angle between the two planes, each of which is determined by one wire (or element) and the line joining the two elements, and the angle that each wire (element) makes with the line joining the elements.[48] He concluded that the forces resulting from the interaction of two infinitesimal current elements were central.[49]

With Ampère, the distinction between static and dynamic electricity became clear for the first time. Although George Simon Ohm finally clarified the relation between electrical tension (or potential difference) and electrical current, Ampère was the first person to explore the question.[50] During the next five years, he worked out the consequences of these observations, which were at this point purely phenomenological, and offered the mathematical theory (mentioned previously) of the interaction between currents based on central forces between small current elements of the wires (see figure 3–2). In a celebrated memoir published in 1825, he described the interaction of parallel currents; and in 1827 he published his "Memoir on the Mathematical Theory of Electrodynam-

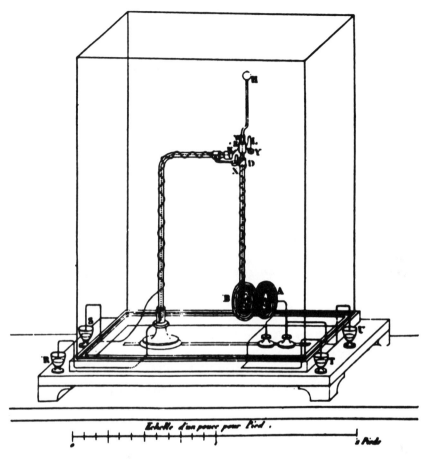

FIGURE 3–2. Ampère's device for measuring the force between two current-carrying spirals. *From* Annales de Chimie et de Physique *15 (1922): 352.*

ics, Uniquely Deduced from Experiment," a work that Pearce Williams has called the *Principia* of electrodynamics.[51]

Oersted's discovery of electromagnetism allowed Ampère to reject Coulomb's theory of two fluids, electric and magnetic, in favor of a single one, which in the case of the interaction between a current and a magnet represented the interaction of two currents of electricity.[52] Ampère had already shown that two currents did exert forces on one another. But perhaps his most profound and far-reaching conclusion was that a magnet is only a collection of electric currents.[53] This view that magnetic forces were due to the motion of the electric fluids in circles was controversial, so his friend Fresnel set out to test the theory of currents circulating about a permanent magnet. By 1821, the negative results of these experiments and Fresnel's failure to find evidence of heating due to the hypothetical currents led Ampère to adopt Fresnel's idea that the currents were molecular rather than circulating around the axis of the magnet.[54] But Ampère never seems to have wavered in his belief that magnetism is in every case the result of aligning electrical currents.[55] This is essentially the assumption that natural permanent magnetism and electromagnetism are the same phenomenon. It was not a popular view: Faraday, in particular, challenged it, as did the French mechanists. Fresnel went further to suggest that the current in a wire should produce a current in an adjacent one and attempted, unsuccessfully, to find such an effect. Nearly a decade later, and only after repeated failures, Faraday did finally see such currents; but they were induced by a *changing* current in the first conductor.[56]

Ampère's metaphysics was a mix of dynamism and mechanism.[57] His model of molecular currents as the source of magnetism was certainly mechanical; but he was strongly influenced by Kant's metaphysics (he believed that metaphysics was the "only really important science"), and he was impressed by the dynamical philosophy of Oersted, Davy, and Faraday.[58] Ampère combined the two threads that led from Newton's work: the mathematical tradition of the great "French" physicists Lagrange and Laplace, and the empirical tradition so honored in Britain. His procedure was to "observe first the facts, varying the conditions as much as possible, . . . in order to deduce general laws based solely on experience, and to deduce therefrom, independently of all hypotheses regarding the nature of the forces . . . the mathematical value of these forces."[59]

Michael Faraday, whose mathematical skills were rudimentary, was put off by Ampère's mathematics: "With regard to your theory, it so soon becomes mathematical that it quickly gets beyond my reach."[60] Given the level at which he could appreciate them, he conceived Ampère's theories to be entirely ad hoc. Nevertheless, Ampère's ideas, especially those concerning the origin of magnetism, greatly influenced Faraday, guiding him in his experiments and constantly serving as a foil against which he tested his own ideas. The process of unifying electricity and magnetism, which was one of the important outcomes of Faraday's work, received its strongest support from Ampère's electrodynamics. In 1847 William Whewell wrote that Ampère had taken the "vague and obscure persuasion that there must be some connection between electricity and magnetism, so long an idle and barren conjecture," and reduced all the phenomena of magnetic and electromotive action "to one single polarity, that of the electro-dynamic current."[61]

In his *Treatise* (1873), James Clerk Maxwell wrote about Ampère's method:

> The experimental investigation by which Ampère established the laws of the mechanical action between electric currents is one of the most brilliant achievements in science. The whole, theory and experiment, seems as if it had leaped, full grown and full armed, from the brain of the "Newton of electricity." . . . We can scarcely believe that Ampère really discovered the law of action by means of the experiments which

he describes. We are led to suspect, what, indeed, he tells us himself, that he discovered the law by some process which he has not shewn us, and that when he had afterwards built up a perfect demonstration he removed all traces of the scaffolding by which he had raised it. . . . Every student should therefore read Ampère's research as a splendid example of scientific style in the statement of a discovery, but he should also study Faraday for the cultivation of a scientific spirit, by means of which the action and reaction which will take place between the newly discovered facts as introduced to him by Faraday and the nascent ideas in his own mind.[62]

Ampère reduced magnetism to the motion of molecular currents and attempted to explain the force that one wire exerted on another as due to vibrations in the aether, which he saw as a neutral fluid formed from the combination of two electrical fluids. This model of matter, with its hypothetical entities, was especially difficult for Faraday to accept; and he kept a respectful distance from Ampère's ideas, recognizing their successes but retaining the strong skepticism and independence of thought that drove his own discoveries. As his health declined, Ampère devoted his later years, from about 1827 on, to philosophy.[63]

Unlike Ampère, who reduced magnetism to electricity, J. J. Prechtl and Jacob Berzelius both proposed purely magnetic explanations of the effect of a current-carrying conductor on a magnet. The theory of Prechtl, an Austrian Naturphilosoph, was especially interesting. It eschewed imponderable fluids and was based on attractive and repulsive powers inherent in matter; these forces, he thought, were generated by small magnets arranged radially in a wire.[64]

Michael Faraday

Michael Faraday (1791–1867) made his way to electromagnetism by a route even more circuitous than Ampère's. The son of a blacksmith, Faraday was primarily self-educated.[65] He was apprenticed to a London bookbinder at the age of fourteen and, in the course of his work, came into contact with some of the ideas of chemistry and electricity, mostly through reading an article on electricity in the third edition of the *Encyclopaedia Britannica* by James Tytler and in Jane Marcet's *Conversations on Chemistry*.[66] In Tytler's article Faraday learned that, "throughout the sciences, as well in most others, few discoveries have been made by reasoning, but many by accident."[67] In 1810 he began attending lectures of the City Philosophical Society and set up a small laboratory in the back of the bookbinder's shop, where he constructed an electrostatic machine. He attended lectures by Humphrey Davy in 1812 and was hired as Davy's assistant at the Royal Institution later that year. During the next ten years Faraday developed his talents in chemistry and began to mature philosophically. By 1818 his determination and experimental ability had already made him a professional chemist.

The fact is sometimes overlooked that Faraday, certainly one of the great experimentalists in the history of physics, was also one of the great theorizers of the nineteenth century. This fact has been obscured by his total lack of mathematical knowledge; to paraphrase Williams, mathematics was a closed book to him. Clearly, however, Faraday possessed extraordinary mathematical intuition, and any reader of his great *Experimental Researches in Electricity* cannot escape the brilliance of his theoretical imagination and the extent to which it guided his experiments: "he bravely theorized on all occasions."[68] On page after page we find him employing analogies between electricity and magnetism and his sense of reciprocity or unity in order to design new experiments and predict their outcome. He was as far as one could be from the undirected experimenter, unencumbered by theory, working blind.[69] Faraday was never cautious or conservative in designing his experiments and generalizing from them. In the final analysis, however, he was thoroughly an experimentalist and had no use for hypotheses that were not fully grounded in the phenomena. Near the end of *Experimental Researches in Electricity,* he wrote: "I feel bound to let experiment guide me into any train of thought which it may justify; being satisfied that experiment like analysis, must lead to strict truth if rightly interpreted." In electricity, and ultimately in magnetism and even gravitation, Faraday was quite willing to challenge Newtonian action-at-a-distance and to base much of his scientific activity on the highly speculative point atomism of Boscovich.

Both Maxwell and Thomson were initially repelled by Faraday's lack of mathematics and the way he "spoke" about physics, but they came to understand his achievements when they saw that the results could be formulated mathematically.[70] Ultimately, Maxwell said of Faraday: "the conjecture of a philosopher so familiar with nature may sometimes be more pregnant with the truth than the best established experimental law discovered by empirical inquirers, and though not bound to admit it as a physical truth, we may accept it as a new idea by which our mathematical conceptions may be rendered clearer."[71]

Faraday was twenty-two years old when he began his career as Davy's assistant, but by 1824 he had been elected a Fellow of the Royal Society and in the following year became the director of the laboratory of the Royal Institution.[72] His first paper in physics, describing electromagnetic rotation (1821) (see figure 3–3), made him an important scientific figure almost overnight.[73] Along the way, he was involved in controversy with Davy (who blocked his election as Fellow of the Royal Society for six months) and William Wollaston, arising in large measure out of petty scientific jealousy.[74] These episodes survive only as footnotes to a remarkable career but probably explain, in part, why Faraday had few close associates.[75] Although he had been Davy's protégé, Faraday himself had no pupils and hardly any disciples.

Religion played a central role in the lives of many nineteenth-century scientists but for no one more than Faraday, whose nonconformist Sandemanian beliefs provided, as Pearce Williams's says, "the firm anchor for his life." But

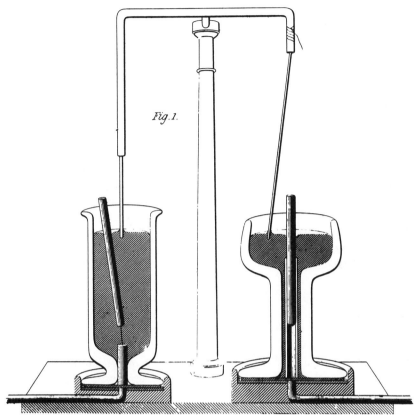

Fig.1.

FIGURE 3–3. Faraday's device that first achieved "magnetic rotation." The wire was made to rotate around the electromagnet, with the circuit being completed through the conducting mercury bath. *From* Experimental Researches in Electricity *2 (1844): plate IV.*

the impact of these beliefs on his scientific work remains largely obscure, in spite of the efforts of Cantor and others.[76]

Faraday's Scientific Work

It is not easy to summarize Faraday's work or clarify for modern readers what he meant by *electrotonic state, magnetic conduction,* and a host of other terms. In the cases of Maxwell, Thomson, and even Helmholtz, we can take refuge in the mathematics when the language seems obscure; but with Faraday there is no mathematics. Furthermore, his career spanned four decades, during which his ideas naturally evolved and changed. He did leave behind his voluminous *Researches* and *Diary,* which offer almost daily insight into the progress of his work.[77] The result is a richness of detail that defies easy summary, that makes one feel glib and superficial if one dispenses with all the texture and leaves the reader with only a distillation. Of course, one has to make this choice and in the end can do little more than refer the reader to Faraday's writings and the journal literature or to biographical material.[78]

By the end of 1821 Faraday, still in a minor post at the Royal Institution, had published a number of papers in chemistry and metallurgy; his analytical work for the Royal Institution helped keep it afloat financially in the 1820s. He first isolated benzene in 1825, became involved in glass research for the Royal Society from 1827 to 1831, and worked with James Stodart on steel.[79] Much of his chemical work was in electrochemistry, which complemented his interests in electricity and magnetism. But Oersted's remarkable discovery in 1819 (that a current could deflect a compass needle) had a galvanizing effect on both Davy and Faraday, as it did on nearly everyone else. Davy first heard of Oersted's discovery of electromagnetism in October 1820, six weeks after the news reached Paris; immediately, he and Faraday set about investigating the effect. By the summer of 1821 Faraday was seriously studying the interaction of currents and magnets. Almost immediately he discovered magnetic rotation, the rotation of a current-carrying wire about a magnet. He described this discovery, in which he used the Oersted effect to construct a forerunner of the electric motor, in an article in the *Quarterly Journal of Science* in September. In this simple experiment, in which a current-carrying wire was made to wrap around a magnet, electricity was converted into mechanical work for the first time.[80]

Also in the fall of 1821, Faraday published his "Historical Sketch of Electromagnetism," in which he described the experimental foundations of electromagnetism, especially the experiments of Oersted and Ampère and the latter's theory of electric current. He was skeptical of Ampère's circular molecular currents, criticizing them as hypothetical entities with no empirical basis. He also expressed doubts about Ampère's model of electrical action, even disputing the reality of electricity.[81] Faraday showed that a hollow cylindrical magnet did not behave like a helix, thus casting doubt on Ampère's early theory of coaxial currents in a magnet.[82]

In that critical year, 1821, Faraday's views were strongly shaped, according to Pearce Williams, by a series of articles in the *Annals of Philosophy* in which the roles of theory and experiment were pitted against one another. This seems also to have sharpened Faraday's views about Ampère's theory of magnetism.[83] Faraday was probably influenced by current ideas about the undulatory nature of light and sound (perhaps even of heat) and gave some thought to the vibratory nature of electricity. Somewhat later he himself said that John Herschel's *A Preliminary Discourse on the Study of Natural Philosophy* (published in 1830) exerted a strong influence on him.

Although Faraday did little further work on electromagnetism until 1831, as early as 1825 he attempted to show that one current could induce another and "came within a whisker of discovering electromagnetic induction."[84] He carried out a series of experiments to see if static currents in one helix could cause currents in another. Many people, including Fresnel, thought that some such ef-

PLATE 6. Faraday's laboratory at the Royal Institution. Illustration by H. Bence Jones. *From* The Life and Letters of Faraday, *2 vols. (Philadelphia: Lippincott, 1870).*

fect must exist—that is, if currents can give rise to magnetic fields, then magnetic fields should give rise to currents.[85] Faraday had a deep-seated belief in a sort of symmetry or reciprocity of action that he later used with great effectiveness into the mid-1840s. In reporting his discovery of electromagnetic induction in 1831, he observed that "it appeared very extraordinary, that as every electric current was accompanied by a corresponding intensity of magnetic action at right angles to the current, good conductors of electricity, when placed within the sphere of this action, should not have any current induced through them, or some sensible effect produced equivalent in force to such a current."[86]

In August 1831, with the aid of large electromagnets, Faraday finally made the discovery that, while static currents or magnetic fields, however strong, did not induce currents, currents were induced when a field was changed or, as Faraday came to see it, when a conductor cut magnetic lines of force. Ampère and De la Rive had actually observed this effect a decade earlier but had passed over it, giving Faraday the chance to make the discovery.[87] While Faraday's discovery is sometimes considered an accident (indeed, it may have been in a very technical sense), it is clear that by 1831 he was actively engaged in a research program that was concerned especially with transient phenomena. Hence, the discovery was less an accident than the outcome of a systematic program of experimentation.[88] The discovery of magnetic induction was among Faraday's greatest. The result was revealed in a paper read to the Royal Society on 24 November 1831 and published in *Philosophical Transactions* in 1832; it became part of the first series in his *Experimental Researches in Electricity.*

Faraday's scientific life between 1831 and 1838 was extraordinarily active;

among other things, he completed the first fourteen series in *Experimental Researches in Electricity*. Much of this period was devoted to electrochemistry as he continued to try to understand the nature of electricity, but he also worked on the question of electrical induction. One result was a theory of induction that pictured electrical action as involving contiguous molecules that transmitted the force. It was an alternative to action-at-distance, of which he was increasingly skeptical. Faraday thought of "induction in curved lines" because, as he saw it, inverse-square action-at-a-distance represented action in straight lines.[89] One of the most important results of this period (especially 1831–33) was his conclusion that all electricities (static, galvanic, or thermoelectric, whether produced by electromagnetic induction or not) were the same. More radically, he rejected the view that electricity was a substance and saw it as a state of strain or polarization.[90]

In 1838 Faraday returned briefly to studies of magnetism. But by 1839 bad health, including severe memory loss, began to plague him; and between 1839 and 1845 he had little energy for experimental work. After a hiatus of nearly six years, he began in 1845 to address seriously the question of how electrical and magnetic forces were transmitted. His theory of action between contiguous particles (through their forces) led him to the idea that matter filled all space, but in the sense of Boscovichean atoms—that is, powers concentrated about point atoms (no extension).[91] Using the analogy of light ("and therefore probably of all radiant action"), he suggested that forces might propagate in time through space, even including gravitation in these speculations.[92]

At this point, he began exchanging letters with twenty-one-year-old William Thomson, a correspondence that had a profound effect on both men. In particular, Thomson's letter in August 1845 led Faraday to a series of experiments in which he discovered the effect of magnetism on the plane of polarization of light.[93] Thomson was at that time trying to cast Faraday's results in mathematical form and suggested that such an effect ought to occur. Moreover, he introduced Faraday to the analogy between lines of heat flow and magnetic lines of force. Later in this chapter we will discuss the reverse side of this interaction when we examine Thomson's attempt to mathematize Faraday's ideas. But there is little question that Faraday's consideration of the analogy between lines of heat flow and magnetism, encouraged by Thomson, helped him conceptualize the field.

In the fall of 1845, stimulated by Thomson's inquiry concerning dielectrics, Faraday discovered the rotation of the plane of polarization of light in a magnetic field and the phenomena of diamagnetism.[94] The interaction between light and a magnetic field gave strong support to the idea that light, electricity, and magnetism were related. The discovery of the magneto-optic effect and diamagnetism allowed Faraday to extend the analogy with electricity: Materials could be polarized by magnetic fields. He arduously investigated the magneto-optic effect in diamagnetic materials subjected to a magnetic field and looked for a "photoelectric current he thought should be produced, by symmetry, when light passed through a substance which rotated its plane of polarization."[95]

Faraday's work between 1845 and 1850 led him slowly toward the idea that lines of force had a physical reality independent of their source. The substance of his work during this time centered on the properties of "a new magnetic condition" that he called diamagnetism, which he discovered at the end of 1845. (In fact, however, the effect had been detected several times earlier, by Brugmans, Coulomb, Becquerel, le Bailiff, and Seebeck.)[96] He chose the name *diamagnet* as the counterpart of *dielectric*.[97] Specifically, Faraday showed that a glass bar would align itself perpendicularly to an applied field rather than along it, as would happen with a ferromagnet. He studied the magnetic properties of diamagnetic materials (such as bismuth), gases, and birefringent crystals in an attempt to understand how they differed from magnetic (that is, ferromagnetic) materials.

By 1850, as the lines of force became increasingly real to him, they played a crucial role in his analysis of electrostatic and magnetic phenomena; somewhat later, they were extended to gravitation. Groping slowly toward the idea of the field, Faraday came to reject action-at-a-distance completely, which he believed to involve straight lines of force. To him, the curved lines of force in electrostatics implied that a force was propagated by contiguous, or adjacent, particles. The Boscovichean model he employed meant that the transmission of force was an intermolecular process rather than action-at-a-distance. As Faraday's ideas evolved (founded, of course, on untiring experimentation), these particles became polarizable (in the modern sense) so that he could explain the properties of dielectric materials.

In the midst of widespread acceptance of the luminiferous aether (and often other fluid or elastic media that pervaded all space), Faraday strove to eliminate any vestige of their cousins, the imponderable fluids. He tentatively proposed that the undulatory vibrations of heat might be vibrations of the lines of force themselves. His theory "endeavor[ed] to dismiss the ether, but not the vibrations." He said: "It seems to me that the resultant of two or more lines of force is in an apt condition for that action which may be considered as equivalent to a *lateral* vibration."[98] At the same time, he offered his solution to the force-matter problem (and the specific problem of electrical conduction) in the form of the interpenetration of the Boscovichean atoms, which provided a plenum that pervaded all matter.[99] Contrasting his theory with aether theory, he wrote: "in the view now set forth, it is the forces of the atomic centres which pervade (and make) all bodies, and also penetrate all space. As regards space, the difference is that the aether presents successive parts or *centers of action*, and the present supposition only *lines of action*."[100]

Faraday and the Field
No aspect of Michael Faraday's scientific career has generated as much controversy as the idea for which he is best known: the concept of the field.[101]

Faraday's understanding of the field evolved over a decade or more and was inseparable from its visualization in terms of lines of force. It certainly cannot be equated with the modern view without considerable qualification. Thus, the answer to "When did Faraday arrive at his conception of the field?" depends, as Nancy Nersessian has pointed out, on what one considers the concept to be.[102] We will take the field concept to imply that forces between two or more bodies are mediated by some influence associated with each of the bodies separately so that, if one body is situated within the range of influence of another, a force will be exerted on the former. Thus, electric or magnetic fields are generated by charges or currents, and a force will be exerted on a charge or current (or magnet) that finds itself in the electric or magnetic field. Considered in this way, the idea is quite different from plenum or vortex theories in which bodies maintain contact with each other through some all-pervading medium.[103] The field provides an explanation for the problem of how distant bodies—bodies that are physically separated—can influence each other but does not invoke the semimystical idea of action-at-a-distance. Today, the field theory lies at the heart of much modern physics.

Some have argued that Faraday possessed the germ of the field theory idea by about 1832 and that he worked it out over the next fifteen years. According to this view, he used experiment to verify theoretical deductions that had grown out of his earlier reading of Kant and Boscovich. This opinion has been vigorously disputed by those who hold that his idea came much later.[104] Nancy Nersessian, however, believes that Faraday arrived at the field concept no later than January 1832, when he suggested that magnetic induction might be due to the cutting of the lines of force: "If a terminated wire move so as to cut a magnetic curve, a power is called into action which tends to urge an electric current through it"; and "if one part of the wire or metal cut the magnetic curves, whilst the other is stationary, then currents are produced."[105] Even if one grants Nersessian's argument, the fact remains that Faraday's concept of the field, and of the role of the physical lines of force, continued to change up to 1848.[106] No one disputes, however, that he believed in the reality of the lines of force—that they could expand and collapse and be cut, that magnetism and even light might involve vibrations of them, and that contiguous particles might act along them. The lines of force provided Faraday with an image of the field that guided his experimentation.[107]

In his later years, Faraday was guided by two great principles: the reality of the lines of force, which was the basis of field theory; and the idea of conservation of forces or powers.[108] He believed that lines of force existed in empty space—that is, had an existence independent of their sources—and effectively denied the existence of the aether. By 1852, he was more than sixty years old and had stopped his experimentation, but he summed up his views in "On the Physical Character of the Lines of Force."[109] Denying action-at-a-distance in the case of gravitation as well, Faraday argued for the existence of a gravitational field associated with a massive object. In his famous paper "On the Conservation of Force" (1859), he analyzed the problem of two bodies acting at a distance and exerting forces on one another. If only one were present, there would be no attraction; if the second were brought up, it would require the creation of force in both particles instantaneously. Faraday's idea of a strain in space, represented by lines of force associated with matter, makes the problem intelligible and lies at the foundation of field theory; to him, the magnetic lines of force existed in a vacuum as well as in matter. He rejected the simple Newtonian inverse-square attraction and repulsion, which even in 1850 continued to constrain the way in which physicists interpreted magnetic forces.[110]

These ideas—the reality of lines of force and the unity of electricity and magnetism—were absorbed by his successor, Maxwell. Born in the same year that Faraday discovered induction, Maxwell developed the mathematical facility that Faraday totally lacked. Using his predecessor's ideas and the example of William Thomson, Maxwell largely created classical electromagnetism as we know it today.[111]

After a long illness, perhaps the result of mercury poisoning, Faraday died on 25 August 1867.[112] Writing in a subsequent issue of *Nature,* Maxwell called him "the father of that enlarged science of electromagnetism which takes in at one view, all the phenomena which former inquirers had taken separately. . . . The way in which Faraday made use of his idea of lines of force in coordinating the phenomena of magneto-electric induction shews him to have been in reality a mathematician of a very high order—one from whom the mathematicians of the future may derive valuable and fertile methods."[113]

William Thomson, Lord Kelvin

By the time of Faraday's death in 1867, two great younger contemporaries, William Thomson (1824–1907) and James Clerk Maxwell (1831–79), had already carried his ideas much beyond him. Thomson (later Lord Kelvin of Largs) was in many respects the foremost figure in the era of transition from classical to modern physics. Like Maxwell, he was raised by a widowed father. In 1840, when he was sixteen years old, Thomson was introduced by his first important teacher, John Pringle Nichol at Glasgow, to Joseph Fourier's *Analytical Theory of Heat* (first published in 1822).[114] Fourier's influence was profound in its implications about the relationship of mathematics to nature; and as Knudsen has written, "the first dozen or so of Thomson's papers are all directly related to Fourier's work."[115] In the words of Peter Guthrie Tait, "Fourier made Thomson."[116]

Dr. James Thomson, William's father, became professor of mathematics at Glasgow in 1832, when William was seven. J. D. Forbes, who taught both William and Maxwell, was appointed to the chair of natural philosophy at nearby Edinburgh the next year; and Nichol joined the Glasgow faculty four years later, taking the chair of practical astronomy, just before William went up to Cambridge. By the age of fifteen, when he received a prize for "Essay on the Figure of the Earth," Thomson had already reached some degree of mathematical maturity. His first professional papers, dealing with the mathematics of Fourier's work and written under the pseudonym PQR, were published when he was sixteen in the *Cambridge Mathematical Journal.* In 1839 he read Laplace's *Mécanique céleste* at the urging of William Meickleham, professor of natural philosophy at Glasgow, and learned of Poisson's theory of electrostatics from Murphy's *Elementary Principles.*[117] Other strong influences on Thomson before he left Glasgow for Cambridge were his brother James, whose interest in engineering and love of machine models provided a strong stimulus for young William, and the 1840 meeting of the British Association for the Advancement of Science (BAAS) in Glasgow, in which both brothers participated.[118] Later BAAS meetings also had important meaning for Thomson.

It is notable that two of the works that most influenced young William Thomson—Laplace's *Mécanique céleste* and Fourier's *Analytical Theory of Heat*—belonged to very different traditions.[119] While neither Fourier nor later George Green invoked hypothetical imponderable fluids, the Laplacian theory of attraction as used by Poisson obtained the necessary differential equations from the forces between the fluids.[120] Jed Buchwald notes that neither Ampère, Arago, nor Fresnel would have simultaneously embraced the approaches of Fourier and Laplace; yet Thomson, even after he came to see the philosophical dichotomy, was able to assimilate both traditions. He soon rejected the idea of an electric fluid for reasons that Hendry would call dynamistic—that is, because it was unobservable and did not possess the attributes of a material substance: extension and impenetrability.[121] By 1850, he was forced to abandon the materiality of heat as well (see chapter 4). At the same time, Thomson's approach was characteristic of the sort of British mechanistic thinking that so distressed Duhem.[122] By the time he reached Cambridge, the mathematical reforms begun by Babbage and Herschel in the 1820s and 1830s had taken hold; and an English school of mathematics had evolved, oriented less toward mathematical rigor and more toward applying mathematics to problems of the natural world.[123] Its models were Fourier and Fresnel rather than Laplace and Poisson, and its strongest adherents in 1845 were Thomson and Stokes.[124]

Thomson went up to Cambridge in 1841 to begin the next stage of his mathematical education, which it was thought he could not get in Scotland. Here he came into contact with the Cambridge Mathematical Tripos system, at that time the route taken by all the best students in natural philosophy (see chapter 2). In this way, a bright young mathematician could quickly rise into the ranks of the elite of British mathematics and science; the Tripos was seen as the process through which the standards of the university and its standing in world were maintained. In the fall of 1841, Thomson was admitted to Peterhouse (St. Peter's College), where he spent three years preparing for the Tripos. The next year he began his studies with renowned private tutor William Hopkins, who also tutored George Gabriel Stokes, Tait, and Maxwell. To his surprise, Thomson emerged as second wrangler (as did Maxwell nine years later), but he ranked first in the Smith's prize examinations held later the same month.

Soon after his arrival in Cambridge, Thomson became editor of the *Cambridge Mathematical Journal* after the death of D. F. Gregory, and in his next three years of undergraduate study he published ten scientific articles.[125] By this time his predilection for applied, or physical, mathematics had already become established.[126] In 1843, in his second year at Cambridge, he published "On the Uniform Motion of Heat in Homogeneous Solid Bodies and Its Connection with the Mathematical Theory of Electricity," which expressed his belief in an analogy between electricity and heat (specifically between potential and temperature or between equipotential and isothermal surfaces). The paper was Thomson's first application of an approach to scientific thinking that became characteristic.[127] In it, he tackled the familiar problems of a surface-charge layer and the potential inside a conductor, using an analogy with thermal equilibrium to obtain, as the sole condition of the equilibrium of the electrical fluid on the surface of a body, that the force be normal to the surface.[128] A quarter-century later, in an 1870 review of Thomson's *Reprint of Papers on Electrostatics and Magnetism* in *Nature*, Maxwell commented that the paper "first introduced into mathematical science that idea of electrical action carried on by means of a continuous medium, which, though it had been announced by Faraday . . . had never been appreciated by other men of science, and was supposed by mathematicians to be inconsistent with the law of electrical action, as established by Coulomb and built on by Poisson."[129] As it turned out, Thomson's results, which shed light on both the problems of electrostatics and equipotential surfaces as well as the problems of heat flow, had already been anticipated by Green and Gauss.[130] Although he quickly became aware of Gauss's results in the same area, Thomson did not learn of Green's "Essay on the Application of Mathematical Analysis to the Theories of Electricity and Magnetism" (1828) until 1845, just as he was leaving Cambridge and only eight days after submitting his paper on the surface charge on conductors to the *Cambridge Philosophical Journal*.[131] It was a revelation to him to find that Green had taken virtually the same path.

After graduating from Cambridge, and at his father's urging, Thomson went to France to make contact with French mathematicians, in part to advance his career and learn the techniques of experimentation, demonstration, and lecturing that might help him obtain the chair of natural philosophy at Glasgow. He worked in Victor Regnault's laboratory in Paris, but his contacts with Cauchy, Chasles, Sturm, and especially Liouville were equally important. Thomson discussed George Green's work, which was generally unknown on the continent, and at Liouville's request summarized the experimental and theoretical aspects of electromagnetism in *Journal de Mathématiques Pures et Appliquées* (1845). He later published the same article in the *Cambridge and Dublin Mathematical Journal* as "On the Mathematical Theory of Electricity in Equilibrium."[132] It synthesized what Thomson had learned from Regnault and the experimental discoveries of Snow Harris and Faraday and the formalism of Fourier, Green, and Thomson himself.[133] In the same year, he read a paper, "On the Elementary Laws of Statical Electricity," at the June meeting of the BAAS, with Faraday in the audience. By the fall of 1846, Thomson, age twenty-two, obtained the chair of natural philosophy at Glasgow, succeeding William Meikleham, who had died in the spring. He occupied the post for fifty-three years.

As we noted earlier, young William Thomson found Faraday's way of "speaking" about electromagnetic phenomena unappealing; indeed, he vehemently rejected Faraday's ideas when he first heard about them.[134] Nor was he at first impressed by Faraday's experiments and the conclusions he drew from them, considering them "more devined than rigorously constituted by quantitative experiments."[135] But by 1845, after putting what he thought were Faraday's ideas into mathematical terms (consistent with Green's electrostatics), he felt less antagonistic.[136] He began corresponding with Faraday that year, and by the following year they had established a relationship of mutual respect. Smith and Wise's *Energy and Empire* details how Thomson surmounted difficulties in his study of electrostatics by using the analogy of the flow of heat and how he found it possible to translate Faraday's ideas into the same terms.[137]

Thomson's next steps were to develop a theory of dynamical electricity and then a theory that would encompass magnetism and magneto-electric, or electromagnetic, phenomena as well. The key to his progress on these fronts may have been his friendship with George Stokes, senior wrangler at Cambridge in 1841, but whom Thomson came to know only after leaving the university. The Thomson-Stokes correspondence is one of the treasures of nineteenth-century scientific communication.[138]

When Thomson learned of Faraday's discovery of the relationship between light and magnetism, he called it the "opening wedge" to a new way of understanding the relation among all forces.[139] Thomson's approach was essentially that of applied mathematics buttressed by a predilection for physical models; he was neither committed to the electrical fluid model of Coulomb or to Faraday's polarization of contiguous molecules.[140] His debt to Fourier is again evident in his indifference to hypothetical entities and predilection for a mathematical description.[141]

In his diary for 31 October 1846, he remarks: "I have this evening . . . after thinking of Faraday's discovery of the effects of magnetism on transparent bodies and polarised light, been recurring to my idea . . . which I had to give up, about magnetism and electricity being capable of representation by the straining of an elastic solid constituted in a peculiar way."[142] Within a month he had completed the paper "On a Mechanical Representation of Electric, Magnetic, and Galvanic Forces."[143] Relying heavily on an analogy with Stokes's work on elastic solids, this paper was written at the very beginning of their long and fruitful correspondence. Thomson adopted the analogy of torsion in an elastic body to replace Faraday's ideas of contiguous particles acting along curved lines of force.[144] In the paper's introduction, he wrote: "Mr Faraday, in the eleventh se-

PLATE 7. William Thomson (1824–1907), age twenty-two, Professor of Natural Philosophy at the University of Glasgow. *From Andrew Gray,* Lord Kelvin: An Account of His Scientific Life and Work.

ries of his *Experimental Researches in Electricity,* has set forth a theory of Electrostatical Induction, which suggests the idea that there may be a problem in the theory of elastic solids corresponding to every problem connected with the distribution of electricity on conductors."

The richness of the interconnections among the theories of heat, fluid flow, and elasticity and their usefulness as a source of analogy as Thomson worked

PLATE 8. George Gabriel Stokes (1819–1903) at the age of fifty-six. Portrait by C. H. Jeens. *From* Nature *12 (1875): facing 201.*

on electromagnetism are illustrated by the titles of the papers he published in the late 1840s: "On a Mechanical Representation of Electric, Magnetic, and Galvanic Forces" (1847), "Notes on Hydrodynamics" (1848), "Note on the Integration of the Equations of Equilibrium of an Elastic Solid" (1848), "On the Theory of Electromagnetic Induction" (1848), and "Notes on Hydrodynamics: On the Vis-viva of a Liquid in Motion" (1849). During the seven years following 1845, he devoted as much time to the theory of heat and its implications for the conservation of energy as to electromagnetism. To discuss these studies separately, as I necessarily must, does violence to his search for unification of forces and especially to the analogy between the mathematical treatment of heat and electromagnetism that was so important to him.

His correspondence and collaboration with Stokes, even though Stokes's interests were primarily in hydrodynamics and elasticity, was critically important.[145] "What an intimate relation there is," Stokes marvels in one letter, "between the mathematical considerations which are applicable to heat, fluid motion, and attraction!"[146] Again, the mathematical analogies proved fruitful, and Thomson received from Stokes mathematical results that he could apply to heat and electromagnetism. One of the products was "On a Mechanical Representation of Electric, Magnetic, and Galvanic Forces." Physical analogy played a crucial role in Thomson's thought, as exemplified by his letter to Stokes on 30 March 1847, asking for help with a model of a magnetized bar represented by fluid flow through a closed tube.[147] Thomson mentions thinking about heat, along with "electricity, magnetism, and especially galvanism; sometimes also water," and goes on to say that he "can strongly recommend heat for clearing the head on *all* such considerations."

Thomson also drew strongly on the theories of Ampère, Gauss, and Green, although he was unwilling to accept Ampère's molecular currents as real; rather, he regarded them as "mathematical metaphor."[148] His goal was to construct a comprehensive and unified dynamical theory, and the path to that theory lay in the evident analogies inherent in the similar mathematical structure of theories of heat, electricity, magnetism, light, and the elastic properties of solids. James Clerk Maxwell, whose program was strongly influenced by Thomson, called this method of analogical reasoning the method of "physical analogy."[149] In the end, Thomson was able to show that Faraday's results were all compatible with action-at-a-distance electrostatics and that one did not have to modify a single result in order to shift to the radically different field theory as a way of thinking about electrical phenomena.[150] Of his analogy between the propagation of electric and magnetic forces and the vibrations of an elastic solid, he wrote to Faraday in 1847: "What I have written is merely a sketch of the mathematical analogy. I did not venture even to hint at the possibility of making it the foundation of a physical theory of the propagation of electric and magnetic forces, which, if established at all, would express as a necessary result the connection between electrical and magnetic forces. . . . If such a theory could be discovered, it would also, when taken in connection with the undulatory theory of light, in all probability explain the effect of magnetism on polarized light."[151]

When Thomson assumed his duties at Glasgow in 1846, he was much concerned with analogies among electricity (static and dynamic), magnetism, elasticity, and fluid dynamics. But his interest in the theory of heat had not waned. Indeed, while he was making an analogy between the flow of heat from a hotter to a colder body (with consideration of the work done thereby) and the flow of electricity from a region of high to low potential, an intrinsic interest of the heat problem must have grown on him.[152] After all, this problem had provided the original model for his theory of electrostatics.

At the 1847 BAAS meeting in Oxford, Thomson talked about the method of images and about terrestrial magnetism.[153] His encounter with Joule at that meeting is described in chapter 4. Perhaps as a result, he sent "An Account of Carnot's Theory of the Motive Power of Heat, with Numerical Results Deduced from Regnault's Experiments on Steam" to *Transactions of the Royal Society of Edinburgh* instead of a memoir on images.[154]

As he matured (although he was only forty in 1864), Thomson devoted much of his time to measurement standards, the problem of the age of the Earth and the cooling of the Earth and the sun, radioactivity, hydrodynamics, and telegraphy. His love of the sea kept him interested in problems arising from navigation, including improving the magnetic compass and in tides and tide prediction; the Atlantic cable occupied him during the latter part of the 1850s. Nearly six years were devoted to *A Treatise on Natural Philosophy,* written with Peter Guthrie Tait for their students at Glasgow and Edinburgh and published in 1867.[155] Some have claimed that "it contributed more than any other book of the nineteenth century to make physics a specialized study."[156] It certainly provided a vehicle through which Thomson and Tait could broadcast their views and was especially important in promoting the energy principle and its importance in physics.

Thomson was offered the position of first Cavendish professor at Cambridge in 1871 but declined, and Maxwell took the post. After Maxwell's premature death in 1879, Thomson was asked again, and again declined. This time Rayleigh was chosen. Five years later he declined for a third time, and J. J. Thomson became Cavendish professor. In 1884 he delivered the famous Baltimore Lectures on the current state of physics to an audience at Johns Hopkins University including Rowland, Michelson, Morley, and Rayleigh; the lectures were finally published twenty years later. Thomson became Lord Kelvin of Largs in 1892 (the Kelvin coming from the river Kelvin in Scotland), in part a reward for his Unionist sympathies, and gave his last lecture at Glasgow in 1899 at age seventy-five.

But he continued to think much about the problem of the aether and corresponded with Stokes about the question of longitudinal waves in the aether. He argued with Fitzgerald about the nature of the aether (again the issue was the solid elastic aether, which Fitzgerald took to be a "mere analogy") and criticized the ideas of Maxwell, whom he considered too much of a positivist.[157] Fitzgerald played an important role in the genesis of the special theory of relativity, which in due course eliminated aether from the vocabulary of physicists.[158] When Kelvin died in 1907, he had for some time been regarded as an obstructionist, especially by Ernest Rutherford and J. J. Thomson. In the Baltimore Lectures, for example, he declared that "the wave theory of light seems to be that branch of physics which is most in want"; and on the occasion of the celebration of his fiftieth year as professor of natural philosophy at Glasgow (1896), he asserted that "I know no more of electric and magnetic force, or of the relation between ether, electricity, and ponderable matter, . . . than I knew and tried to teach to my students of natural philosophy fifty years ago in my first session as professor"—this more than two decades after Maxwell's *Treatise* (1873).[159]

James Clerk Maxwell

Arguably, there is only one physicist whose name is likely to be mentioned in the same breath as Newton's and Einstein's: James Clerk Maxwell (1831–79).[160] Though born forty years after Faraday, he survived him by only twelve. Yet almost without thinking, we now call classical electromagnetic theory "Maxwell's theory." By the time of his death in 1879 at the age of forty-eight, he had revolutionized, if not created, two of the major areas of classical physics: electromagnetic theory and the kinetic theory of gases. The direct heritage of his work was the statistical mechanics of Boltzmann, Gibbs, and Einstein, one of the great pillars of quantum theory, and relativity theory, which grew directly from his theory of electromagnetism.

On the surface, the contrast between Faraday and Maxwell could hardly have been greater. While Faraday had little education and no mathematical training, Maxwell was born into landed gentry, educated at Edinburgh and Cambridge, and became a product of the Tripos system. Faraday was the supreme experimentalist, if also a remarkable intuitive theorist. Although Maxwell, too, was a skilled experimenter (note especially his work on color), he is renowned for his theoretical physics—much like Newton in that regard.

Although the Scottish Enlightenment was past its prime in the 1830s and 1840s, when Maxwell was being raised by his widowed father, the boy profited from a climate that still attached great value to intellectual gifts and learning. During his schoolboy days at Edinburgh Academy, he began a lifelong relationship with Peter Guthrie Tait, a friendship that continued at the University of Edinburgh and Cambridge and lasted until Maxwell's death.[161] He attended Edinburgh for three years and then went up to Cambridge (initially Peterhouse, Thomson's college, and then Trinity) in the autumn of 1850.[162] At Edinburgh, his wide-ranging interests were stimulated by William Hamilton's moral philosophy (which was Kantian but infused with Scottish commonsense philosophy), James Forbes's natural philosophy, and the mathematics of Phillip Kelland.[163] Among other things, Forbes introduced him to probability theory.[164]

By the middle of the nineteenth century, the estrangement between moral and natural philosophy was well developed in Britain. Nevertheless, while William Hamilton had little use for mathematics, he had an impact on Maxwell at Edinburgh that was second only to Forbes's. At Cambridge, William Whewell became an even greater influence on Maxwell's philosophical development, especially through his *History of the Inductive Sciences* and *Philosophy of the Inductive Sciences*. One could say that he was responsible for Maxwell's remaining a natural philosopher in the fullest sense of the term, for Maxwell was deeply thoughtful about metaphysical and epistemological questions. The young Maxwell

PLATE 9. James Clerk Maxwell (1831–79). *From W. D. Niven, ed.,* The Scientific Papers of James Clerk Maxwell *(Cambridge: Cambridge University Press, 1890).*

was also strongly influenced by Thomson, seven years his senior, who had already assumed the chair of natural philosophy at Glasgow in November 1846, when Maxwell was sixteen, and whom the young man first met four years later. There was a rich correspondence between the two extraordinary scientists.[165]

At Cambridge Maxwell attended lectures by G. G. Stokes, Lucasian professor of mathematics, and was prepared for the Tripos by renowned coach William Hopkins. He finished as second wrangler to E. J. Routh in the 1854 Tripos and shared with Routh the more meaningful Smith's Prize.

In 1855, twenty-four-year-old Maxwell became a Fellow of Trinity College and read his first paper on electromagnetism to the Cambridge Philosophical Society. He had learned about the subject from Thomson's papers and through correspondence with him. Over the next decade, Maxwell developed his theory with frequent stimulation from Thomson, especially from the latter's 1856 paper "Dynamical Illustrations of the Magnetic and the Helicoidal Rotary Effects of Transparent Bodies on Polarized Light." In a sense, this represented a changing of the guard; for by 1864 Maxwell presented his "Dynamical Theory of the Electromagnetic Field" to the Royal Society, thus completing a program that Thomson had initiated more than two decades earlier. Thomson introduced Maxwell to Faraday's work, or at least convinced Maxwell of the value of studying Faraday, a fact that Maxwell acknowledged in his *Treatise on Electricity and Magnetism:*

> I was aware that there was supposed to be a difference between Faraday's way of conceiving phenomena and that of the mathematicians, so that neither he nor they were satisfied with each other's language. I had also this conviction that this discrepancy did not arise from either party being wrong. I was first convinced of this by Sir William Thomson, to whose advice and assistance . . . I owe most of what I have learned on the subject.
>
> As I proceeded with the study of Faraday, I perceived that his method of conceiving the phenomena was also a mathematical one, though not exhibited in the conventional form of mathematical symbols. I also found that these methods were capable of being expressed in the ordinary mathematical forms, and thus compared with those of the professional mathematicians.[166]

In the winter of 1855–56 Maxwell read to the Cambridge Philosophical Society his first great scientific paper on electromagnetism, "On Faraday's Lines of Force," which he had been working on for at least two years.[167] In this paper, Faraday's field theory had, for the first time, a mathematical foundation (although Thomson had already provided a mathematical basis for much of Faraday's work in electrostatics). Maxwell wrote that his own methods "are generally those suggested by the processes of reasoning which are found in the researches of Faraday, and which, though they have been interpreted mathematically by Prof. Thomson and others, are very generally supposed to be of an indefinite and unmathematical character. . . . By the method which I adopt, I hope to render it evident that . . . the limit of my design is to shew how, by a strict application of the ideas and methods of Faraday, the connexion of the very different orders of phenomena which he has discovered may be clearly placed before the mathematical mind."[168]

Maxwell began the paper by adopting an analogy with heat conduction, which, of course, Thomson had done before him.[169] Then he developed a further analogy between lines of force and the motion through a porous medium of an incompressible fluid in tubes. With this foundation, he discussed permanent magnets, dielectrics and paramagnets, and electromagnetic induction. In the mathematical parts of the paper, he relied heavily on results obtained by

Thomson and others, including Green and even Helmholtz in "On the Conservation of Force" (see chapter 5).[170] Maxwell used the symmetry that exists between electrical and magnetic phenomena ("since the mathematical laws of magnetism are identical with those of electricity") to obtain differential and integral expressions for various quantities.[171] These quantities included the spatial variation of the current as a function of the electric density (charge density), which is essentially the continuity equation; the relation between the magnetic intensity of a closed curve (an integral statement, but he wrote down the differential form—the spatial variation of that intensity); and the electric quantity passing through the curve.[172] In the absence of current, the change in the magnetic tension, or potential, was given in terms of components of the magnetic intensity (in modern terms, the magnetic intensity given as the gradient of the magnetic potential: $H = \nabla V_m$). He then considered work and energy, arriving at an expression for the work of sources on a current as related to the time rate of change of a quantity that involves components of the electrotonic intensity multiplied by corresponding components of the current. This resulted in an expression for the electric intensity or electromotive force as the time derivative of the electrotonic intensity. Maxwell then stated six laws summarizing the theory of the electrotonic state, which he said did not contain "even the shadow of a true physical theory." It was "a temporary instrument of research" that "does not, even in appearance, *account* for anything."[173] He went on to say that Weber's theory is an example of what a physical theory should be and gave additional credit to Thomson for having stimulated his interest in the mathematical formulation given in the paper.[174] The remainder of the paper was devoted to application of Thomson's method of images to a number of problems in magnetostatics.[175]

It is interesting to note how Maxwell treated Faraday's electrotonic state:

> In this outline of Faraday's electrical theories, as they appear from a mathematical point of view, I can do no more than simply state the mathematical methods by which I believe that electrical phenomena can be best comprehended and reduced to calculation, and my aim has been to present the mathematical ideas to the mind in an embodied form, as systems of lines or surfaces, and not as mere symbols, which neither convey the same ideas, nor readily adapt themselves to the phenomena to be explained. The idea of the electro-tonic state, however, has not yet presented itself to my mind in such a form that its nature and properties may be clearly explained without reference to mere symbols, and therefore I propose in the following investigation to use symbols freely, and to take for granted the ordinary mathematical operations. By a careful study of the laws of elastic solids and of the motions of viscous fluids, I hope to discover a method of forming a mechanical conception of this electro-tonic state adapted to general reasoning.[176]

In 1856, Maxwell's father died. That same year Maxwell took the chair of natural philosophy at Marischal College, University of Aberdeen, which he lost four years later when the two colleges of Aberdeen merged. Two years later, after the retirement of James Forbes, Edinburgh passed him over in favor of Tait for the chair of natural philosophy (Maxwell was not regarded as a remarkable teacher) and took the chair of physics and astronomy at King's College in London.[177] Soon after finishing his first great paper on the kinetic theory of gases (1859), he returned to the problem of electromagnetism, which he had put aside in early 1856.

As Maxwell approached thirty, his life was complicated by his marriage in 1858 and a near fatal case of smallpox acquired when he visited Glenlair in 1860.[178] He was also burdened by the demands of his new post at King's College, which he had assumed in the fall of that year. In 1861, the year he met Faraday, Maxwell was made a Fellow of the Royal Society and began writing the first of his two great works on electromagnetism, "On Physical Lines of Force."[179] In this paper he introduced the model of molecular vortices, which provided a basis for and a visualization of the mathematics. During the next five years he published "A Dynamical Theory of the Electromagnetic Field" (1864) and "On the Dynamical Theory of Gases" (1866).[180] These three celebrated papers established Maxwell's field theory of electromagnetism and enormously advanced the kinetic theory of gases.

Although Maxwell's use of analogy was perhaps his most powerful weapon in arriving at a description of a problem, his commitment to the resulting mechanical models depended on the problem at hand and on the way in which his treatment and understanding of a problem matured. This particular issue has been the source of much controversy among Maxwell scholars. Some may prefer the Maxwell of the *Treatise* (or even of modern textbooks), in which the wheels and vortices have largely disappeared, and thus claim that he was never committed to the models' reality. In truth, however, Maxwell did believe in the model, particularly in the case of "On Physical Lines of Force," and in the molecular vortex model.[181] Siegel shows that, while Maxwell considered the fluid analogy he used in "On Faraday's Lines of Force" to be imaginary, his studies of the rotatory nature of magnetism required the existence of a rotatory cause that he found in Thomson's molecular vortex model. Thomson was not the only scientist tempted by the explanatory power of vortices; Hermann von Helmholtz and W.J.M. Rankine had employed them as well.[182] Eventually, by the late 1860s, Maxwell's model of wheels and vortices had virtually disappeared, and only the mathematical theory remained.[183] Referring to the mathematical theory we call "Maxwell's equations" (which appear in the two great papers mentioned in this paragraph), Ludwig Boltzmann (whose statistical mechanics generalized Maxwell's kinetic theory) asked if "it was God who wrote these lines"?[184]

Maxwell and Electromagnetism

Maxwell's stimulus in the years 1859–60 seems to have come from Thomson's attempts to understand some of Faraday's experiments (including the magneto-optic, or "Faraday," effect) in terms of the molecular motions that underlay the dynamical theory of heat, which Thomson had just come to embrace, and from the conservation and interconvertibility of energy, which had become established only in the previous decade. Thomson had helped found the modern science of thermodynamics in the early 1850s and now saw the possibility of relating electromagnetic induction and the effects of magnetic fields on light propagating through matter to "the matter of which the motions constitute heat."

The first two parts of Maxwell's "On Physical Lines of Force" were called "The Theory of Molecular Vortices Applied to Magnetic Phenomena" and "The Theory of Molecular Vortices Applied to Electric Currents." Using Thomson's "Mechanical Representation of Electric, Magnetic, and Galvanic Forces" as a starting point, Maxwell developed a theory of electromagnetism based on "an imaginary system of molecular vortices."[185] In a note appended to part 2, he commented that Helmholtz had written a paper on fluid motion that showed that the lines of fluid motion behave in a way similar to those of magnetic force. Maxwell remarked: "This is an additional instance of a *physical analogy,* the investigation of which may illustrate both electro-magnetism and hydrodynamics." (In "On Faraday's Lines of Force" Maxwell had explained what he intended by the term *physical analogy.*)[186]

"On Physical Lines of Force" was a highly speculative attempt to find a mechanical model of a medium that mediated all electric and magnetic phenomena. In a passage that provides deep insight into how he used such models, Maxwell admitted:

> The conception of a particle having its motion connected with that of a vortex by perfect rolling contact may appear somewhat awkward. I do not bring it forward as a mode of connexion existing in nature, or even as that which I would willingly assent to as an electrical hypothesis. It is, however, a mode of connexion which is mechanically conceivable, and easily investigated, and serves to bring out the actual mechanical connexions between the known electro-magnetic phenomena; so that I venture to say that any one who understands the provisional and temporary character of this hypothesis, will find himself rather helped than hindered by it in his search after the true interpretation of the phenomena.[187]

Maxwell is here referring to the device of *idle wheels,* small particles that couple adjacent vortices (see figure 3–4). It seems clear that, while he made no claim for the reality of these objects, he did not entertain similar doubts about the rotatory motion that was represented by the molecular vortices.[188]

The second and third parts of the paper were published in 1862 and applied the same theory to electrostatics and the effect of magnetism on polarized light. In part 3, Maxwell undertook to calculate the speed of propagation of transverse waves in the elastic medium in which the cellular vortical motion took place. Noting that the undulatory theory of light "requires us to admit this kind

Fig: 2.

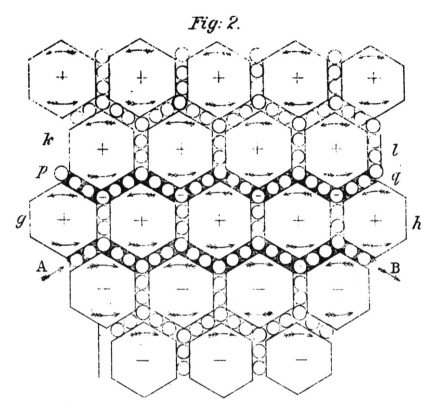

FIGURE 3–4. Maxwell's mechanical model of hexagonal molecular vortices with rolling idle wheels between them. *From J. C. Maxwell, "On Physical Lines of Force," in W. D. Niven, ed.,* The Scientific Papers of James Clerk Maxwell *(Cambridge: Cambridge University Press, 1890), facing 488.*

of elasticity in the luminiferous medium, in order to account for transverse vibrations," he added that "we need not then be surprised if the magneto-electric medium possesses the same property."[189] His calculation of the speed of propagation, while not entirely correct, yielded a value consistent with the speed of light.[190] He concluded that "we can scarcely avoid the inference that *light consists in the transverse undulation of the same medium which is the cause of electric and magnetic phenomena.*"[191] Concealed in this modest sentence was his announcement of one of the most profound discoveries of the nineteenth century: the possibility of the existence of electromagnetic waves. Maxwell elaborated upon this discovery two years later.

However mechanistic the vortical motions of "On Physical Lines of Force" are, we must keep in mind that Maxwell was not totally a mechanist and that the vortex model was a heuristic device entirely consistent with his ideas about physical analogy. Later he wrote: "[The] theory of molecular vortices applied to magnetism, electricity, &c . . . though rough and clumsy compared with the realities of nature, may have served its turn as a provisional hypothesis."[192] One of the most remarkable aspects of the great paper "A Dynamical Theory of the Electromagnetic Field," which he read to the Royal Society in December 1864, is that the unwieldy mechanical model has mostly disappeared: Only the equations remain (like the smile of the Cheshire cat). Nevertheless, Maxwell's route to the electromagnetic theory of light was inseparable from the vortex model, and it was this model that allowed him to unify electromagnetism and optics. One might suppose that Maxwell first discovered that his equations could be combined to yield a wave equation and then set about to understand how to model it mechanically, but this was apparently not the case.[193]

Maxwell finally rejected Weber's action-at-a-distance theories and committed himself fully to the aether, a medium filling space and permeating bodies, and which transmitted not only transmitted light but electric and magnetic forces as well. In "Dynamical Theory of the Electromagnetic Field," he explicitly declared that electricity and magnetism are in some sense one electromagnetic field. He made full use of the energy principle, by this time well over a decade old, and ultimately arrived at what he called the "General Equations of the Electromagnetic Field," in part 3 of the paper. Because he did not use vector notation, an expression such as $\int \mathbf{E} \cdot \mathbf{dl}$ was written $\int (P\ dx/ds + Q\ dy/ds + R\ dz/ds)ds$, where P, Q, and R were components of the electromotive force.[194] Surely the most revolutionary and controversial (even notorious) feature of this paper was the introduction of the displacement current, which, according to Maxwell, resulted from the change in time of the electric displacement and contributes to the spatial change of the magnetic field.[195] Yet today the paper is remembered as much for the introduction of electromagnetic waves as for the displacement current.[196] Maxwell wrote: "it seems we have strong reason to conclude that light itself (including radiant heat, and other radiations if any) is an electromagnetic disturbance in the form of waves propagating through the electromagnetic field according to electromagnetic laws."[197] In other words, "light consists in the transverse undulations of the same medium which is the cause of electric and magnetic phenomena."[198]

In early 1865, tiring of the demands of his academic position, Maxwell retreated from London to pastoral Glenlair. The nearly six years he spent there were fruitful but perhaps not entirely fulfilling; for when he was offered the chair of experimental physics at Cambridge (after Thomson and Helmholtz declined it), he accepted the position, if somewhat reluctantly.[199] He became the first Cavendish professor and presided over the creation of the Cavendish Laboratory, funded by William Cavendish, Duke of Devonshire, whose great-uncle had been Henry Cavendish.[200] Much of Maxwell's time thereafter was devoted to teaching, working on the laboratory, and editing the Cavendish papers.[201] It was a measure of his selflessness that he undertook this task of editor. Knowing now that he had only a few years to live, we might wish Maxwell had not assumed it; for he died within a month of the book's publication.

In 1873 Maxwell's great *Treatise on Electricity and Magnetism* was published. It is one of the first great broad applications of the energy concept to physics and the first application of William Rowan Hamilton's new dynamics.[202] Maxwell had worked on the *Treatise* at Glenlair and, perhaps because he had finished it, could again face the demands of a busy professional and social life. Within four years, however, he showed the first signs of his fatal illness (cancer of the stomach); and he passed serenely from this world in 1879.[203]

Electromagnetism in Germany before Helmholtz

In 1832 the great German mathematician Carl Friedrich Gauss (1777–1855), having read Biot's *Traité de physique* and Poisson's 1812 memoirs on the distribution of electricity on the surface of a conductor, showed that the volume distribution of magnetic sources could be replaced by a surface distribution. During the next few years he devoted considerable effort to the problem of terrestrial magnetism, publishing a large paper about it in 1838. The following year he generalized these results in a work in which he developed potential theory that could be applied to magnetism, electrostatics, and gravitation.[204] As noted earlier, he had anticipated Thomson in these discoveries but, in turn, had himself been preceded by Green, although the discoveries were independent. Typical of the great but neurotic Gauss, much of his work on electromagnetism was never published. Nor did he learn of Green's 1828 work himself until Thomson revived it in 1846 and it appeared in Crelle's *Journal für de Reine und Angewandte Mathematik* in 1850–54.[205] Although many of us today know Gauss as one of the premier mathematicians of his age, he also began the move toward more precise instrumentation with the construction of a magnetometer in 1833, two years after Weber came to Göttingen.

Wilhelm Weber (1804–91), an older contemporary of Thomson and Maxwell, did theoretical and experimental work on acoustics, electricity, and magnetism, concentrating on the latter subjects after he became ordinary professor of physics at Göttingen, where he worked with Gauss. He continued his work even after he was driven away from the university for political reasons. Working outside the field tradition of Faraday, Weber (1846) treated electromagnetic phenomena in terms of an inverse-square force acting at a distance, mediated by the aether.[206] As we have mentioned, Maxwell admired Weber's theory in spite of its lack of intersection with his own ideas.

George Simon Ohm, who had even greater problems than Weber did in securing a position that would support his research, also was as much a theorist as an experimenter. Like Thomson, Ohm used the analogy between Fourier's heat conduction and the conduction of electricity. He claimed in his *Galvanic Circuit* that mathematics had proven its utility in a new field of physics from which it had been previously excluded.[207]

Beginning in 1832 with a paper on electromagnetic induction, Faraday's work was regularly translated in *Annalen der Physik*. As a result, Gauss and Weber immediately began a series of experiments on induction, Ohm's law, the velocity of currents, and so on. But these experiments ended when Weber was dismissed.[208]

In the 1840s Franz Neumann, motivated by Fresnel's wave theory of light, extended the work of Navier, Poisson, and Cauchy, especially on the problem of wave propagation in crystals, and double refraction.[209] He also published work on Ampère's law and electrodynamics, an area in which both he and Weber were working, at Königsberg and Leipzig, respectively (1845–46).[210] At Königsberg, Neumann became Kirchhoff's mentor; but Kirchhoff went to Breslau in 1850 and eventually to Heidelberg.

Gauss was succeeded at Göttingen first by G. L. Dirichlet and then by Bernhard Riemann, both men continuing a tradition of strong interaction between mathematics and physics. Each was interested in the theory of partial differential equations and potential theory. In 1858 Riemann presented a paper to the Royal Society of Göttingen in which he proposed adding a d^2V/dt^2 term to Poisson's equation, which resulted in a propagation of the potential V through space with a velocity near the known speed of light. Although this took place six years before Maxwell's paper on electromagnetic waves, it was not published until 1867, the year after Riemann's death.[211]

Hermann von Helmholtz

Hermann von Helmholtz (1821–94) was in many ways Maxwell's German counterpart. A physiologist as well as a physicist, he was also a student of color and an opponent of Weber's action-at-a-distance theory of electromagnetism. In 1838, to advance his studies in physics, he had to take advantage of an opportunity for free medical study at the University of Berlin (Royal Friedrich-Wilhelm Institute), which obliged him to serve as an army physician. He took his doctorate in 1842, having written his dissertation on the physiology of nerves under Johannes Müller, and received his M.D. the same year.[212] His circle consisted mostly of Müller's students, who were interested in reducing life processes to physics and chemistry; Helmholtz himself was especially interested in physiological heat.[213] By early 1847, he had completed a draft of his seminal work on the conservation of force, which he read to the Berlin Physical Society on 23 July (see chapter 5). As Jungnickel and McCormmach remark, this paper is one of the most remarkable first publications in the history of physics.

Helmholtz was professor of physiology at Königsberg for six years. In 1855 he moved to Bonn as professor of anatomy and physiology; and in 1858, at Kirchhoff's urging, he assumed the professorship of physiology at Heidelberg. At Bonn Helmholtz worked on physiological optics and acoustics, but in 1857 he published an important work in mathematical physics: "On the Integrals of the Hydrodynamic Equations Which Express Vortex-Motion."[214] He made a number of trips to England, including one in 1853 when he managed to see Faraday and another ten years later when he visited Thomson in Glasgow. That same year he met Joule in Manchester, describing him as the "chief discover of the conservation of energy."[215] In the late 1860s Helmholtz published several profound works on the foundations of geometry. When Magnus died in 1870, the opportunity arose for Helmholtz to assume the chair of physics at Berlin. The move was a watershed for him, even though he was now fifty years old, for it marked his turn away from physiology to physics.[216] Just at this point, Maxwell's papers on electromagnetic theory were translated into German, and Helmholtz spent the next five years developing his own theory of electrodynamics and comparing it with Maxwell's.[217] Much of that work was experimental, as were subsequent studies on the galvanic pile and the nature of electricity, which led him to the conclusion that electricity was corpuscular. He expressed that view in the 1881 Faraday Lecture.

By the time of his death in 1894, Helmholtz had been for thirty years an influential figure in German and international science. It is no coincidence that during his lifetime German science gained ascendancy, although the contributions of Rudolph Clausius (who died in 1888) and Kirchhoff were equally influential. The movement toward establishment of the first professorships of theoretical physics in German universities had begun; and Max Planck became one of the first scientists to hold such a position, at Kiel in 1885.

The reception of Maxwell's theory was far from effusive; Thomson often expressed his misgivings, and many continental (especially German) physicists were still committed to action-at-a-distance electrodynamics.[218] But Oliver Heaviside (1850–1925), as much an engineer as a physicist, wrote a series of influential and practical articles based on the Maxwell theory, aiding its acceptance in Britain (recasting it in the familiar form of "Maxwell's equations").[219] To Heinrich Hertz (1857–94), "Maxwell's theory *is* Maxwell's system of equations" (my emphasis).[220] By the end of the century they would be expressed most commonly in the language of vector calculus. Hertz played an important role in the reception of Maxwell's theory, especially in Germany, through his verification of Maxwell's prediction of electromagnetic waves.[221] Yet even though he was a protégé of Helmholtz, Hertz represents a rare example of unguided experimentation—to such an extent that, when he did generate electromagnetic waves, he believed that his results contradicted Maxwell's theory.[222] Gustav Kirchhoff, who made profound contributions to so many areas of physics in the later nineteenth century, formulated a theory of diffraction of light in the 1880s that put the Huygens-Fresnel theory on firm mathematical ground.[223] This formed the basis for widespread application to the propagation and scattering of electromagnetic and sound waves.

At the centenary of Maxwell's birth in 1931, Albert Einstein said: "Since Maxwell's time, Physical Reality has been thought of as represented by continuous fields, governed by partial differential equations, and not capable of any mechanical explanation."[224] Maxwell's equations described the propagation of electromagnetic waves through the aether with a fixed velocity determined by the properties of the aether. The theory incorporated a wide range of well-known electromagnetic phenomena and predicted the existence of other phenomena, including electromagnetic radiation. A century later, it is easy to see what had been accomplished by 1873, yet Maxwell's achievement was not lost on his contemporaries. The battle that had raged over the wave versus particle nature of light for nearly two centuries had already been largely settled by Young and Fresnel in favor of the wave, and in 1890 Hertz wrote, "For all practical purposes, the wave theory of light is a certainty."[225] Still, there was no unanimity on the question of whether light and the newly proposed and detected electromagnetic waves were one. Ironically, although consensus was essentially achieved by the turn of the twentieth century, within five years that simple view was shaken anew; and within a quarter-century light was understood in a substantially different way. This new understanding of the dual nature of light did not, however, destroy the unification accomplished by Faraday and Maxwell. The battle moved on to other forms of electromagnetic energy, especially X rays.

CHAPTER 4

Heat and Thermodynamics

$\mathcal{T}$he earliest idea of heat was undoubtedly of a substance that could be moved from place to place, a view embodied in the *caloric* principle of the late eighteenth century. The contrary view of heat as motion also has an ancient, if somewhat dubious, pedigree, beginning with the speculations of the ancient Greek atomists and their successors Epicurus and Lucretius. But it was only in the seventeenth century that the first attempts at a quantitative theory of the effects of heat on gases (mainly air) were made; and thermometry, one of the essential bedrocks of the experimental science of heat, dates from the eighteenth century.[1] In spite of earlier speculation, the concept of heat as the result of internal motion, usually the agitation of unseen small parts of bodies, was also essentially a seventeenth-century idea founded upon the atomism of Gassendi and others and advanced by Galileo, Bacon, Descartes, and Huygens.[2] Most important in the reception of this idea was Isaac Newton, who in the queries of book 3 of his *Opticks* attributed heat to the motion of the small parts of bodies.[3] John Locke echoed Newton, saying that "heat is a very brisk agitation of the insensible parts of the object . . . so that what in our sensation is *heat,* in the object is nothing but *motion.*"[4] Robert Boyle clarified that "heat seems principally to consist in that mechanical property of matter we call motion."[5]

This early form of the dynamical theory of heat rested on the insecure foundation of atomism—that is, the postulate of atoms.[6] While this conception of heat was not universally embraced in the late seventeenth and early eighteenth centuries, it nonetheless had wide support. Once it became established, it was not seriously challenged until almost a century later, when Joseph Black discovered latent heat in 1757.[7] His discovery, made at a time when imponderable fluids were widely discussed as providing an underlying explanation for a range of physical phenomena, led to the ascendancy of the caloric theory of heat. Thus, the theory that heat was an imponderable fluid—the *caloric*—that passed from one body to another was actually new and controversial in 1783 when Lavoisier and Laplace equivocated on the issue of the nature of heat: "Other physicists think that heat is only the result of indetectable movements of the molecules of matter. . . . we shall not decide between the two . . . hypotheses; several phenomena appear to be favorable to the latter; such as, for example, that of the heat which the rubbing together of two bodies produces."[8] In short, although the atomic theory of heat prevailed from the early seventeenth century until the late eighteenth century, it was superseded for nearly a half-century by the caloric theory before once again being resurrected in the 1820s.

Much of the mathematical structure of classical thermodynamics (what we now call the theory of heat) had to wait for the discovery of energy conservation near the middle of the nineteenth century. Thus, it may have been fortunate that a good part of that structure did not depend on whether heat was an imponderable fluid or a manifestation of motions of the small constituents of matter, which the observations of Benjamin Thompson (Count Rumford) and Humphrey Davy seemed to show about 1800. Nor, therefore, did it depend on the existence of atoms. Indeed, all of classical physics developed vigorously throughout the nineteenth century while the atomic theory was still in dispute.

Thermodynamics has a peculiar ontological status. Founded on the practical applications of heat, especially the steam engine, during the last half of the eighteenth century, and advanced by experiments from Boyle to Joule, it has nonetheless been given an axiomatic formulation in "rational thermodynamics" in which its gritty beginnings are almost entirely hidden.[9] Maxwell may not have had that in mind when he declared that it was "a science with secure foundations, clear definitions, and distinct boundaries."[10] But many years later Albert Einstein expressed a similar view of thermodynamics as a perfect example of what a physical theory should be—founded on a limited number of axioms from which the entire theory could be elaborated mathematically yet applicable to a wide range of phenomena.

The Early Gas Laws of Boyle, Mariotte, and Gay-Lussac

Although thermodynamics became a mature science in the middle of the nineteenth century, its origins lie in the seventeenth century with the speculations of Galileo and the careful experiments of Evangelista Torricelli (1608–47). Torricelli showed that the weight of air caused a column of water or mercury to rise, thus dismissing Galileo's force of the vacuum and laying the foundation for the practical barometer. He also first detected changes in the ambient air pressure. At almost the same time, Blaise Pascal (1623–62), Marin Mersenne (1588–1648), Gilles Roberval (1602–75), and Valeriano Magni (1586–1661), carried out similar investigations.[11] From this early work emerged the concept of air pressure as analogous to the pressure exerted on an object immersed in a fluid—for example, water. Otto von Guericke (1602–86) demonstrated the efficacy of the air pump and the power of the vacuum in his 1654 experiment with the evacuated Magdeburg hemispheres, which could not be pulled apart by a team of eight horses. Guillaume Amontons (1663–1705) investigated the increase in a gas's pressure with temperature and discovered that, when water boils, it does so at constant temperature. This discovery led, nearly a century later, to the idea of latent heat. Cardwell suggests that the significance of the discovery was not lost on Amontons and that he realized its importance for thermometry: that it provided a fixed point on the temperature scale.[12]

In 1660–61, two Englishmen, Henry Power (1623–68) and Richard Towneley (1629–1707), showed that at constant temperature the pressure and volume of a gas are inversely related: that is, pV = constant. Their work was known to Robert Boyle (1627–91), who mentioned it in the second edition of *New Experiments Physico-Mechanical, Touching the Spring of the Air, and Its Effects* (1662), acknowledging Towneley but not Power. Robert Hooke mentioned their work in *Micrographia,* which in turn Newton acknowledged.[13] Edmé Mariotte (ca. 1620–84) published the law in *Essai de la nature de l'air* (1679) without giving credit to Boyle, of whose work he was apparently aware.[14] The law is known by Boyle's name in English and by Mariotte's in France but ought to be known as Power and Towneley's.

The law known variously as Charles's law and the law of Gay-Lussac, which describes the thermal expansion of gases, was first noted by J.A.C. Charles (1746–1823) about 1787.[15] It was never published by him, however, and his understanding was incomplete.[16] The measurements of Joseph Louis Gay-Lussac (1778–1850), made in 1801–2 with the encouragement of Laplace and Berthollet, were more precise and systematic.[17] They showed not only the equal expansion of gases for equal temperature changes but that, for every degree Celsius of temperature rise, the fractional change in volume for all gases is 1/266.66.[18] John Dalton of Manchester (1766–1844) was only slightly anticipated by Gay-Lussac in these measurements.

Jacob Hermann (1678–1733) and Daniel Bernoulli (1700–82) were the first to attempt to understand the known gas laws in terms of the motion of the constituent atoms of the gas.[19] The ultimate result of this line of thought was the kinetic theory of gases, developed in the 1860s by Krönig, Clausius, and Maxwell (see chapter 7). In *Hydrodynamica* (1738), Bernoulli introduced the idea of an interval between collisions (analogous to the modern mean free path, defined more than a century later by Clausius) and derived the Boyle-Mariotte law.[20]

The Caloric Theory

In the late eighteenth century the phlogiston theory of combustion enjoyed brief popularity, and the important and influential Claude-Louis Berthollet (1748–1822) advocated it into the 1780s. But by 1772 the great French chemist Antoine Laurent Lavoisier (1743–94) showed that the theory was untenable. Nonetheless, the idea that heat is a substance that can be passed from one body to another was not an isolated or singular hypothesis: Imponderable fluids were inherent to the Laplacian mechanistic physics of the late eighteenth and early nineteenth centuries, and many thermal phenomena seemed to be most easily understood in terms of an imponderable fluid. Thus, the caloric theory soon strode the stage from which the phlogiston theory had been driven.

Joseph Black (1728–99) taught medicine and chemistry at Glasgow and Edinburgh after receiving his M.D. at Edinburgh in 1754. He was a contemporary of Adam Smith and David Hume at Glasgow and Edinburgh and a friend and advocate of James Hutton, whose revolutionary uniformitarian ideas in geology helped transform that science. Arguing from the experiments of George Martine, Black concluded that water and mercury have different heat capacities: Water rises in temperature less rapidly than mercury does. Black's most famous contribution, however, was his notion of latent heat: the heat absorbed by a body when changing state (that is, melting or boiling) while not changing temperature.[21] Black noted that, without the existence of latent heat, only a small rise in temperature would catastrophically melt a snowfield.[22] J. C. Wilcke's 1772 and 1781 discoveries of the same phenomenon were entirely independent of Black's, who never published his results.[23] It was this discovery of latent heat, which at the time seemed impossible to reconcile with the dynamical theory of heat, that gave impetus to the caloric theory. Theories of the materiality of heat (or fire) were also aided by existing theories of an imponderable electrical fluid.[24]

The nature of heat—whether it was an imponderable fluid that passed from one object to another and resulted in the heating or cooling of a body or whether it was simply a manifestation of the motion of the particles internal to a body— was by no means the most vital question by the 1780s. Nevertheless, it was a critical one. Newton's authority was important and persuasive (as in the more general question of atomism) but not decisive. The hypothesis that heat was a substance, embodied in the caloric theory of Lavoisier, Laplace, and Gay-Lussac, had a direct appeal that gave it an advantage over the less concrete theory of vibrations of small, unseen parts of a body. Many experiments were easily interpreted in terms of this paradigm of the materiality of heat. Members of the Society of Arcueil, especially Laplace and Poisson, strongly championed the caloric hypothesis and, more generally, the importance of imponderable fluids in explaining electrical and thermal phenomena. In the end, it is likely that the decline of Laplace's authority after his death in 1827 was a major factor in growing disenchantment with the caloric theory.

Quite apart from the ultimate success or failure of the caloric theory, the 1783 memoir by Lavoisier and Laplace organized and clarified the confused science of heat; and their synthesis of theoretical and experimental results formed the basis for further development of thermodynamics, whether based on caloric or *vis viva* theories.[25] Because Laplace had not yet converted to the caloric theory, their memoir straddles the fence on the issue. But by 1789 Lavoisier's *Elementary Treatise of Chemistry* clearly set forth the new chemistry and the material theory of heat, and both Lavoisier and Berthollet assumed that the caloric was an element.[26]

Robert Fox has admirably chronicled the story of the rise and fall of the caloric theory; therefore, we shall not summarize it in detail.[27] But after Joseph

Black's discovery of latent heat (1757) until the second quarter of the nineteenth century, the theory was hardly challenged, which shows that the influence of the experiments of Thompson and Davy has sometimes been overstated. Only after 1815 did a group of physicists, most of them recent graduates of the Ecole Polytechnique and led, at least in spirit, by Fresnel's criticism of the corpuscular theory of light, rose in opposition to the caloric theory espoused by Laplace, Biot, and Poisson. The Newtonian theory that light is a substance—a stream of corpuscles—went hand in hand with the materiality of heat; the plausibility of the latter was buttressed by the fact that both light and heat traveled through a vacuum from the sun, which seemed impossible if light were an undulatory phenomenon. On the other hand, the general acceptance of the wave theory of light after about 1820 supported an undulatory or wave theory of heat.[28] The connection was tenuous but sufficient to provide a transitional undulatory theory between the caloric and the final dynamical theories of heat.

Experiments showing that radiant heat behaved very much like light and could be reflected and refracted were offered in support of the wave theory of light or an undulatory theory of heat. But they were difficult to reconcile with a dynamic or kinetic theory of heat because they implied that heat was not only molecular motion. The eventual, crucial, step was the realization that heat and what we now call thermal radiation are distinct. Beginning with the work of the Swedish chemist and apothecary Carl William Scheele in 1777 and leading eventually through Maxwell's development of the electromagnetic theory of light almost a century later, it came to be understood that radiant heat is a form of energy different from heat although convertible into it. In "On Air and Fire," Scheele showed the distinction between heat transported by what we now call convection (in his words, "radiant heat"), which had properties similar to light's; and De Saussure and Pictet made further important studies.[29] By the first decade of the nineteenth century, William Herschel found from heat measurements that the sun's spectrum extended far beyond the red and that heat obeyed Snell's law of refraction.[30] In about 1811, François Delaroche carried out a series of experiments on radiant heat, showing that it was the same as light.

As we shall see, the caloric theory reached a dead end by about 1830, after the death of Laplace. It stalled partly because of pressure from Fresnel, Arago, Petit, Dulong, and even Ampère and partly because of the electrochemistry of Berzelius. Its demise was a reaction to Laplacian mechanistic science and the Arcueil Circle, with their fondness for unobservable, imponderable substances. Although the experiments of Thompson and Davy played a role in the retreat of the theory, they were not decisive: The caloric theory survived them by a quarter-century.

The material theory of heat had lost most of its supporters by about 1850, but the death knell may have been Clausius's paper "On the Mean Length of the Paths," which appeared in the *Annalen der Physik* in 1858, and Maxwell's 1860 paper "Illustrations of the Dynamical Theory of Gases," which firmly established the microscopic basis for the observed properties of gases.[31]

The Steam Engine and Thermodynamics

The first modern heat engine was developed by Thomas Savery in 1698; this pumping engine cannot have worked well because of air logging that resulted from air dissolved in the water that was turned into steam. This difficulty was removed in 1712 in the Newcomen engine, which ushered in the era of British technological domination. In about 1765 James Watt (1736–1819) invented the true steam engine, with both a hot cylinder and a cold condensing cylinder or reservoir; previously a single cylinder had been alternately heated and cooled, producing large heat losses. Implicit in Watt's invention is the notion that, to do work, an engine must operate between two reservoirs, one hot and one cold, an idea that was not given formal expression until Sadi Carnot's "Reflections on the Motive Power of Fire" (1824). Watt's patents effectively strangled the attempts of others to improve on his engines.[32] But when the patents expired in 1800, Cornish engineer Richard Trevithick invented the high-pressure steam engine, which exhausted steam directly into the air, obviating the need for a condenser and making the steam engine much more mobile. Near the end of the eighteenth century, Davies Gilbert, mathematician, member of Parliament, president of the Royal Society, and Cornish friend of Humphrey Davy, theoretically related the work done by the steam engine to the area under the p-V curve.[33]

As others have often noted, this period was typical of those in which technology drives scientific discovery. Watt was a Glasgow instrument maker whose interest in steam engines seems to have been aroused by his friend John Robison and a model of a Newcomen engine brought for repair to the instrument shop at the University of Glasgow, where Watt was working in 1764.[34] Trying to understand the principle of the Newcomen engine, Watt consulted fellow Glaswegian Joseph Black in the summer of that year and learned of Black's theory of latent heat.[35] Watt developed his own version of the steam engine largely through physical intuition and practical insight, for what little theory he had to guide him was wrong. The correct theory of the action of the steam engine came after the fact and was in part the result of squabbling over patent rights.[36]

The Newcomen engine fostered a widely held view of the operation of the steam engine: that work was produced by atmospheric pressure against a vacuum left by condensing the steam. But by 1797 an article by John Robison in the *Encyclopaedia Britannica* emphasized the importance of the expansive power of steam, an idea that had become increasingly influential.[37] Important technical issues in this period included the relative merits of standard and high-pressure steam engines and understanding why the latter seemed more efficient. One of the chief deficiencies in the theoretical arguments advanced in the first two decades of the century was the failure to consider a complete cycle in which the working fluid returns to its original state. (See the discussion of the Carnot cycle later in the chapter.)

During the fifty years up to 1825, much of the activity in thermodynamics

PLATE 10.   Newcomen steam engine of the type used to pump mines. *From Jean Theophilus Desaguliers,* A Course of Experimental Philosophy *(London, 1744). Reprinted by permission of the Houghton Library, Harvard University.*

centered on the steam engine (and, to a lesser extent, the air engine). This was especially true in England; but Cardwell notes as well that, "with the possible exception of Fourier, virtually every distinguished French physicist, theoretical or experimental, concerned himself with the problems of the steam-engine."[38] It has been argued that science owes more to the steam engine than the steam engine owes to science and that all of the basic principles of thermodynamics except for latent and specific heats were discovered by studying the steam engine.[39]

In sum, the modern theory of heat had these important antecedents: (1) the experiments and observations of Power and Towneley, Hooke and Boyle, Charles, Mariotte, and Gay-Lussac, which determined the approximate relation between the pressure, volume, and temperature of a gas; (2) the discussions surrounding the caloric theory of heat and the dynamical theory of heat; and (3)

PLATE 11.   James Watt (1736–1819). *Portrait by John Partridge after Sir William Beechey. Courtesy of the National Galleries of Scotland, Picture Library.*

the need to understand and perfect the steam engine, which raised the issue of the interconvertibility of heat and mechanical work.[40]

Benjamin Thompson (Count Rumford) and Humphrey Davy

The names of Benjamin Thompson (1753–1814) and Humphrey Davy (1778–1829) are inseparably linked, not only through their experiments, which showed that heat is not a fluid, but equally through their association with the Royal Institution of Great Britain: Thompson as founder, Davy as lecturer in chemistry and an enormously popular public speaker. Although Davy was primarily a chemist (and a critic of Lavoisier's chemistry), he possessed a Cornish appreciation of technology and had invented the mine safety lamp.[41] Thompson was a military engineer, an urban planner, and a designer of lighting and heating apparatus.

The observations of Thompson (1798), Davy (1799), and others that mechanical work could be converted into heat through friction contributed to the demise of the caloric theory and the resurgence of the dynamical theory of heat.[42] Thompson claimed to have taken his understanding of the relation between heat and vibrations from Shaw's edition of Boerhaave's *A New Method of Chemistry* (1741). He was apparently also influenced by Boerhaave's earlier *Elements of Chemistry* (1732), although the latter's metaphysical view of heat (or fire) was very different from the one Thompson adopted.[43]

Thompson was an expatriate American whose royalist sentiments during the American Revolution forced him to settle in England in 1781.[44] His title, Count of Rumford, was bestowed on him by Carl Theodor, Duke of Bavaria, in 1792, and he served the Bavarian court until his death in 1814, in spite of spending only 1785–93 in Munich. (Thompson was also knighted by England's George III.) He read his first paper to the Royal Society of London in 1786 and received the society's Copley Medal in 1792 for his work on heat conduction and insulation. Thompson's scientific work always had practical goals, as indicated by his interest in stoves, lighting and photometry, silk, steam engines, cannon boring, and calorimetry. He also reorganized the Bavarian army and established a system of poorhouses to which he applied some of his technological ideas. Thompson founded the Royal Institution of Great Britain in 1799 with the goal of making the fruits of science available to common people and even the poor. In 1801 he hired Thomas Young as professor of natural philosophy and editor of the Institution's journal and Humphrey Davy as assistant lecturer in chemistry.

Thompson read his now-famous paper on heat evolved in cannon boring to the Royal Society on 25 January 1798.[45] In it he concluded that heat "cannot possibly be a material substance" and that it was nearly impossible to imagine heat's being excited and communicated as it was in his experiments "except it be motion." It was common knowledge that heat was generated by friction; but Thompson's cannon-boring experiments seemed to show that the amount of heat that could be generated was unlimited, suggesting that it was not a material substance. Thompson also tried to weigh the caloric. When he got a null result, he took this as an argument against its existence. Nevertheless, one can agree with Cardwell, who says that "Rumford's famous paper was not so much a piece of science, in the proper sense of that word, as an instance of special pleading; a mustering of evidence for a particular point of view."[46] Thompson clearly did not hold views that foreshadowed energy conservation: In 1801 he claimed that the heat emitted by bodies consisted of "undulations in the Etherial fluid."

Thompson spent the last thirteen years of his life mainly in France, and he left England forever in 1802. He made the acquaintance of Berthollet, Lagrange, and Laplace; and although he possessed little understanding of the mathematical details of Laplace's physics, he criticized them nonetheless, especially Laplace's theory of capillarity, thus creating an antagonism between the two men. Thompson's practical style of physics raised little interest in France, and the papers he delivered to the Institute of France often had a hostile reception. In 1813, a year before his death, he dined in Paris with Davy and Davy's scientific secretary, Michael Faraday. The meeting took place shortly after Faraday began working for Davy, and it made an enormous impression on the young man.

Davy's earliest writings rejected the caloric theory and returned to the ideas of Robert Boyle.[47] His friend and patron, Dr. Thomas Beddoes, who published the volume in which Davy's first work appeared (1799), pointed out the similarity between Davy's ideas (the melting of ice by friction) and Thompson's.[48] Fox discounts the influence of Thompson's and Davy's experiments, noting that the caloric theory flowered in the first decade of the nineteenth century on the heels of these experiments. While he is correct, the fact that a number of other influential scientists rejected the static caloric theory in favor of the dynamical theory shows the direction of the tide.[49] Nevertheless, other influential scientists, including William Thomson, were not converted until about 1850.

John Herapath (1790–1868) was led to the dynamical theory of heat through his esoteric speculations on the cause of gravity, and as early as 1820 he attacked the caloric theory of gases.[50] His views on the source of the thermodynamic properties of gases derived from his theory of the collision of infinitely hard but perfectly elastic atoms with each other and the walls of the containing vessel. Herapath, who was self-taught and whose speculations were largely ignored by his empirically disposed contemporaries, was among the first to attempt the transition from the qualitative dynamical theory of heat to a quantitative kinetic theory from which the properties of gases could be calculated. Rebuffed by the British scientific establishment, he eventually retreated to the editorship of a railway journal, where he freely indulged his speculative bent.[51] We explore Herapath's role in the development of the kinetic theory in chapter 7.

Although his work in thermodynamics was less important than the foundation he provided for the kinetic theory of Clausius and Maxwell, Herapath implicitly defined an absolute Fahrenheit temperature scale. Starting from the formula used by Poisson to relate pressure and density, $p = a\rho(1 + b\theta)$, where θ is the temperature and ρ the density, Herapath arrived at a formula that was essentially $p = a\rho (F + 448)$.[52] He thus located absolute cold (absolute zero) at -448°F, not far from the modern value of -459°F. He wrote in 1821: "I conceive heat to consist in motion; and that the temperature of a body is the intensity of the intestine motion of its particles estimated, when you compare the temperature of different bodies, not by their velocity but by their momentum. The degree of absolute cold is where the particles have no motion."[53]

Fourier to Joule and Mayer

The science of thermodynamics can be thought to have sprung from Sadi Carnot's memoir of 1824, and its principal foundations (the laws of energy conservation and entropy) date from no earlier than 1850. But the theory of heat as an independent mathematical subject arose in Fourier's great *Théorie analytique de la chaleur* (1822), translated as *The Analytical Theory of Heat.*[54] Fourier showed that one can go very far in the theory of heat without committing oneself on the issue of the *nature* of heat. He was explicit in his rejection of unnecessary assumptions; his theory was macroscopic and nonhypothetical. In describing heat conduction and static problems in the theory of heat, it postulated neither atoms nor imponderable fluids.[55] It was not a mechanical theory because it ventured no explanation of heat conduction, only a description.[56] Yet in spite of this tension between the mathematical or analytical theory of heat and empirical questions that could be answered only by experiment, Fourier's work served as a lifelong guide for William Thomson, who was born two years after its publication and read it at age fifteen or sixteen. In a sense, Thomson's entire career was built upon Fourier's work and its implications. The very fact that the theory was not based on hypothetical entities allowed it to serve easily as a model for electric and even optical phenomena, especially for Thomson.

Fourier did most of his work while he was a public administrator for Napoleon, between 1801 and 1815.[57] Earlier he taught at the new Ecole Polytechnique as a junior to Lagrange. At Grenoble he applied the analytical techniques of French mathematical physics to the problem of heat diffusion or conduction; and when the Academy proposed heat conduction as the subject of the 1811 competition, Fourier's submission won the prize. Nevertheless, he was reproached by the jury; and even though he won the support of Laplace, he was opposed by Lagrange and Poisson.[58] In particular, Lagrange, whose *Analytic Mechanics* certainly influenced Fourier's great work, objected to his use of the infinite trigonometric series now known as the Fourier series.[59]

Fourier's importance is at least threefold. He was the first to treat heat conduction seriously as a boundary value problem in the theory of partial differential equations. He also introduced new and powerful techniques for solving these problems, and his theory served as a model for applying the techniques to other physical problems.

Among the unsolved problems in 1815 were the specific heats of gases; the precise relation between pressure, temperature, and volume of a gas (what we call the equation of state); and the temperature scale.[60] Alexis Thérèse Petit and Pierre Louis Dulong were interested in these problems. Their experimental studies of the laws of cooling by radiation and convection gave them a tool for measuring specific heats, and they made the watershed discovery in 1819 that the specific heat per gram multiplied by the atomic weight was the same for all gases.[61] This discovery supported Dalton's ideas about atomic weight; and the Petit-Dulong law stayed at the center of the controversy over atoms through the creation of the kinetic theory in the 1860s to the genesis of quantum theory in the first quarter of the twentieth century.

Laplace had shown in 1816 that Newton's calculation of the speed of sound in a gas was in error by a factor $\gamma = c_p/c_v$ due to the adiabatic heating and cooling of the air during compression and expansion.[62] Consequently, the problem of specific heats was often posed in terms of the value of γ.[63] Although it was successful in this case, Laplace's theory predicted wrongly that the specific heat of a gas should increase with pressure or density.[64] This problem of adiabatic expansion or compression had both practical and theoretical interest. Since the invention of the air pump, scientists had noticed that when a gas is suddenly compressed it becomes heated and that it cools on expansion; and after 1806 adiabatic heating was used in the fire piston to generate enough heat to ignite tinder. Such a phenomenon was readily explained using the material theory of heat.[65] Between 1807 and 1823, Poisson contributed significantly to this problem, and in his memoir "On the Heat of Gases and Vapors" he obtained the formula $p'/p = (\rho'/\rho)^k$, where ρ is the density. For adiabatic processes, this is the equivalent of the modern expression $pV^k = $ constant, where $k = \gamma = c_v/c_p$.

Sadi Carnot

One of the great documents in the history of thermodynamics is Sadi Carnot's 1824 memoir "Reflections on the Motive Power of Heat and on Engines Suitable for Developing This Power."[66] In it we find a nearly complete statement of what we now call the second law of thermodynamics, a statement made two decades before the first law was formulated and more than a quarter-century before the final statement of the second law.[67] Carnot (1796–1832) stated that, whenever work is done by a thermodynamic system in a cyclic process, heat must flow from a hotter to a cooler reservoir—in other words, that a temperature difference is required to do work. This is the first clear, if incomplete, statement of the law of entropy. Although he employed the caloric theory, he asked in a footnote, "Is it possible to conceive the phenomena of heat and electricity as due to anything else than some kind of motion of the body?"[68] And in another place he mused, "We may say, in passing, [that] the fundamental principles on which the theory of heat rests require the most careful examination. Many experimental facts appear almost inexplicable in the present state of this theory."[69] Nevertheless, he believed wrongly in the conservation of heat by itself—that no heat was lost when work was done in a heat engine—a position that became untenable by the early 1840s.[70] Carnot had undergone a conversion to the motional theory of heat by the time he died in 1832, but that conversion came to light only in a posthumous note that his brother finally submitted for publication in 1878, more than forty years later. In the note, Carnot committed himself clearly to the idea that "heat is then the result of motion" and "is simply motive power, or rather motion which has changed its form. It is a movement among the particles of bodies."[71]

PLATE 12. Sadi Carnot (1796–1832), age thirty-four.

A military engineer, Carnot was the son of the renowned Lazare Carnot, engineer and revolutionary. He attended the Ecole Polytechnique where he was exposed to the ideas of Poisson, Gay-Lussac, Ampère, and Arago, all of whom were faculty members.[72] Carnot's most important work began in 1821, when he first became interested in the steam engine, and continued until his death just over a decade later from complications of scarlet fever.[73]

When fully and correctly elaborated, the 1824 paper contains what later became known as the Clausius-Carnot statement of the second law of thermodynamics.[74] It includes the discoveries that (1) the work done by a heat engine is a function only of the temperature differential between the hot and cold reservoirs between which it operates and (2) the difference between the specific heats at constant volume and pressure is a constant.[75] The paper is a watershed in the history of science, representing such a bold departure from the past that we often think of it as the foundation of the science of thermodynamics.[76] The Carnot cycle (figure 4–1), which Carnot introduced in this paper and which was a product of his study of the steam engine, provided a clear and graphic example of a closed cycle that could be used to model the actual operation of a heat engine. Today this is illustrated using the p-V, or indicator, diagram, which Carnot did not use.

Emile Clapeyron (1799–1864), a classmate of Carnot's at the Ecole Polytechnique, first used the p-V diagram to provide a graphical representation of Carnot's closed cycle and first expressed Carnot's ideas mathematically.[77] Especially important was his recognition that ideal engines produce equal amounts of mechanical work from equal amounts of caloric when they operate between the same temperatures. Nevertheless, neither his nor Carnot's ideas received much notice until after 1844, when Karl Holtzmann (1811–65) and William Thomson became interested.[78] Rudolph Clausius, who has the strongest claim of priority in the matter of the second law, learned of Carnot's paper through Clapeyron.

James Prescott Joule

William Wollaston (1766–1828) proposed in 1805 that the amount of heat generated by combustion in a steam engine should be measured by the work that it can do (for example, the height to which a weight can be raised), thus giving one of the earliest expressions of the idea of the interconvertibility of heat and mechanical work. Although we address this issue at length in chapter 5, it is germane here because it represents a process that led to the first law of thermodynamics. Nevertheless, scientists such as Holtzmann and Mayer were willing to compute the mechanical equivalent of heat (the amount of mechanical work that could be generated from a given amount of heat)—not committing to the idea of interconvertibility but adhering to Carnot's idea of conservation of heat alone.[79] Equivalence did not necessarily imply conversion.

Like Faraday, James Prescott Joule (1818–89) was a dedicated and creative experimental scientist with relatively little mathematical training. He was

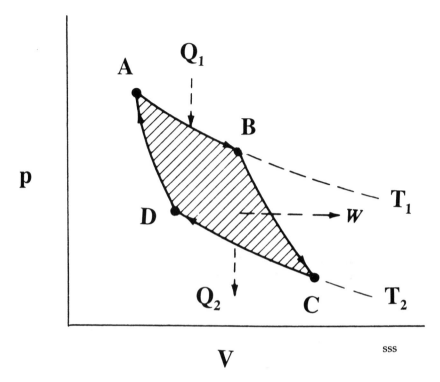

FIGURE 4–1. Carnot cycle, displayed on a p-V, or indicator, diagram. This reversible cycle consists of an isothermal expansion from A to B, an adiabatic expansion from B to C, an isothermal compression from C to D, and an adiabatic compression from D to A. Heat (Q_1) is absorbed at the higher temperature T_1 along the path AB, and work is performed by the system. In BC no heat is exchanged; work is done by system as it expands. In CD, at the lower temperature T_2, heat (Q_2) is given up as work is done on the system; and in DA, no heat is exchanged and work is done on the system.

a student of the aging Dalton and a product of the Manchester school that included Dalton and Peter Ewart. Joule not only played a crucial role in reestablishing the dynamical theory of heat but also, more effectively than any of his contemporaries, advanced the idea that heat could be converted into mechanical work and vice versa. Indeed, Hermann von Helmholtz credited Joule with the discovery of conservation of energy (see chapter 5).

For Joule, the interconvertibility of heat and mechanical work initially meant the conversion of electrical energy into heat.[80] Eventually, however, he came to see it as the heat generated by friction in liquids, and finally he generalized it to all processes. The idea of the conversion of heat into mechanical work became commonplace after the invention of the steam engine and the rapid growth of steam power; yet, as we have seen, Carnot believed that work could be done by transferring heat from a warm body to a cold one with no loss of

PLATE 13. James Prescott Joule (1818–89), age sixty-four. *Portrait by C. H. Jeens. From* Nature *26 (1882): facing 616.*

heat. Even into the 1840s, William Thomson could accept the conversion of mechanical work into heat but not the contrary.

Joule was educated at home and at age nineteen became a protégé of William Sturgeon, one of the inventors of the electromagnet.[81] Joule was guided by a belief in the simplicity of physical laws, an approach he seems to have inherited from Dalton.[82] But even though his experiments were simple, he generalized boldly from them. In 1841, when he was twenty-three years old, he discovered what we now call Joule heating, the I^2R heat loss due to a current I flowing through a resistance R, by placing a current-carrying coil into water and carefully measuring the temperature rise.[83] By rotating the armature of an electric motor in the field of an electromagnet in a water bath, he not only discovered that mechanical energy was converted into heat in the water but used the induced current as a measure of the mechanical work by determining the heating produced. In his paper describing the experiment, he wrote: "The mechanical power exerted in turning a magnetoelectric machine is converted into heat evolved by the passage of the currents of induction through its coils; and, on the other hand . . . the motive power of the electro-magnetic engine is obtained at the expense of the heat due to the chemical reactions of the battery by which it is worked."[84]

Further experiments between 1845 and 1847 involved direct measurements of the heat evolved (or absorbed) in mechanical processes, including the heat generated in water from the friction of a paddle wheel, and in the expansion and compression of air.[85] Joule concluded that, when a gas (in this case, air) expands without doing work, no heat is lost.[86] By this time he had read the papers of Carnot and Clapeyron, which motivated him in part to carry out his experiments on the expansion and compression of gases.[87] But he was critical of the Carnot-Clapeyron view that mechanical effect could be generated by moving heat from a hot to a cold body without the discharge of heat; and most important, he was committed to the dynamical theory of heat—that is, heat as vis viva—unlike either Carnot or Clapeyron. (Carnot was dead by this time.) As Cardwell and others have shown, the clarity of Joule's ideas about converting heat into an equivalent amount of work derive from not only Dalton and Ewart (whom Dalton admired) but also J.N.P. Hachette (1769–1834), perhaps through William Whewell's writings.[88] When Joule read a paper to the Manchester Society in October 1848, he invoked the kinetic theory of John Herapath as a way of using the dynamical theory of heat to explain the properties of gases.[89]

In short, there were two issues: the nature of heat and the interconvertibility of heat and work (which we address more fully in chapter 5). Joule, perhaps unfettered by theoretical biases, was ahead of Thomson on both questions. By 1847 he was ready to announce his conclusions—in a public lecture in the reading room of St. Ann's Church, Manchester, on 28 April and at the Oxford meeting of the BAAS in June.[90] In the audience at the Oxford meeting was twenty-two-year-old William Thomson. Under ordinary circumstances Thomson would not have heard the paper. But the program of the chemical section, in which Joule's paper would naturally have been placed, was full; therefore, it was placed in the mathematics and physics section. This encounter had profound consequences for both men: For Joule, it led to the recognition he deserved; for Thomson, it led to his rejection of the caloric theory and Carnot's postulate of the conservation of heat and eventually to his adoption of Joule's theory of energy conservation.[91] Joule's actual route to scientific recognition was an 1850 letter from Thomson in the *Philosophical Magazine* supporting his ideas.

Thus, 1847 marked the beginning of Thomson's slow and agonized conversion to the dynamical theory of heat. Three years later he finally "consider[ed] it thus established that heat is not a substance but a dynamical form of mechanical effect."[92] But in 1848 he was still unconvinced. He thought that the conversion of heat was "probably impossible" but admitted that Joule had a contrary opinion based on experiments "seeming to indicate an actual conversion of mechanical effect into caloric."[93] In a letter written in October 1948, commenting on Thomson's "On an Absolute Thermometric Scale," Joule noted (with surprise?) "that you still adhere to Carnot's theory of the Motive Power of Heat." And in another letter later that month, he said: "There is one point mentioned in your paper especially interesting to me, namely that the property of converting heat into mechanical effect is not admitted."[94] Thomson acknowledged the force of Joule's objections but still did not yield on the critical point.

Ultimately, it took William Rankine (1820–72) and his work on steam to convince Thomson of the value of Joule's experimental work. Rankine attempted the final step—the reconciliation of Carnot's theory with Joule's results. Although Clausius and Thomson finally accomplished the task, both were influenced by Rankine—Thomson, crucially. The conflict between Joule's work and Carnot's caloric-based theory was still critical for Thomson as late as 1849; but stimulated by Rankine and Clausius, he arrived at his own formulation of the second law of thermodynamics within two years.[95]

Joule was by no means the only scientist thinking about the interconvertibility of work and heat in the 1840s. German scientists Julius Robert Mayer and Hermann von Helmholtz and Danish scientist Ludwig Colding joined Joule in advocating the conservation of energy; so did at least seven other contemporaries who argued for some form of the idea of the indestructibility of force or the interconvertibility of heat and work.[96] Because most of this work is only incidently related to the development of thermodynamics, we will discuss it more fully in chapter 5.

Clausius and Thomson

Rudolph Clausius (1822–88) was born in Pomerania, Prussia (now in Poland). He entered the University of Berlin at age eighteen and received his doctorate at Halle in 1847. In April 1850 he published his most famous paper, "On the Motive Power of Heat, and on the Laws Which Can Be Deduced from It for the Theory of Heat," in *Poggendorff's Annalen,* which presented the second law of thermodynamics in complete form for the first time.[97] Clausius began teaching at the Royal School of Artillery and Engineering and in 1855 became professor of mathematical physics at the Polytechnic in Zurich. In 1867 he moved to the University of Würzburg and in 1869 became professor of physics in Bonn, where he remained until his death. In 1870 he was wounded while leading an ambulance corps in the Franco-Prussian war and never fully recovered.[98]

By August 1850 Thomson had read Clausius's "On the Motive Power of Heat," which he quickly followed with his own "On the Dynamical Theory of Heat" (discussed later in this chapter).[99] Clausius's clear priority in the second law resulted from his greater willingness to embrace Carnot and Joule simultaneously, which meant rejecting Carnot's theory that no heat was lost in a cycle.[100] Clausius began his 1850 paper with a tribute to Carnot, whose ideas he learned from Clapeyron and Thomson because he had not been able to obtain a copy of Carnot's book.[101] He quoted Carnot as saying that the conservation of heat is "not to be doubted; it was assumed at first without investigation, and then established in many cases by calorimetric measurements. To deny it would overthrow the whole theory of heat, of which it is the foundation." But quickly he departed from Carnot:

> I am not aware, however, that it has been sufficiently proved by experiment that no loss of heat occurs when work is done; it may, perhaps, on the contrary, be asserted with more correctness that even if such a loss has not been proved directly, it has yet been shown by other facts to be not only admissible, but even highly probable. If it be assumed that heat, like a substance, cannot diminish in quantity, it must also be assumed that it cannot increase. It is, however, almost impossible to explain the heat produced by friction except as an increase in the quantity of heat. . . . To this it must be added that other facts have lately become known which support the view, that heat is not a substance, but consists in a motion of the least parts of bodies.[102]

Thomson's ambivalence toward Carnot's theory was clear—his enthusiastic support for it clashed with his grudging acceptance of Joule's contradictory results: "If we [abandon Joule's fundamental axiom], however, we meet with innumerable other difficulties—insuperable without farther experimental investigation, and an entire reconstruction of the theory of heat from its foundation. It is in reality to experiment that we must look—either for a verification of Carnot's axiom, and an explanation of the difficulty we have been considering; or for an entirely new basis of the Theory of Heat."[103]

But Clausius was not so easily deterred:

> I believe that we should not be daunted by these difficulties, but rather should familiarize ourselves as much as possible with the consequences of the idea that heat is a motion. . . . Then too I do not think the difficulties are so serious as Thomson does. . . . It is not necessary to discard Carnot's theory entirely, a step which we certainly would find hard to take, since it has to some extent been conspicuously verified by experience. A careful examination shows that the new method does not stand in contradiction to the essential principle of Carnot, but only to the subsidiary statement *that no heat is lost.*[104]

Clausius titled the first section of his paper "Consequences of the Principle of the Equivalence of Heat and Work." He proceeded from what we now call the equation of state of a gas, which he described as the combined laws of Boyle-Mariotte and Gay-Lussac and which he applied to a Carnot cycle, a closed figure on the p-V (indicator) diagram (see figure 4–1). It represents an isothermal (constant temperature) expansion from *a* to *b,* during which the gas extracts heat Q_1 from a high-temperature reservoir A, followed by an adiabatic expansion from *b* to *c* in which no heat is transferred ($\Delta Q = 0$) but pressure, volume, and temperature all change. Next there is an isothermal compression from *c* to *d,* during which heat Q_2 is expelled to a low-temperature reservoir B and finally an adiabatic compression from d to a. The work in each step is computed from $\int$ pdV—that is, the area under each curve a-b, b-c, and so on. It is not hard to see that the net work is the area enclosed by a-b, b-c, c-d, d-a. In this discussion, Clausius noted the importance of infinitely small changes of state of the gas.

Next Clausius obtained an explicit form for the first law of thermodynamics in a familiar form, dQ = dU + pdV, in which U is simply an arbitrary function of V and t (the temperature)—what Thomson called intrinsic energy and we call internal energy.[105] Clausius considered adiabatic and isothermal processes and obtained results that are now standard: pV^k = constant, for adiabatic, and Q = AR (a + t_o) ln V/V_o = A $p_o V_o$ log V/V_o for heat evolved or absorbed, for isothermal processes.[106]

Section 2 of the paper is titled "Consequences of Carnot's Principle in Connection with the One Already Introduced." In it, without much fanfare, Clausius expresses what we know as the second law of thermodynamics. The paper asks us to consider two processes in which an amount of heat Q is transferred from a hot body A to a cold body B, but that in the first case less work, W_1, is done than in the second, where the work is W_2. Clausius asks us to assume that the two processes (engines) are connected, with the second running in reverse so that heat Q_1 is transferred from A to B, doing work W_1, which is then used to transfer heat Q_2 (> Q_1) from B to A. Thus, "by repeating these two processes alternately it would be possible, without expenditure of force or any other change, to transfer as much heat as we please from a *cold* to a *hot* body, and this is not in accord with the other relations of heat, since it always shows a tendency to equalize temperature differences and therefore to pass from *hotter* to *colder* bodies." A detailed analysis then leads to an analytical expression of the second law involving a universal function of temperature.[107]

The following year (1851), Thomson gave a different statement of the second law, which he referred to as the law of Carnot and Clausius, but noted that the two forms were equivalent. Thomson's statements of the two versions (axioms) of the second law are as follows:[108]

> It is impossible, by means of an inanimate material agency, to derive mechanical effect from any portion of matter by cooling it below the temperature of the coldest of the surrounding objects. (Thomson)

> It is impossible for a self-acting machine, unaided by any external agency, to convey heat from one body to another at a higher temperature. (Carnot-Clausius)

Declaring their equivalence, Thomson said: "It is easily shown, that, although this and the axiom I have used are different in form, either is a consequence of the other."[109]

In 1854, Clausius developed the concept of entropy to describe the transformational content of a body, although he did not introduce the term *entropy* for another decade: "I prefer going to the ancient languages for the names of important scientific quantities, so that they may mean the same thing in all living tongues. I propose, accordingly, to call S the *entropy* of a body, after the Greek word τροπή, 'transformation.' I have designedly coined the word *entropy* to be similar to 'energy,' for these two quantities are so analogous in their physical significance."[110] In the same work (1854), he offered his famous statement of the two laws of thermodynamics: (1) the energy of the universe is constant; (2) the entropy of the universe tends to a maximum.

The 1854 paper may have been influenced by Thomson's conclusion that, in a cyclic process "of a perfectly reversible kind, the quantities of heat which it takes in at different temperatures are subject to a linear equation, of which the coefficients are the corresponding values of an absolute function of the temperature." Clausius took this function to be $1/T$ and was able to show that the sum of the terms Q_i/T_i equaled 0 ($\Sigma Q_i/T_i = 0$).[111] Using a Carnot cycle, Clausius was able to establish this result himself as well as its generalization to all reversible processes, which was $\int dQ/T = 0$. Finally, he considered *irreversible* changes, in which more heat flows from the hot reservoir A to the cold reservoir B for the same amount of work than in the case of the perfect engine.[112] Here $\int dQ/T > 0$ because the reversed cycle, with heat converted into work, does not compensate for the transformation of a larger amount of heat at high temperature into low-temperature heat. If $\int dQ/T < 0$, then heat could be made to flow from a cold body to a hot body by simple conduction. Note that for a *reversible* process, if $\int dQ/T > 0$, reversing the cycle would make $\int dQ/T < 0$, violating the second law; therefore, the equal sign must hold for reversible processes and $dQ/T > 0$ for irreversible ones.

Clausius defined the internal energy as consisting of one part that represented conversion of mechanical energy into heat (the vis viva of the body) and of another part, the disgregation, that depended on the internal arrangement of the particles. The entropy change of a body, then, consisted of two parts—one representing the change in the heat content, the other a change in molecular arrangement.[113]

Clausius was awarded the Royal Society's Copley Medal in 1879 and published his papers on thermodynamics in a two-volume work, *Die Mechanische Wärmetheorie* (translated as *The Mechanical Theory of Heat*), in 1865–67, revising it in 1875.

Thomson and the Second Law

Although Joule's paper at the 1847 BAAS meeting was critical to the development of William Thomson's ideas about heat, his brother James's interest in applications of steam engineering and the discussions between the brothers about related problems during the early 1840s had an equally important influence, making Thomson more receptive to Joule's ideas. William had come across Carnot's work in 1844, presumably through Clapeyron's paper, but had not moved much beyond Carnot by 1847.[114] Following the ideas of Carnot and Clapeyron, Thomson thought of heat as falling from one degree of intensity to another, which was the source of the mechanical effect; work was done as heat was transferred, without loss, from a hot to a cold body. To Joule, of course, such a process was the conversion of heat into work. At the end of the BAAS meeting, Joule gave Thomson copies of his 1843 and 1844 papers, which both William and James examined carefully.[115] Thomson was particularly occupied with the problem of the loss of mechanical effect in conducting heat from a high to a low temperature even without operating a machine.

After his encounter with Joule in 1847, Thomson faced the problem of how to reconcile Joule with Carnot.[116] During the next three years, James and William applied the Carnot-Clapeyron theory: that to do work, an engine had to operate between two temperatures. In particular, they tried to understand the operation of the Stirling air engine.[117] Together they also predicted (correctly) that the freezing point of water must be lowered under pressure. An outcome of these considerations and William's experiments was the absolute temperature scale divorced from any particular substance and based on the idea that temperature intervals should be related to specific quantities of heat and the equivalent mechanical effect produced (1847–48). This became the Kelvin scale. More to the point, however, they tried to reconcile their own view (that if the state of a body remained unchanged, then no heat was gained or lost) with Joule's (that in a steam engine, operating in a cycle, the heat lost by the steam is converted into work).[118] In his paper on the absolute temperature scale, Thomson acknowledged that work might be converted into heat, but not the reverse.[119]

A critical element in Thomson's attempts to reconcile Carnot and Joule was Fourier's theory of heat conduction, which Thomson had learned as a schoolboy. Heat conduction was a process in which the flow of heat produced no discernable mechanical effect and which was irreversible, contrasting strongly and dramatically with the reversible fall of heat through a temperature difference, in which mechanical work could be done. When he finally realized that heat was wasted or dissipated in irreversible processes such as conduction and not available to do work, he was able to concede that Carnot was wrong and almost simultaneously resolved the questions of the nature of heat and the interconvertibility of heat and mechanical effect. He had arrived at Clausius's second law of thermodynamics.[120]

But in 1850, after three years of searching correspondence between the

Plate 14. Sir William Thomson, 1870. *Photograph by Emery Walker. From Silvanus P. Thompson,* The Life of William Thomson, Baron Kelvin of Largs *(London: Macmillan, 1910).*

Thomson brothers and Joule, the problem was still unresolved. Nevertheless, James was writing, "*if* Carnot's principles are admitted," and Joule had written to William, "as your brother said, there ought to be some connecting link between the results I have arrived at and those deduced from Carnot's theory."[121] It took a paper by W.J.M. Rankine delivered to the Royal Society of Edinburgh on 4 February 1850 to break the impasse.[122] On 15 October Thomson wrote to Joule, saying, "your discovery alone can reconcile Mr. Rankine's discovery with known facts." The letter, with a foreword by Joule, was published in both the *Philosophical Magazine* and *Poggendorff's Annalen* in late 1850.[123]

Rankine's discovery was his conclusion that, if saturated steam is allowed to expand in what we call an isothermal process (that is, at constant temperature), then the work done by the expanding steam must be at the expense of heat supplied from without. Otherwise, some of the vapor would have to condense, supplying the heat from the latent heat of condensation. He offered this conclusion as an explanation for what Joule called "the nonscalding property of steam issuing from a high pressure boiler."[124] During the 1840s Rankine, who had been educated at the University of Edinburgh, had worked on engineering problems associated with the Scottish and Irish railway systems; and from 1855 to 1872 he held the chair of engineering at Glasgow, to which James Thomson ultimately succeeded. In the paper based on his vortex hypothesis, Rankine wrote, "No mechanical power can be given out in the shape of expansion, unless the quantity of heat emitted by the body in returning to its primitive temperature and volume is *less* than the quantity of heat received."[125] To Joule Thomson admitted, "If we are contented to take Regnault's result as an experimental fact, and if we adopt your mechanical equivalent for a thermal unit . . . , we may demonstrate Mr. Rankine's remarkable theorem without any other hypothesis than the convertibility of heat and mechanical effect."[126] Finally, Thomson alluded to Clausius's "On the Motive Power of Heat," in which "a similar conclusion to that which I have quoted of Mr. Rankine's . . . is announced."[127]

For Thomson, Rankine's insight quickly released a flood of work, and by February 1851 he had satisfied himself that his theory was correct.[128] In March he presented his results to the Royal Society of Edinburgh in the first part of his seminal work, "On the Dynamical Theory of Heat."[129] In it he offered his own version of the second law, as an axiom, claiming that he had "obtained [it] about a year ago." He also stated Clausius's version and, as I have already noted, declared that "either is the consequence of the other."[130] (I quoted both versions earlier in the chapter.)

Thomson's slow conversion to Joule's ideas meant that he was overtaken by both Clausius and Rankine while he spent six years mulling over the problem of conservation of heat. On the matter of priority, he said, "It is with no wish to claim priority that I make these statements, as the merit of first establishing the proposition upon correct principles is entirely due to Clausius, who published his demonstration of it in the month of May last year, in the second part of his paper on the motive power of heat." He went on to say "that I have

Plate 15. W.J.M. Rankine (1820–72). *Portrait by H. Allard. From W.J.M. Rankine,* Miscellaneous Scientific Papers *(London, 1881).*

given the demonstration exactly as it occurred to me before I knew that Clausius had either enunciated or demonstrated the proposition."[131] Commenting on Thomson's paper, Joule said, "You seem to have completely solved the difficulty which before seemed to render the results of Carnot's theory and what we must consider the true theory irreconcilable."[132]

While never far from the surface, Thomson's ideas about dissipative effects in the universe are hidden in the final version of the 1851 paper.[133] In an early draft, however, he wrote:

> I believe the tendency in the material world is for motion to become diffused, and that as a whole the reverse of concentration is gradually going on. . . . the difficulty which weighed principally with me in not accepting the theory so ably supported by Mr. Joule was that the mechanical effect stated in Carnot's Theory to be *absolutely lost* by conduction, is not accounted for in the dynamical theory otherwise than by asserting that *it is not lost.* . . . The fact is, it may I believe be demonstrated that the work is *lost to man* irrecoverably; but not lost in the material world.[134]

The full implications of the dissipation of mechanical energy in all physical process were more fully explored in "On a Universal Tendency to the Dissipation of Mechanical Energy," published during the following year. A decade later, in "On the Age of the Sun's Heat," Thomson gave an explicit picture of the final state of the universe as "a state of universal rest and death," a condition popularly labeled "heat death."[135]

In the 1860s Thomson worked with Tait to complete their *Treatise on Natural Philosophy* and responded to ideas about the origin of the Earth that circulated after the publication of Lyell's *Principles of Geology* and Darwin's *Origin of Species.*[136] In 1862 he read a paper to the Royal Society of Edinburgh titled "On the Secular Cooling of the Earth"; his anti-uniformitarian views put him at odds with T. H. Huxley. In "On the Age of the Sun's Heat," Thomson adopted a theory of the sun's energy that Helmholtz had earlier offered—namely, gravitational contraction—but great uncertainty about the resultant age of the sun made the entire enterprise indecisive.[137] Thomson himself favored an age of 10 to 15 million years, far less than the estimate given by some geologists and paleontologists. Nevertheless, he did allow an upper limit of 500 million years, which Huxley remarked would serve the purposes of geologists.

After the death of Clausius, Josiah Willard Gibbs contributed an article to the American Academy of Arts and Sciences claiming that with Clausius's 1850 paper "the science of thermodynamics came into existence."[138] Nevertheless, in his study of the Gibbs's article, Martin Klein shows that the demonstrated equivalence of Thomson's and Clausius's statements of the second law and their different but overlapping views of entropy and dissipation by no means settled the issue. In particular, Thomson's schoolmate and collaborator, Peter Guthrie Tait, disputed Clausius's stature in the creation of thermodynamics. The controversy was sparked by the publication of Maxwell's *Theory of Heat,* which in the first edition barely noted Clausius's work, and Tait's *Sketch of the History of Thermodynamics.*[139] Tait even claimed that Thomson was mistaken in his demonstration that Clausius's and his own versions of the second law were logically equivalent. The meaning of the second law had been clouded by these polemics, but in 1873 Maxwell wrote to Tait: "It is only lately under the conduct of

Professor Willard Gibbs that I have been led to recant an error which I had im-
bibed from your [work], namely that the entropy of Clausius is unavailable
energy. . . . The entropy of Clausius is . . . only Rankine's Thermodynamic
Function."[140]

In the last third of the nineteenth century, thermodynamics evolved in several different ways. As we shall see, the kinetic theory of gases, created by
Clausius and Maxwell in the 1860s, provided a microscopic foundation for the
theory of heat. Its offspring was statistical mechanics, founded by Ludwig
Boltzmann but elaborated on greatly by American physicist Josiah Willard Gibbs
and eventually Albert Einstein. The problem of the thermal properties of electromagnetic radiation, which had to be formulated in terms of statistical mechanics, drew increasing attention and stimulated work by Kirchhoff, Wien,
Stefan, Boltzmann, Lord Rayleigh (William Strutt), James Jeans, and ultimately
Max Planck and Einstein. Similarly important, and also depending on statistical mechanics, was the problem of the discrete radiation emitted by excited gases.
As we shall see in chapter 8, these problems, especially the intractability of the
latter one, helped usher in the quantum revolution. Of course, there were widespread practical applications of thermodynamics to a host of problems in physics, including thermal processes in stars and the cooling of the Earth, as well as
to technology, particularly the internal combustion engine.

CHAPTER 5

Energy and the Energy Principle

One of the greatest accomplishments of nineteenth-century physics was the discovery of the energy principle in the years just before 1850. Supreme among the global principles of physics, energy conservation has been a sine qua non of virtually every area of physical science in the twentieth century.[1] Although its roots are ancient, its foundations as a concrete mathematical principle can be traced to the seventeenth century in the form of the conservation of vis viva (kinetic energy).[2] Yet energy conservation as we now know it (in the context of classical physics) dates from only the 1840s. The discovery was twofold, involving (1) the concept of energy as an essential property of physical systems and (2) the conservation of energy—in other words, that the energy of an isolated system is constant. As we shall see, its conservation is what made energy a critically important quantity historically.

The ontological status of energy conservation is different from most of the discoveries of the nineteenth century. While often broad and far-reaching, many of these discoveries were nonetheless concrete and relative to assumptions about the detailed structure of matter. Because they relied on hypotheses about atoms or molecules, the existence of the aether, and so on, they have fared more or less well depending on the assumptions that underlie them. Energy conservation, on the other hand, is a global generalization from the particulars of nineteenth-century experimental physics to an abstract principle assumed to have universal validity independent of the physical systems whose study led to its discovery or of the models used to interpret those systems. It has, of course, been generalized in the twentieth century, a fact that simply underscores its universality.

The conservation of energy is often treated as an axiom—assumed to be inviolably true despite the fact that it can be verified (like any other physical law) only in a limited or approximate sense and hence only to a high degree of probability.[3] Thus, the idea of conservation of energy provides a good case study in the history of scientific discovery as the story of the establishment of a principle that by extrapolation is taken to be far more general than any of its empirical or deductive foundations. It is an excellent example of what separates physics and mathematics. In the end, however, it is an empirical law subject to continual experimental test.[4]

Vast breadth or scope is no guarantee of the validity or usefulness of an idea; many such ideas have been imposed on a scientific discipline, often from the outside, to the detriment of that science. This is especially true of philosophical structures or predispositions that have found their way into science but have no empirical foundation at all. One thinks, for example, of the belief in final causes or of nineteenth-century Naturphilosophie, which was friendly to the idea of conservation but could offer neither rigorous theoretical plausibility nor experimental basis. Other examples, while historically interesting, had even less to offer as guides to scientific progress.

Because the question of the conservation of force, or energy, is inseparable from the evolution of mechanics, thermodynamics, and electromagnetic theory, this chapter will only summarize the disparate paths that led to a general understanding of the law of conservation of energy. In previous chapters we have alluded to some of the problems that first raised the issue.[5]

Conservation in the Late Eighteenth Century

Serious consideration of energy, in the form of living force or dead force, and its conservation, first arose in the eighteenth century in the wake of Newton's *Principia*. Nevertheless, as early as 1669, Huygens, Wren, and Wallis had provided essentially correct descriptions of elastic and inelastic collisions.[6] As Wilson Scott emphasizes, however, studies of vis viva loss, primarily those growing out of the study of machines, formed an important part of the discussion.[7] Each of these ideas—that vis viva was conserved in hard body impact and that in real machines, which operated by impact, loss of vis viva did occur—were important in the evolution of the idea of conservation of energy into the 1840s. Huygens, Leibniz, and the Bernoullis were among those who first noted the importance of mv^2 or vis viva (living force) in collisions and raised the question of its conservation.[8] Johann Bernoulli wrote: "Where *vis viva* disappears, the power to do work . . . is not lost, but is only changed into some other form."[9] Descartes assumed *vis motiva,* or force of motion (our momentum), to be constant; and subsequent discussions of collisions showed that it was indeed a conserved quantity in this context.[10] Also providing some impetus toward a generalization of the concept of conservation of motion was the issue of perpetual motion, which I shall not pursue here.[11] Laplace's introduction of what we now call potential energy led to the idea of conservation of kinetic plus potential energy in mechanics, and the Lagrangian formulation of mechanics encouraged scientists to think of the conservation of the scalar quantity $T + V$ (the first integral of the equation of motion) rather than the (vector) momentum $\underline{p}$, whose conservation emerges immediately in the Newtonian formulation based on force.[12]

In this chapter, we will not recount the history of conservation of force in mechanics or in collision theory.[13] What I wish to examine is the more general idea, the energy principle, which ultimately contains the assertion that all forms of energy are in principle interchangeable or interconvertible and that the total amount of energy in the universe is constant. The clarity of the Newtonian idea of "force" (what we now mean by force or impulse [Fdt]) was eroded in the eighteenth century: That which we call kinetic energy became living force (vis viva), and momentum was force of motion.[14] Energy associated with position was dead force. Thus, Faraday described force as the "*cause* of all physical action" and spoke of its conservation.[15] Even after Helmholtz's 1847 paper, the conceptual confusion between force and energy continued.[16]

By the 1820s the concept of work (*travail*) was understood in something approaching the modern sense ($\int F \cdot dx$).[17] But even by the 1780s and 1790s, scientists were making use of the concept. The interconvertibility of mechanical energy and heat is another matter, however; and if not strictly a nineteenth-century idea, it was not fully accepted until the late 1840s. As we have noted, the concept of energy became truly meaningful only in the context of discussions of its conservation. The eighteenth-century idea of the conservation of mechanical energy in elastic collisions, and even the idea of interconvertibility, were simply first steps toward the full generalization, which in turn would give energy its exalted place. Perhaps the most important reason why these progenitors bore no fruit for at least a half-century was the need to arrive at some understanding that all phenomena are, at bottom, mechanical.[18] This was by no means obvious at the beginning of the nineteenth century.

The Fallow Years: 1800–31

Modern use of the term *energy* derives from Thomas Young, who in *Lectures on Natural Philosophy* wrote: "The term Energy may be applied, with great propriety, to the product of the mass or weight of a body into the square of the number expressing its velocity. . . . This product has been denominated the living force."[19] Yet Young evidently held no general concept of the conservation of this quantity.[20] The final discovery of the law of conservation of energy was a complex interaction of ideas and speculations in the 1840s that we can only summarize here. Scientists involved include Julius Robert Mayer, Ludwig Colding, Marc Seguin, James Joule, and Hermann von Helmholtz, with the last two playing the most important roles.[21] For the moment, however, we will concentrate on the first third of the nineteenth century, the period in which the mechanical theory of heat slowly began to gain adherents and when the first crucial empirical discoveries of Oersted, Ampère, Faraday, and others added a new dimension to classical physics—the years just before Thomson, Maxwell, Joule, Helmholtz, and their contemporaries stepped onto the stage.

As we have seen, Benjamin Thompson (Count Rumford) was a vigorous adherent of the dynamical theory of heat at a time when the caloric theory was in vogue, especially in France. Furthermore, he believed that "the sum of the active forces in the universe must always remain constant."[22] Kuhn is probably right in considering this statement to be no more than an echo of the much older idea of conservation of vis viva.[23] But even if Thompson intended a wider application, his work shows that his thoughts remained imprecise and vague: He clearly did not realize the importance of the idea.

It is neither useful nor enlightening to try to search every statement of general or metaphysical commitment to the idea that the quantity of motion or force in the universe must remain unchanged for the germ of the idea of conservation of energy. There was a widely held, intuitive view that the total quantity of force (or motion) might be somehow constant, but it did not necessarily imply an ability to distinguish the vector nature of force or momentum from the scalar energy. Nor, until Helmholtz, was there really anything resembling a *proof* of energy conservation. We might distinguish, as Kuhn does, between those who arrived at energy conservation by patient, quantitative means and those for whom it was essentially a leap of faith.

In the early nineteenth century many scientific figures were groping toward energy conservation, prompted by experiments such as Thompson's and Davy's, the technology of the steam engine, speculations about processes in living organisms, the conservation of mechanical energy, and the conversion of mechanical energy from one form to another or the extraction of mechanical effect from heat. The rapid acceptance of energy conservation was due in considerable measure to a preexisting sympathy for such modes of thought, which had been gradually gaining strength during the preceding fifty years. The steam engine—Thomas Savery's invention in 1698, Thomas Newcomen's improvement in 1712, and especially James Watt's true (condensing) engine in the mid-1760s—were a watershed in the history of energy conservation. By the eighteenth century, conversion of one form of mechanical effort into another was easily exhibited in mechanical devices such as the waterwheel and other hydraulic machines. Of course, conversion does not necessarily imply conservation; indeed, in practical machines, the loss of mechanical effect was also obvious. Furthermore, the problem posed by the steam engine was very different: An apparently nonmechanical quantity, heat, was transformed into mechanical effort through the expansive effect of steam. But the early steam pioneers came to measure the mechanical effect by the distance a weight could be raised—in effect, force times distance, or what we know as work. The importance of this realization (or its generalization, $\int \mathbf{f} \cdot d\mathbf{s}$) became apparent first in the works of Euler, Lagrange, and Laplace and in Lazare Carnot's "Essay on Machines in General" (1782).[24]

The dynamical theory of heat finally began to take hold in the 1820s; and the implications of experiments—which began about 1800—on the interconvertibility of heat and mechanical work (Thompson, Davy, Joule) and on the

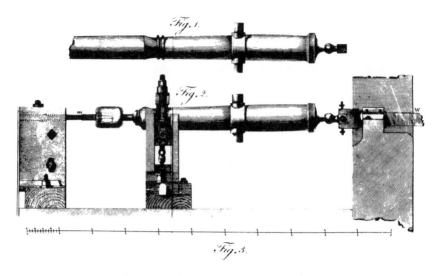

PLATE 16.   Thompson's illustration of cannon boring. *From B. Thompson (Count Rumford), "Inquiry Concerning the Source of Heat Excited by Friction,"* Philosophical Transactions of the Royal Society *88 (1798): 80.*

generation of heat by currents really only began to become clear in the 1840s. But these discoveries did not necessarily imply conservation, nor did they require a motional theory of heat. Indeed, the material theory of heat may have made it easier to think of its conservation, and Thompson's idea that "the heat generated by friction is inexhaustible" explicitly denies conservation.[25] Sadi Carnot's great work of 1824 and Thomson's writing in the late 1840s (based on Carnot's work) offered the idea of heat as a substance falling from one degree of intensity to another to explain the extraction of power from a heat engine rather than interconvertibility. It is certainly possible to formulate a theory of complete interconvertibility of heat and mechanical effect in either the caloric or the mechanical theories of heat.

But the case of Carnot is more complicated. Although he was still committed to the caloric theory in 1824, he asked, "Do we not know . . . *a posteriori* that all the attempts to produce perpetual motion by any means whatever have been fruitless?"[26] And in his "Posthumous Remarks," written no later than 1832 (the year in which he died), Carnot committed himself fully to energy conservation, crediting the influence of Davy and Rumford:

> Heat is simply motive power, or rather motion which has changed its form. It is a movement among the particles of bodies. Wherever there is destruction of motive power, there is at the same time production of heat in quantity exactly proportional to the quantity of motive power destroyed. Reciprocally, wherever there is destruction of heat, there is

production of motive power. We can establish the general proposition that motive power is, in quantity, invariable in nature; that it is, correctly speaking, never either produced or destroyed. It is true that it changes its form—that is, it produces sometimes one form of motion, sometimes another—but it is never annihilated.[27]

Unfortunately, this commitment to energy conservation came to light only in 1878 and thus had no influence on ensuing events.[28] We might be pardoned for considering it a lost opportunity, for in this work Carnot proposed some of the same experiments that Joule used a dozen years or more later to determine the mechanical equivalent of heat (including the Joule-Thomson experiment). Yet Carnot's "Reflections" of 1824 passed almost without notice; his later work might also have faded into obscurity.

The Discovery of the Law of Conservation of Energy

The early progenitors of energy conservation, who came to a conclusion about the idea in the six years surrounding 1840, were all part of an older generation of scientists born before 1815.[29] In contrast, the figures who dominated classical physics at midcentury and beyond were just beginning their scientific careers in the early 1840s. Energy conservation was part of their intellectual inheritance; or, at the very least, it was in the air.[30]

By the mid-1830s Michael Faraday was making much of the conversion of one power from and into another; by 1840 he was firmly committed to its conservation. His London counterpart William Grove had arrived at energy conservation by 1842 and wrote and lectured widely about it. By 1839, Marc Seguin, as a result of his work on steam engines, had concluded that heat and motion were interconvertible.[31] As we have seen, Carnot expressed a similar view in "Posthumous Remarks" (written by 1832). Julius Robert Mayer (1814–78) arrived at a firm commitment to energy conservation by 1842, particularly through his observation of animal heat; and Ludwig Colding (1815–88) reached the same conclusion by 1840.[32] In 1837 Friedrich Mohr (1804–79) expressed his belief in interconvertibility of forces in "On the Nature of Heat," rejected by *Poggendorff's Annalen* but published in the then obscure *Zeitschrift für Physik*.[33] Mohr himself did not learn of its publication until nearly thirty years later, thus suffering a fate similar to Carnot's: His far-reaching ideas had no influence on the evolution of the energy principle. Mohr wrote: "Besides the known fifty-four chemical elements there exists in nature only one agent more, and this is called force [kraft]; it can under suitable conditions appear as motion, cohesion, electricity, light, heat, and magnetism."[34]

Colding and Mayer, as well as Justus von Liebig (1803–73), who led a movement to subject living organisms to quantitative scrutiny, focusing especially on the nature of animal heat, and Gustav Adolph Hirn (1815–90) were influenced by Naturphilosophie. (Colding, an engineer, was a Danish protégé of Oersted, and Liebig studied with Schelling.) Even Helmholtz's father was a friend of Fichte.[35] So strong was this influence that Kuhn claims that Carnot and Joule stand almost alone in having their ideas about interconvertibility and conservation free of metaphysical predispositions.[36] Mendoza observes that Mayer's famous paper of 1842 "is almost unreadable today, the physical ideas being obscured under layers of philosophical verbiage."[37] Nevertheless, Elkana believes that Mayer was no "hunch" philosopher: "There is . . . no doubt as to the universality and depth of his genius." Furthermore, it is interesting that Oersted considered Colding "too exact" and that Holtzmann, Seguin, and Hirn were steam engineers and thus far from armchair philosophers. In short, the claim that the early progenitors of energy conservation scarcely went beyond the stage of metaphysical enchantment is groundless.

Mayer was a physician on a Dutch merchant vessel on a trip to the East Indies. In the process of trying to understand the nature and origin of animal heat, he concluded that there was a fixed relationship between heat and work and that they were two manifestations of a single indestructible quantity. With an original mind but no training in mechanics, Mayer expressed his ideas personally and idiosyncratically, especially in his analyses of collisions and gravitation. But he was firmly dedicated to the conservation of force and believed it to be the foundation for physiology. He argued that forces are causes; therefore, one has a chain of causes and effects in which no term can vanish, meaning that all causes must be indestructible.[38]

Mayer's claim to priority rests on his determination of the mechanical equivalent of heat in 1842, based in part on the experiments of Thompson.[39] For the height that a 1 gram mass could be raised against gravity, assuming that the heat required to elevate the temperature of 1 gram of water 1°C (the calorie) was converted into mechanical work, he obtained a value of 365 meters.[40] A year later Colding obtained a value for the "mechanical equivalent of heat" based on heat generated by friction, almost simultaneously with Joule. Yet as late as 1864 Colding wrote that "the forces of nature are something spiritual and immaterial, entities . . . [that] . . . must of course be very superior to everything in the material world; and . . . it is obvious that it is through them only that the wisdom we perceive and admire in nature expresses itself." He concluded: "It is inherent in the nature of things that the forces which sensibly disappear must again appear as active in some other way."[41]

By 1822 French engineer Marc Seguin (1786–1875) had written to John Herschel about the conservation of vis viva; Brewster published the letter in the *Edinburgh Philosophical Journal.*[42] Seguin traveled frequently to England and was acquainted with Davy and Faraday as well as with Herschel. His ideas and those of his uncle, Joseph Montgolfier of ballooning fame, may have influenced Joule, who might have read Seguin's letter to Herschel and did read Seguin's book, published in 1839.[43] William Grove (1811–96), the first lecturer at the London Institution in 1840, gave two celebrated lectures in 1842 and 1843, the latter titled "Correlation of Physical Forces," in which he expressed his faith in the interconvertibility of all forces and lamented the lack of a quantitative measure of their "equivalent of power." Mayer, Colding, and Joule were in the process of providing just that quantitative measure in 1842–43.

Chapter 4 mentions the experiments of Joule that led to his defense of the dynamical theory of heat, his determination of the mechanical equivalent of heat between 1843 and 1850, and his clear statement of energy conservation in 1847. During the seven years before 1850, Joule published descriptions of ten experiments that yielded values for the mechanical equivalent of heat.[44] These studies by the consummate experimentalist not only showed that mechanical work could be converted into heat through friction (for example, a paddle vigorously stirring water) but went further to demonstrate that electrical processes evolved heat and that one could relate, quantitatively (through the mechanical equivalent of heat), the heat evolved and the mechanical effort or effect.

But in 1847 Joule was still virtually unknown; at the Oxford BAAS meeting in June, the chairman of his session asked him to give only a brief summary of his remarks rather than read them at length. Joule recalled: "This I endeavored to do, and discussion not being invited, the communication would have passed without comment if a young man had not risen in the section, and by his intelligent observations created a lively interest in the new theory. The young man was William Thomson, who had two years previously passed the University of Cambridge with the Highest honour, and is now probably the foremost scientific authority of the age."[45]

Thomson himself remembered: "I can never forget the British Association at Oxford in the year 1847, when in one of the sections I heard a paper read by a very unassuming young man who betrayed no consciousness in his manner that he had a great idea to unfold. I was tremendously struck with the paper. I at first thought it could not be true because it was different from Carnot's theory, and immediately after the reading of the paper I had a few words of conversation with the author of the paper, James Joule, which was the beginning of our forty years' acquaintance and friendship."[46]

We have already noted the importance of this meeting, which contributed materially to the establishment of the law of energy conservation and to Thomson's work on thermodynamics and especially dissipation and the second law.

Joule offered his clearest statement of energy conservation in his St. Ann's Reading Room lecture in Manchester on 28 April 1847: "Wherever living force is apparently destroyed, whether by percussion, friction, or any similar means, an exact equivalent of heat is restored. The converse of this proposition is also true, namely, that heat cannot be lessened or absorbed without the production of living force or its equivalent attraction through space. . . . all three therefore—namely heat, living force, and attraction through space (to which I might also add *light,* were it consistent with the scope of the present lecture)—are mutually convertible into one another."[47]

In spite of the clarity of Joule's statement, some scholars (Elkana, in particular) consider Helmholtz to be the most important of all the scientists who

confirmed the energy principle because of the profound generality of his conclusions. In a sense, however, he was to Joule what Thomson or Maxwell were
to Faraday (although the analogy is weakened chronologically by the fact that
Helmholtz delivered his seminal "Die Erhaltung der Kraft," his first scientific
paper, to the Physical Society of Berlin on 23 July 1847, only a month after the
BAAS meeting at which Joule spoke). Helmholtz himself called Joule the "chief
discover of the conservation of energy." In the end, the two men were complementary, and the reception of the law of conservation of energy was hastened
because its empirical and mathematical foundations were offered simultaneously.

Privately published, Helmholtz's "Die Erhaltung der Kraft" was a fifty-
three-page manuscript divided into six sections: (1) the conservation of vis viva,
(2) the conservation of force, (3) application to mechanics, (4) the force-
equivalent of heat, (5) the force-equivalent of electric processes, and (6) the
force-equivalent of electromagnetism.[48] Helmholtz showed that conservation of
force (energy) was a generalization of the idea of conservation of vis viva in
analytical mechanics and that the phenomena of heat and electromagnetism, and
even living organisms, could be brought under this general principle. He concluded that for a collection of interacting bodies "the sum of their *vires vivae*
and tensions must be constant" and that "in systems for which the principle of
conservation of force can be applied in all its generality, the elementary forces
of the material points must be central forces."[49]

Again, it was important that, beginning in 1847, the different contributions of Joule and Helmholtz became known almost simultaneously. Those of
Helmholtz, while deeply fundamental (and immediately recognized as such by
Maxwell and Thomson), could be seen as a priori; therefore, they could be rejected by scientists who wished to eliminate all nonempirical elements from
physics. Because Joule's experiments could not be so readily dismissed, the two
sets of discoveries together had far more impact than either could have had alone.
But while that consensus was achieved rapidly after 1847, it had been retarded
by the fact that *Annalen der Physik,* the leading German physics journal, rejected
not only the papers of Mohr and Mayer but also the work of Helmholtz, evidently because of the absence of original experiments and the hostility toward
theoretical deductions that had been engendered by the speculations of the
Naturphilosophs.[50]

Route to Consensus

According to Kuhn, there are at least a dozen scientists who deserve some
credit for contributing, during the ten years after 1837, to the realization that
energy in all its manifestations is conserved.[51] They did not all have the same
thing in mind, and we must resist the temptation to lump them together or assume that everyone who spoke of conservation of force in the 1840s had a clear
idea of energy and energy conservation. Furthermore, technical historical priority is of little note; pioneers such as Mohr, Mayer, and Sadi Carnot (in the

PLATE 17. Hermann von Helmholtz (1821–94). *Portrait by C. H. Jeens. From* Nature *15 (1887): facing 389.*

matter of energy conservation) had little influence on those who came later or arrived independently at the same idea and carried it forward.

By 1845 Grove, Mohr, Holtzmann, Mayer, Colding, and Joule had all gone beyond philosophical speculation and obtained a value for the mechanical equivalent of heat.[52] Yet not all of them embraced the concept of conservation. So when examining the history of the idea of conservation of energy in the nineteenth century, we can draw a distinction between those like Colding and Mayer, who were committed to energy conservation for reasons that fell far short of mathematical or empirical proof, and Joule and Helmholtz, the former carefully demonstrating the conversion of one form of energy into another, the latter understanding the principle thoroughly enough to offer the soundest proof possible at the time.

By about 1850, rapid consensus meant that the energy principle (that is, energy and its conservation) became the foundation for a new methodology, a new way of approaching physical problems. Thomson and Tait's influential *Treatise on Natural Philosophy* (1867) took almost as its raison d'être the introduction of students to the energy program.[53] Force was relegated to second place—something derivative, to be defined in terms of work. With no comparable textbook in French or German, the *Treatise* became one of the principal vehicles through which the importance of the energy principle was broadcast in the 1870s and 1880s. The generalization of the energy principle by Einstein, who in 1905 showed that the conservation of energy was only an approximation, merely served to elevate the status of this fundamental law.

CHAPTER 6

Atomism

*I*n the philosophy of nature, certain issues stand apart from mathematical details, specific models of physical systems, or empirical results, even though they may be generalizations from experiment or a priori notions ultimately subjected to experimental test. Examples include the homogeneity of space and time and various symmetry and invariance (conservation) arguments—including energy conservation—all of which, in the end, have some sort of mathematical expression.

This issue in this chapter is somewhat different. We are concerned here with the ultimate structure of matter—specifically, whether it is continuous (and therefore essentially a fluid or an elastic solid) or whether it is made up of elementary discrete structures. This well-known problem has a long history, originating, as far as we know, with the Greek atomists of the fifth century B.C. The confrontation of two polar opposites—the atomistic and continuum theories of matter—is one of the major themes of the eighteenth and nineteenth centuries and lies behind much of the development of classical physics in the nineteenth century.

Until at least the seventeenth century, the discrete structure of matter was strictly a hypothesis unsupported by any empirical evidence. In the seventeenth century, considerations about the nature of heat moved speculations about atomistic structure from the realm of philosophy into physical theory, but only in the nineteenth century did Daltonian chemistry begin to provide the long-sought empirical evidence of atoms. The atom may be a commonplace of modern life, but we should not forget that even at the beginning of the twentieth century controversy still raged over the reality (as opposed to the theoretical efficacy) of atoms.

It is important to distinguish atomism as a philosophical position (that is, as an a priori hypothesis about the underlying structure of matter) from what we might call chemical atomism, which around 1800 first gave the atomic theory an empirical foundation—just before John Dalton published *A New System of Chemical Philosophy.* For more than 150 years, many philosophers had assumed that matter on the small scale was corpuscular and that the properties of the small particles that composed it somehow determined the gross structure of matter. But nothing resembling empirical evidence for this view existed until Daniel Bernoulli offered the first kinetic theory of gases in 1738. It was more than seventy years before chemical stoichiometry began to turn the tide in favor of atomism; and nearly another century elapsed before the evidence, this time from physics, became inescapable.

I will not discuss at length the relationship between ancient Greek atomism and modern atomic theory, with its origins in early nineteenth-century chemistry and the kinetic theory of gases.[1] But are we entitled to consider modern atomic theory a direct descendent of Greek atomism? Edward MacKinnon remarks that "the eventual success of atomism provides an abiding temptation to overemphasize the philosophical worth of the original doctrine."[2] Surely the idea that matter is divisible, even indefinitely divisible, is an idea that must have occurred independently to many thinkers as an alternative to the conception that matter is uniform. Thus, it is easy to scorn the idea that modern atomism owes anything to the Greeks, who had no empirical evidence of the underlying structure of matter. Alexandre Koyré, who held this view, not only dismissed Epicurus's atomism but declared Gassendi's revival of it in the seventeenth century "perfectly sterile."[3] But this attitude ignores the critical importance of the fact that an idea may be absorbed and accommodated by simple repetition. This process of preparing the ground for an idea, of making it plausible by continued discussion, is neither easily quantifiable nor attractive to methodologists. Nonetheless, it has played an important role in the reception of ideas—certainly in this case.

Moreover, despite a hiatus of sixteen centuries, we can connect indeed Greek and Roman atomism with that of modern times through the recovery of Lucretius's *De rerum natura* in the fifteenth century and trace the evolution of atomistic ideas into the seventeenth century, where they flourished in the hands of French humanist Pierre Gassendi (1592–1655) among others. In Francis Bacon's *On the Philosophy of Democritus, Parmenides, and Telesius,* "the atomic theory is selected upon and dwelt upon by him as the chain which connects the best parts of the physical philosophy of the ancient and the modern world."[4] Although Isaac Newton rarely mentioned atoms, and little of his work depended on the detailed structure of matter, his influence was crucial in giving rise to the corpuscular philosophy; and there is a direct causal chain linking Newton, Boyle, Dalton, and nineteenth-century atomism. As we shall see, John Dalton is important to both two threads of nineteenth-century atomism, but the chemical atomism that prevailed after 1810 had little in common with physical atomism until almost 1870.

It is true that from antiquity until the time of Newton—indeed, until the end of the eighteenth century—atomism had no experimental foundation, although it was employed as a conceptual framework for interpreting certain phenomena—for example, heat and, by 1738, the kinetic theory. Nonetheless, between 1646 and 1691 at least seven European natural philosophers either proposed or attempted to determine the sizes of atoms, employing dust particles, smoke, gold leaf, and so on. Typically, they arrived at a dimension on the order of 1 micron (micrometer [10^{-6}m]). Thus, Whyte calls the period between 1640 and 1700 the birth of experimental quantitative atomism.[5]

Eighteenth-Century Atomic Theory

The reason for the sterility of atomism in the eighteenth century—despite Newton's authority and a general acceptance of the idea of an underlying corpuscular structure to matter—is fairly clear.[6] It lies in the fact that, with chemistry and physics of matter in their infancy, there were few observations or experiments upon which the atomic theory could shed any light at all, and there was no theoretical basis for interpreting them. The most notable exception is the kinetic theory of Daniel Bernoulli (1738), which made a credible attempt to explain the observed properties of gases in terms of atoms. Euler's success in formulating hydrodynamics in terms of partial differential equations describing the state of a continuous medium pushed the problem of microscopic structure into the background as an unnecessary metaphysical assumption.[7] As we noted in chapter 2, the ideas of Immanuel Kant dominated the discussion of the metaphysics of natural philosophy during the second half of the eighteenth century. He, better than anyone else, drew a clear distinction between mechanical explanations, which depended upon hypotheses of atoms and the void and forces of repulsion and attraction, and dynamical natural philosophy, which made use of "the moving forces of attraction and repulsion originally belonging to them."[8] The dynamical philosophy did not entirely reject atoms; rather, it rejected the method that explained the properties of matter by making detailed assumptions about atoms interacting in a void. To call this view positivistic is perhaps to indulge in labeling at the expense of understanding because the dynamical position was anything but free of metaphysical assumptions.

The atomic theory of Jesuit natural philosopher Roger Boscovich (1711–87) was very much in this dynamical tradition.[9] Although his theory retained atoms, they were sources of inherent powers or forces that were actually the primary entities: "Matter is unchangeable, and consists of points that are perfectly simple, indivisible, of no extent, & separated from one another; . . . each of these points has a property of inertia, & in addition a mutual active force depending on the distance."[10] Boscovich's theory owed much to Leibniz and his monads, especially the idea of nonextended primary elements. To Newton's forces of mutual attraction Boscovich added repulsive forces that kept the point force centers from being in contact: "The primary elements of matter are in my opinion perfectly indivisible & non-extended points; they are so scattered in an immense vacuum that every two of them are separated from one another by a definite interval; this interval can be indefinitely increased or diminished, but can never vanish altogether . . . for I do not admit as possible any immediate contact between them."[11] Boscovich's atomism appealed strongly to certain natural philosophers with dynamical inclinations, including Humphrey Davy and his protégé, Michael Faraday.

Chemistry and Atomism: 1770–1830

To anyone with even a tenuous interest the history of chemistry, the names Priestley, Lavoisier, Dalton, and Avogadro symbolize the birth of chemistry, which took place in the decades surrounding 1800. There were, of course, others who participated in the rise of chemistry as a scientific discipline based on sound experimental practice and theoretical structures. The most fundamental of these theories—the sine qua non of chemistry as we know it—is the atomic theory. As we have already noted, atomism in the nineteenth century existed in two different forms, which Scott, Rocke, Knight, and others have separated into physical and chemical atomism.[12] While the origins of physical atomism go back to Boyle, Gassendi, Newton, Bernoulli, and Laplace, chemical atomism really began with John Dalton and William Higgins, who based it on the "new" chemistry of Priestley, Lavoisier, Richter, and others. The idea that the elements had fixed or constant properties was advocated by Boyle, Georg Stahl (who founded the phlogiston theory), and Lavoisier, leading to the conclusion that the chemical properties of matter were the result of its discreteness.[13] Of course, some scientists whom we label as chemical atomists, such as Davy and Wollaston, were not strangers to the idea of physical atomism.[14] Some vacillated between one view and the other, some were merely cautious, and others declined to speculate at all.

Jeremias Benjamin Richter (1762–1807), a porcelain chemist, and Joseph Louis Proust (1755–1826) independently established the law of combining or definite proportions—the law of equivalents—which lies at the heart of chemical stoichiometry and was a foundation of Dalton's atomism.[15] Richter's German countryman E. G. Fischer consolidated this work, which showed that elements could combine in only *one* proportion or ratio, and published the first table of equivalent weights.[16] Equivalent weights are equal to the atomic weight or a submultiple of it and depend on valence. In *Essay on Chemical Statics* (1803), Berthollet disputed the discreteness of this law of combination, without effect.

Thackray characterizes the late 1770s and the 1780s as "the heroic days of Newtonian chemistry," especially in France: "The long line of speculation and debate deriving from Newton's own work was crystallized in the rival systems of Boscovich and Buffon. The faltering inquiries of Taylor and Hauksbee, and the empirical affinity tables of a variety of chemists were available to experimenters. Thus there seemed to be both a sound theoretical base, and a smattering of technique to guide the Newtonian investigator. With such men as Guyton, Fourcroy, Kirwan, Bergman, and Wenzel engaged in the struggle, progress seemed assured and the quantification of chemical force-mechanisms imminent."[17]

Priestley and Lavoisier

The case of Joseph Priestley (1733–1804) is especially interesting. Like Dalton, he was a religious dissenter (Presbyterian, in this case), and he eventually became an Arian; he also was of humble origin.[18] In *History of Electricity* (1767), Priestley deduced (from the observation that a pith ball inside a hollow conductor did not experience any electrical force when an electrified object was brought near it) that the electrical force, like Newtonian gravity, must be inverse-square, thus anticipating and perhaps inspiring Cavendish's experiment.[19] In 1780 Priestley moved to Birmingham and became a member of the Lunar Society, whose members supported him through subscriptions. Although he wrote eleven volumes of religious history between 1781 and 1791, he devoted the last thirty-five years of his life largely to the study of gases—"pneumatic chemistry," in the language of the time.

Priestley's connection to atomism is tenuous. But he was one of the founders of late eighteenth-century chemistry (especially pneumatic chemistry); and with Cavendish, Scheele, Lavoisier, Hales, and Black, he provided much of the raw material that was used to build atomic chemistry. Furthermore, he enthusiastically embraced the point atomism of Roger Boscovich, not least because it could easily be reconciled with theological concerns.[20] Nevertheless, this issue caused a break between the Jesuit Boscovich and Priestley, who, according to Thackray, saw point atomism as a way of eliminating matter-spirit dualism.[21] Priestley's *Disquisitions Relating to Matter and Spirit* (1777) met suspicion, even hostility, from continental Jesuits, commonsense Scottish philosophers, and even Quakers such as Richard Price and John Whitehead. According to Thackray, Whitehead "gave utterance to the common fear that 'materialism'—by which he meant the denial of solid atoms—'must terminate in atheism.'"[22]

Ironically, while Priestley's discovery of oxygen (independently of Swedish chemist Carl Wilhelm Scheele) helped lead to the demise of the phlogiston theory, he never relinquished it, partly because of the firm conservatism of James Watt, a fellow member of the Lunar Society of Birmingham. In the words of Robert Schofield, Priestley "is particularly remembered in the history of science for the peculiar stubbornness with which he defended a theory of chemistry that he had done much to destroy."[23]

Members of the Lunar Society (Priestley, in particular) were favorably disposed toward the French Revolution. In general, however, English reaction against its potential for social disruption made such views unpopular. Angered by his religious views and his support of the revolution, a mob destroyed Priestley's laboratory in the Birmingham riots of 1791. Shortly thereafter, he moved to America, where he died in 1804. The Lunar Society was effectively dissolved by disagreement over support for Priestley and his notorious views.[24]

Anton-Laurent Lavoisier (1743–94) was educated in the Collège des Quatre Nations, which d'Alembert had also attended, and studied under the astronomer Lacaille. At the age of twenty he received a law degree, and throughout the next thirty years he was a distinguished statesman and scientist.[25] Although he played an important role in the French Revolution, he fell victim to the Reign of Terror, which suppressed the Royal Academy of Sciences, and he was guillotined in May 1794. Lavoisier's contributions to chemistry are too numerous and profound for us to describe here; for the most part they are also beyond the scope of this chapter. He was, however (with Priestley and others), instrumental in establishing chemistry as an empirical science with theoretical underpinnings.[26] If we couple this work with Dalton's chemical atomism, we have the main ingredients of the revolution that created modern chemistry between the 1770s and 1810.

In 1777, Lavoisier took the young Laplace under his wing; and in 1783 they presented to the Academy their famous "Memoir on Heat," which developed Lavoisier's theory of combustion. An element of this process was fire (or caloric). Two years later Lavoisier wrote "Reflections on Phlogiston," which sounded the death knell for the phlogiston theory, even though Priestley adhered to it to the end. Lavoisier made the case for his new chemistry in *Elementary Treatise of Chemistry,* published in 1789.

Like Priestley, Lavoisier made no direct contribution to atomism. Indeed, it is interesting to note that such a pioneer of chemistry was still grappling with the problem of the four elements (earth, air, fire, and water) inherited from Greek antiquity. But by the end of the eighteenth century, Lavoisier, Priestley, and their contemporaries had shown the compound nature of many substances; identified more than fifty "elements" (in Lavoisier's words "substances non decomposés"), which included hydrogen, carbon, oxygen, and nitrogen as well as light and caloric; and greatly elucidated the nature of chemical reactions.[27] It remained for John Dalton to provide the theoretical insight that would complete the revolution and make these reactions intelligible.

John Dalton

It is to John Dalton (1766–1844) that we give the credit for establishing chemistry on the principles of atomism. Newton's atomism was a strong influence—in particular, Newton's derivation of Boyle's law.[28] In fact, the atomic model used by Boyle and Newton was a static one; a model of the properties of a gas based on random motions of atoms was not proposed until 1738.

Dalton was the son of nonconforming middle-class Quaker parents and grew up in rural Cumberland, where he attended Quaker schools. He became a teacher by the age of twenty-one, other careers being largely closed to him because of his humble origins and his religious beliefs.[29] Dalton's mentors included Elihu Robinson and John Gough, and he heard lectures by Joseph Banks in 1782, when he was sixteen years old. The young Dalton profited from the Quaker emphasis on natural philosophy; from itinerant natural philosophers and other visitors to Kendal Boarding School in the English Lake District, where he taught in the 1780s; and from an excellent library at the school. In 1792 he became professor of mathematics and natural philosophy at New College, Manchester, which had been established by the Unitarians (Socians). He held this post for eight years and thereafter remained in Manchester as an independent teacher until his death. He lectured at New College, at the Royal Institution in London, and in Edinburgh during the years when he was developing his theory of chemical atoms. Dalton was the shining light of the Manchester Literary and Philosophical Society, which he joined in 1793. Although he was elected to the Royal Society in 1822, he had little use for it, preferring instead the British Association for the Advancement of Science (BAAS), which he helped to found in 1831 and believed to be more seriously scientific.

Dalton wrote *Meteorological Observations and Essays* in 1793 while teaching at the Kendal School.[30] But he accomplished his most renowned work after he moved to Manchester and then left New College in 1800. One might expect that the father of modern atomic theory would have turned early to the idea that heat was merely a manifestation of the motion of atomic particles. On the contrary, he was devoted to the caloric, which he pictured as surrounding each atom in an envelope:

> All bodies of sensible magnitude ... are constituted of a vast number of extremely small particles, or atoms of matter bound together by a force of attraction, which is more or less powerful according to circumstances. ... Whether the ultimate particles of a body, such as water, are all alike ... is a question of some importance. From what is known, we have no reason to apprehend a diversity in these particulars: if it does exist in water, it must equally exist in the elements constituting water, namely hydrogen and oxygen. ... In other words, every particle of water is like every other particle of water; every particle of hydrogen is like every other particle of hydrogen. ... Besides the force of attraction, which, in one character or another, belongs universally to ponderable bodies, we find another force that is likewise universal ... namely a force of repulsion. This is now generally, and I think properly, ascribed to the energy of heat. An atmosphere of this subtle fluid constantly surrounds the atoms of all bodies and prevents them from being drawn into actual contact. This appears to be satisfactorily proved by the observation, that the bulk of a body may be diminished by abstracting some of its heat.[31]

Dalton's earliest work, beginning with "Meteorological Investigations" (1793) and elaborated on during the next decade, concerned mixtures of gases, including their diffusion, and led to the law of partial pressures that bears his name. In particular, he was interested in the idea that the absorption of gases in water or air did not involve chemical affinity. He appended to "Absorption of Gases by Water" (read in 1803 and published in 1805) a list of chemical equivalents as well as what Thackray calls "a disconsolate footnote," saying that "subsequent experience renders this conjecture less probable."[32] The major difficulty in arriving at atomic weights from combining proportions was the need to assume the number of atoms of each element involved. For example, assuming that water contains one atom each of hydrogen and oxygen, it follows that an oxygen atom is about eight times as heavy as a hydrogen atom (according to Dalton, six times as heavy). By assuming a ratio of two atoms of hydrogen to one of oxygen, Dalton's result would be that oxygen is about twelve times as heavy.[33] Hence, he preferred the entirely empirical equivalents rather than atomic weights. In *New System of Chemical Philosophy* Dalton presented his models of spherical atoms that combined chemically due to attraction of the unlike parts and repulsion of the like parts (see figure 6–1). In this model, the atoms *are* the elements, which have no internal structure. He also gave a rationale for determining the structure of each substance, a "law of simplicity" that generally assumed, somewhat arbitrarily, that only one atom of each element would combine with one of another, reducing the mutual repulsion of like atoms. Dalton thus established the chemical basis for atomism and provided a plausible integration of caloric into his picture by showing how the repulsive forces between the hard particles of caloric and the attractive forces among the atomic particles led to the structure of matter.[34] It is ironic that his pupil, James Prescott Joule, played a vital role in eliminating caloric from physics and chemistry by championing the mechanical theory of heat and taking one of the first steps toward a kinetic theory of gases (see chapter 7).[35]

From the observed stoichiometry of chemical reactions, Dalton found that elements combined by weight in definite proportions that were the ratios of integers. In a lecture to the Royal Institution in 1810, he described his discovery: "This idea occurred to me in 1805. I soon found that the sizes of the particles of elastic fluids *must* be different. . . . The different sizes of the particles of elastic fluids under the circumstances of temperature and pressure being once established, it became an object to determine the relative *sizes* and *weights,* together with the *relative number* of atoms in a given volume. . . . Thus a train of investigation was laid for determining the *number* and *weight* of all chemical elementary particles which enter into any sort of combination one with another."[36]

Even this result failed to convince many scientists of the existence of atoms; contradictory results seemed to suggest that these ratios were not fixed, and it was not clear how such simple ratios emerged from the combination of (presumably) untold numbers of atoms. At the same time, Gay-Lussac was showing that the volumes of gases also combined in simple ratios, although it proved difficult to reconcile in detail his results with Dalton's. According to Scott, these difficulties prevented Gay-Lussac from interpreting his results in terms of the atomic theory, which he was inclined to do; thus, he became a supporter of the law of multiple proportions but not of atomism.[37] William Wollaston and Humphrey Davy (and others, both British and French) wanted to use a term with less philosophical baggage than *atom* and opted for *chemical equivalent* or *proportion numbers.*[38] In 1810 Amadeus Avogadro proposed what could have been the solution—namely, that every atom or molecule of whatever gas occupied

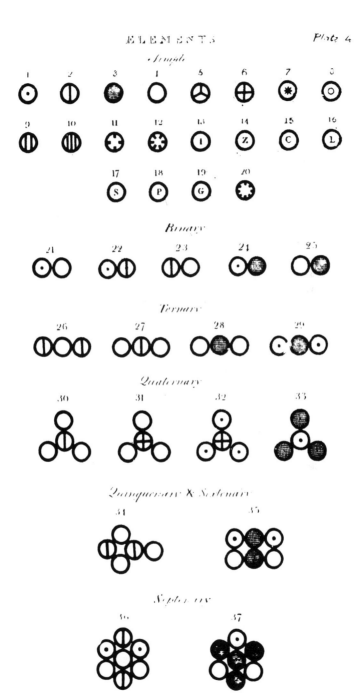

FIGURE 6–1. Dalton's models of spherical atoms. *From John Dalton,* New System of Chemical Philosophy *(London: Dawson, 1808).*

the same volume. This "discovery" attracted little attention, however; and although there continued to be considerable interest in atomism in its general aspects during the next few decades, not much progress was made. Indeed, the discovery of organic molecules introduced a substantial complication into chemistry.

Dalton and Humphrey Davy represented the two strands of Newtonianism that we have already discussed in chapter 3: that of the *Opticks* (Dalton) and that of the *Principia* (Davy). Scott speculates about how we might view matter today if the pictorial character of Dalton's theory had not prevailed, although it is hard to agree with him that atoms as visualizable entities might have vanished permanently from science in favor of empirical proportions or equivalents. Scott quotes W. V. Farrar as saying: "John Dalton had a pictorial imagination. There was never anything shadowy or metaphysical about his atoms; they were (in Newton's phrase, which he often quoted) 'solid, massy and hard'; too small to see but very real. This concreteness of the imagination proved to be Dalton's great strength as a chemist; for it so happened that chemistry thrived in the nineteenth century when it was naive and pictorial, and languished when it tried to be abstract and subtle."[39]

Until almost midcentury, "real" atoms were primarily the province of chemists; physicists employed them only in general terms, often reflecting a philosophical position rather than a commitment growing from the work of Dalton and Avogadro.[40] But what Dalton had accomplished—and not alone, by any means—was to remove the atom from the realm of speculative philosophy to the laboratory. Initially, this was only the chemistry laboratory, where its manifestations seemed irrefutable. But in due course the implications of atoms for crystal structure, electricity, and magnetism (Faraday and Ampère); the theory of heat; and atomic spectra became clear.

Between 1805 and 1810 Dalton's ideas began to gain acceptance. First, he won over Wollaston and Thomas Thomson and then others; his *New System of Chemical Philosophy* was widely read.[41] Still, these scientists often did not accept atoms per se but instead the law of proportions, of equivalents, which was "divested of all hypothesis."[42] Knight's image of Dalton as the "Kepler of chemistry, which still awaited its Newton," is apt; for a true atomic theory was not offered until the early years of the twentieth century.[43] Although chemistry underwent great changes in the nineteenth century, those who thought it would soon be transformed from its empirical stage into a deductive, mathematical science were frustrated for a full century. In his posthumously published *Consolation in Travel: The Last Days of a Philosopher,* Humphrey Davy wrote: "A young man who has performed the ordinary course of college studies, which are supposed fitted for common life and refined society, has all the preliminary knowledge necessary to commence the study of chemistry. . . . All the implements absolutely necessary may be carried in a small trunk; and some of the best and most refined researches of modern chemists have been made by means of an apparatus which might with ease be contained in a small travelling case, and the expense of which is only a few pounds."[44] But his death ended a period of much popular excitement over chemistry. The new science became more quantitative, highly specialized, and beyond the reach of most amateurs.[45]

Davy and Faraday

During the first third of the nineteenth century, atomism advanced on several fronts. One of the eighteenth-century bases of nineteenth-century atomic theory was the point atomism of Roger Boscovich. His *Theory of Natural Philosophy,* published in Venice in 1763, attempted to explain the properties of matter in terms of point atoms, which were surrounded by forces that were alternately attractive and repulsive; matter itself essentially disappeared because the atoms had no extension. Boscovichean point atomism was at the heart of Davy's and Faraday's theoretical descriptions, and Davy's last written words analyzed how point atoms could be arranged to give matter its observable properties.[46]

Perhaps we shall never fully understand the extent of Davy's commitment to atoms or to Boscovichean rather than Daltonian atoms. We know that he employed atoms in his discussions, and we also know that he considered them too hypothetical and speculative. In lectures to the Royal Institution in 1802, he said: "Though the corpuscular philosophy supposes the existence of bodies composed of similar particles, yet we are not certain that any such bodies have been yet examined. The simple principles of the chemists are substances which have not been hitherto composed or decomposed by art; and they are elements, only in relation to other known substances."[47] Later he told Berzelius that Dalton's ideas were "more ingenious than correct."[48] In 1812 Davy regarded atoms as "mere creatures of the imagination."[49] And in *Elements of Chemical Philosophy,* published that year, he wrote more philosophically, "Our numerical expressions ought to relate only to the results of experiments."[50] John Davy claimed that his brother never accepted Dalton's atoms, but Coleridge thought otherwise, lamenting Davy's commitment to atoms. Davy seems to have been a major influence on Faraday's gradual acceptance of Boscovichean atomic theory.[51]

It is clear that Davy did make use of definite or combining proportions after about 1808. In *Elements of Chemical Philosophy,* he argued from the fact that two volumes of hydrogen combined with one of oxygen to produce water. Thus, he adopted the 2:1 ratio of hydrogen to oxygen atoms and concluded that "the specific gravity of hydrogene is to that of oxygene as 15 to 1."[52] In *History of Chemistry,* Thomas Thomson argued that all Davy did was substitute the word *proportion* for *atom* and touted Davy as a "strenuous supporter" of atomism.[53] But in *Consolations in Travel,* Davy hedged, saying that because we "never can see the elementary particles of bodies, our reasoning upon them must be founded upon analogies from mechanics, and the idea that small indivisible particles follow the same laws of motion as the masses that they compose."

Although Faraday did not declare his support for Boscovich's atoms until 1844, his belief in a corpuscular structure of matter, inherited from Davy and

Boscovich, is evident in his writing during the 1820s.[54] Faraday also absorbed the atomic theory from Thomas Thomson, who introduced Dalton's atomism into the third edition of *A System of Chemistry* in 1807.[55] The Daltonian atom was much like a billiard ball; rearrangements and splittings due to collisions were responsible for chemical changes. The atoms were attracted by gravitational forces but possessed a repulsive force due to caloric. The atom of Boscovich was, on the other hand, a point atom; and the properties of matter resulted from the arrangement of these point atoms, an arrangement due in turn to the attractive and repulsive powers that surrounded each atom.

Like Davy, Faraday worked to keep hypotheses at a minimum. To him, Boscovich's theory, while not necessarily superior to Dalton's in chemistry, explained crystalline structure and electrical conductivity, which the latter did not. Feeling that simple Daltonian atoms were inadequate, Faraday offered a theory in 1816 that, "supposing a polarity to exist in the particles and that certain points attract whilst others repel," tried to explain how such regular arrangements of atoms might form. Fourteen years later, his views were still maturing. Nevertheless, his commitment to point atomism may have cemented as early as 1823, with his work on liquefaction of chlorine and its implications.[56]

Faraday only became explicit about his commitment in 1844, when, in a paper read to the Royal Institution in January, he drew a distinction between what he called theory and hypothesis and "that which is the knowledge of facts and laws."[57] After summarizing the problems involved in understanding conductors and insulators according to the usual model of material atoms separated by empty space, he declared:

> If we must assume at all, as indeed in a branch of knowledge like the present we can hardly help it, then the safest course appears to be to assume as little as possible, and in that respect the atoms of Boscovich appear to me to have a great advantage over the more usual notion. His atoms, if I understand aright, are mere centers of forces or powers, not particles of matter, in which the powers themselves reside. . . . Hence matter will be *continuous* throughout, and in considering a mass of it we have not to suppose a distinction between its atoms and any intervening space. The powers around the centres give these centres the properties of atoms of matter; and these powers again, when many centres by the conjoint forces are grouped into a mass, give to every part of that mass the properties of matter. In such a view all the contradiction resulting from the consideration of electric insulation and conduction disappear.[58]

Just as Davy's most important discoveries were in electrochemistry, much of Faraday's work concerned the effects of galvanic currents on chemical compounds, especially electrolytes. This technique provided a different way of obtaining relative atomic weights that was a basis of Berzelius's theory of chemical proportions. The implications of Faraday's 1833 discovery of the second law of electrochemistry, which asserts that the amount of a substance deposited at an electrode in an electrochemical process is proportional to its equivalent weight, are revolutionary (as Williams points out in hindsight) because they imply a quantum of electricity—in effect, the electron.[59] Faraday, however, saw chemical affinity in terms of the force fields of point atoms rather than an electrical fluid or electrical matter. He wrote that he was "jealous of the term *atom;* for though it is very easy to talk of atoms, it is very difficult to form a clear idea of their nature."[60]

For Faraday, the problem of electrical conductivity was an especially clear example of a phenomenon that required Boscovichean point atoms. Daltonian billiard-ball atoms were not in contact even in solids, and certainly not in gases, meaning that electrical conductivity seemed impossible unless space were a conductor. But if space were a conductor, how could one explain insulators? Faraday said: "It would seem, therefore, that in accepting the ordinary atomic theory, space may be proved to be a non-conductor in non-conducting bodies, and a conductor in conducting bodies, but the reasoning ends in this, a subversion of that theory altogether. . . . Any ground of reasoning which tends to such conclusions as these must in itself be false."[61] In Faraday's Boscovichean theory, the forces filled all space; and because, in this model, matter was nothing but force, it was coextensive with space.[62] Above all, the theory was a denial of action-at-a-distance.

Faraday had thought since 1833 that force was transmitted by contiguous particles, and in groping his way toward the field theory he sought to understand magnetic lines of force as lines of contiguous-acting particles as he had earlier viewed the electrostatic force. But by the early 1850s, he was thoroughly convinced of the reality of the magnetic lines of force and had moved away from explaining them in terms of the action of contiguous particles: "If they exist, it is not by a succession of particles, as in the case of static electric induction, but by the condition of space free from such material particles. . . . these lines exist in . . . a vacuum as well as where there is matter."[63] Thus, while Faraday continued to believe in the point atoms of Boscovich, he had dispensed with the contiguous particles as the means of transmitting magnetic force in favor of the independently existing lines of force in empty space.

Chemical Atomism: 1830–70

During the forty years before 1830, France was peerless among the scientific nations of the world. The names Laplace, Poisson, Berthollet, Ampère, Fourier, Carnot, Fresnel, Biot, Cauchy, Gay-Lussac, Lavoisier, de Morveau, Fourcroy, Dulong, and Petit do not exhaust the roll of scientific luminaries working in France during this period.[64] Despite the inclinations of scientists such as Fresnel and Fourier to eschew untestable hypotheses, French discoveries proved crucial in propagating the atomic theory. Notable among them was the work of Pierre Louis Dulong (1785–1838) and Alexis Thérèse Petit (1791–1820), who showed that the specific heat of a body was directly related to its atomic weight—in particular, that the product of specific heat and atomic weight was a constant.[65] The Dulong-Petit law (mentioned in chapter 4 and reexamined in chapter 7) provided an absolute determination of atomic weight.

Elsewhere, in about 1815, Swedish chemist Jakob Berzelius (1779–1848) accurately determined atomic weights, thus clarifying many of the issues surrounding the atomic theory. He attempted to explain chemical affinity in terms of electric polarity or polarization.[66] Although he was a disciple of Richter and Davy, Berzelius was a confirmed atomist and made a strong case for atomism through his prolific writing.[67] He opened new horizons by extending ideas of chemical affinity to organic compounds, which unified chemistry by showing that these substances also obeyed the law of multiple proportions.[68] Rocke says that Berzelius may have been the most skilled analyst of his day and notes that he was critical of the precision of the work of both Dalton and Davy.[69] He was also responsible for introducing the now standard nomenclature for chemical elements, using the first letter of the Latin name of the element.[70] At about the same time (1815), Thomas Young was attempting to deduce the size or separation between molecules by observing water in a capillary tube. The value he obtained, somewhat less than 10^{-9} inches (or 10^{-9} cm), is within an order of magnitude of the correct value.[71]

It would take us far afield to trace the evolution of inorganic and organic chemistry during this period, when analytical techniques improved and physical chemistry was founded. But we would be remiss if we did not pause to mention the genesis of the periodic table of the elements, which arose from the discovery of the laws of combining proportions and measurements of atomic weights, including those of newly discovered elements. This task was accomplished by Newlands, Meyer, Mendeleev, and Thomsen between 1864 and 1869, when they came to understand that the elements could be arranged in families with similar chemical properties. In a sense, they realized the old idea of British physician William Prout, who in 1815 had attempted to show that the elements were all multiples of hydrogen. His attempt immediately raised the issue of whether the elements were simple or instead compounded of smaller, more fundamental parts. The fact that atomic weights (as measured by Berzelius, for example) were not simple integers cast doubt upon Prout's thesis, which was laid aside for the time being.

William Whewell, with whom Faraday often corresponded, was not enthusiastic about any atomic hypothesis. In 1837 he wrote in *History of the Inductive Sciences:*

> So far as the assumption of such atoms as we have spoken of serves to express those laws of chemical composition which we have referred to, it is a clear and useful generalization. But if the Atomic Theory be put forwards (and its author, Dr. Dalton, appears to have put it forwards with such an intention,) as asserting that chemical elements are really com-

posed of *atoms,* that is, of such particles not further divisible, we cannot avoid remarking, that for such a conclusion, chemical research has not afforded, nor can afford, any satisfactory evidence whatever. The smallest observable quantities of ingredients, as well as the largest, combine according to laws of proportions and equivalence which have been cited above. How are we to deduce from such facts any inference with regard to the existence of certain smallest possible particles?[72]

Whewell epitomized the confusion that clouded the debates over atoms and the use of the term *atom,* which, in fact, implied indivisibility. Were chemical elements atoms themselves, or were they comprised of smaller, indivisible entities? Whewell was correct in saying that the answer was beyond the realm of chemistry. He thought that "the grounds of the assumption of the atomic structure of substance are to be found rather in the idea of substance itself, than in the experimental laws of chemical affinity."[73] Furthermore, after quoting Newton's preface to the *Principia* and recounting Laplace's success in explaining capillarity, elasticity, heat conduction, and so on in terms of particles interacting through attractive and repulsive forces, Whewell remarked: "The actual constitution of bodies as composed of distinct and separate particles is by no means proved by these coincidences. The assumption, in the reasoning, of certain centers of force acting at a distance, is to be considered as nothing more than a method of reducing to calculation that view of the constitution of bodies which supposes that they exert a force at *every* point. It is a mathematical artifice of the same kind as the hypothetical division of a body into infinitesimal parts, in order to find its center of gravity; and no more implies a physical reality than that hypothesis does."[74]

By the 1860s, the issue was far from decided. In the first place there had long been evidence that atoms divide in chemical reactions or that they form molecules that divide in such processes. Dumas had written in 1837 that "it is my conviction that the chemists' equivalents . . . which we term *atoms,* are nothing but molecular groups."[75] The utility of the atomic doctrine (using the term to describe the combining of elements into constant, discrete ratios or proportions) was indisputable; but as a picture of the underlying structure of matter, it was still a subject for debate. The two sides of the issue were joined at the Chemical Society meetings in 1867 and 1869. In his presidential address in 1869, Alexander Williamson said: "I am not overstating the case when I say that, on the one hand all chemists use the atomic theory, and that, on the other hand, a considerable number of them view it with mistrust, some with a positive dislike."[76] An 1860 chemical conference in Karlsrühe, Germany, helped organize a consensus on the new idea of valence, which was built directly on the assumption of atoms and their weights. To a considerable extent, this consensus vindicated an old idea of Berzelius's; he had long since argued that the atomic constituents of molecules were held together by electric forces.[77]

Atomism and Physics: 1840–70

Outside chemistry, the issue of atomism arose most immediately in the theory of heat (see chapter 4). For more than a century, until the 1780s, heat was generally assumed to be a manifestation of atomic motions. Then, after a period during which the caloric theory dominated (roughly 1780–1825), scientists entered a time of transition between 1824 and 1847 as the dynamical theory of heat began to return to ascendancy. Nonetheless, the reality of atoms was still under debate. In the late 1850s, the kinetic theory of gases arose from these considerations in an attempt to give the dynamical theory of heat a microscopic basis, creating a field of physics that required the existence of atoms; treated atoms quantitatively; and yielded information about atomic or molecular sizes, speeds, and so on.[78] This began the era of quantitative physical atomism in which, for the first time, predictions of the atomic theory could be subjected to experimental test. It led to early measurements or estimates of the diameters of molecules—Loschmidt's in 1865 (the first reliable estimate of atomic size), Stoney's in 1868, and William Thomson's in 1870.[79] The story is largely told in chapter 7.

By the time Thomson went up to Cambridge in 1841, the reforms instituted by Babbage, Herschel, and the Analytical Society had taken effect. One of the consequences (at least, for the aether and light) was that mathematical description sufficed and microscopic theories based on point atoms attracted less interest. The triumph of the wave theory of light was a further blow to those who favored a corpuscular point of view; Smith and Wise note that adherents of particle theories were mainly experimentalists.[80] The problems that occupied Thomson's interest during the early 1840s were those in electrostatics and heat flow, both of which yielded readily to continuum techniques. Using the caloric theory of heat and the concept of electricity as a fluid, scientists as diverse as Gauss, Green, Ampère, and Thomson were able to make great strides. For Thomson, this nonhypothetical viewpoint was a deeply ingrained predisposition, which, according to Smith and Wise, "should be understood to involve his deepest emotional commitments and family ties, along with his social and political commitments, and not merely a rational view of what constitutes good science."[81]

But atoms and other discrete microscopic structures did not arise only in the theory of heat. As the wave theory of light gained adherents, attempts to understand the nature of the aether led to mechanical models of it, some requiring point atoms, others employing vortex atoms.[82] It might seem that no one could go far in electrostatics without the notion of a point charge, but that did not itself imply any commitment to the real existence of particles. Rather, to someone like Thomson, it was an idealization, an equivalent to using infinitesimal elements of volume or mass in applications of the calculus. Thomson did, however, attempt to devise an atomic theory based on the properties of vortex rings as Helmholtz used them; and Maxwell thought enough of the theory to

discuss it at length in his *Encyclopaedia Britannica* article on atoms, even defending it as being remarkably free from adjustable parameters.[83] Perhaps surprisingly, Pierre Duhem gave it a nod in *The Evolution of Mechanics,* published
in 1903.[84] Thomson himself had little patience for the uses to which atoms had
been put, and in 1870 he wrote:

> The idea of an atom has been so constantly associated with incredible
> assumptions of infinite strength, absolute rigidity, mystical actions at a
> distance, and indivisibility, that chemists and many other reasonable
> naturalists of modern times, losing all patience with it, have dismissed
> it to the realms of metaphysics, and made it smaller than "anything we
> can conceive." Chemistry is powerless to deal with this question, and
> many others of paramount importance, if barred by the hardness of its
> fundamental assumptions, from contemplating the atom as a real por-
> tion of matter occupying a finite space, and forming a not immeasur-
> ably small constituent of any palpable body.[85]

Nevertheless, in "The Size of Atoms," he grudgingly proceeded to calculate an
average size for molecules of somewhat less than 10^{-8} cm, comparable to modern values. He also defended his favorite vortex atoms against the ideal of rigid
billiard-ball–like atoms. Thomson had developed this theory in the 1860s and
was surprised to find that Helmholtz shared his interest in vortex motion.[86]

In 1847 Joule gave his famous St. Ann's reading room lecture, and at the
1848 BAAS meeting in Swansea he presented for the first time his version of
the kinetic theory of gases. On the basis of this theory he was able to calculate
the specific heat of several gases and obtain the speed of hydrogen atoms at
60°F.[87] Joule was clearly influenced by Herapath, who had published *Mathematical Physics* early in 1847.[88] Characteristically, however, Joule was not explicit
about his view of the motion of the atoms on which his kinetic theory was based.

Thus began a process that had its genesis in the late 1840s and reached
fruition in the late 1860s. It was initiated by Joule's championing of the dynamical theory of heat and the introduction of the kinetic theory by Clausius and
Maxwell, which eventually led to the conversion of many reluctant scientists.
Neither Joule, Clausius, and Maxwell nor those who preceded them (namely,
Waterston, Herapath, and Krönig) could have developed their ideas without a
thorough commitment to atoms or to molecules as collections of atoms. By the
late 1860s the dynamical theory of gases—in particular, Maxwell's theory of
viscosity—made it possible to calculate the diameters of atoms. The physical
atom, which had been the subject of speculation for 2,000 years, finally entered
the realm of experimental physics.[89] This made the chemical atom real as well;
for as we have seen, chemists were also debating the reality of atoms as opposed to the empirical combining proportions or equivalents. Maxwell commented on these physical and chemical atoms (in the context of a discussion of
molecules):

The definition of the word molecule . . . as employed in the statement of Gay-Lussac's law [that the weights of the chemical equivalents of different substances are proportional to the densities of these substances in gaseous form] is by no means identical with the definition of the same word as in the kinetic theory of gases. The chemists ascertain by experiment the ratios of the masses of the different substances in a compound. From these they deduce the chemical equivalents of the different substances, that of a particular substance, say hydrogen, being taken as unity. The only evidence why substances combine in definite ratios is that the molecules of the substances are in the ratio of their chemical equivalents, and that what we call combination is an action which takes place by a union of a molecule of one substance to a molecule of another. . . . This kind of reasoning, when presented in a proper form and sustained by proper evidence, has a high degree of cogency. But it is purely chemical reasoning; it is not dynamical reasoning. It is founded on chemical experience, not on the laws of motion.[90]

In a lecture to the BAAS at Bradford in 1873, Maxwell drew a clear distinction between the indivisible atoms of the pre-Socratics and his modern term *molecule;* and in a lecture to the Chemical Society two years later, he attempted close the gap between the chemical and dynamical views of molecules.[91]

Helmholtz adopted what he called the hypothesis of Ampère—that bodies are composed of atoms "which themselves are composed of subordinate particles (chemical elements, electricity, &c)"—and argued that "the conservation of force in these motions [in the atoms] will hold good in all cases where hitherto the conservation of caloric has been assumed."[92] Yet positivists such as chemist Wilhelm Ostwald and physicist Ernst Mach were still able to oppose the use of the atom concept. According to Mach, "the atomic theory has in physical science a function which is similar to that of certain auxiliary concepts in mathematics; it is a mathematical *model* for facilitating the mental reproduction of facts."[93] Ostwald had been involved in framing the science of energetics, in which energy was all-important and matter superfluous. But finally in 1909 he converted to atomism in the face of evidence from cathode rays, radioactivity, and Brownian motion.[94]

The quantum of electricity—the electron—was implicit in much of the work in electrochemistry dating back to Faraday and in the ionic theory of Arrhenius.[95] Thus, a crucial event in the history of atomism was H. A. Lorentz's introduction of the hypothesis of an elementary electrical charge.[96] Johnstone Stoney gave this charge the name *electron* even before its actual "discovery" by J. J. Thomson in 1894–97. The availability of good vacua and high voltages in the last two decades of the nineteenth century made possible the first experiments of Crookes and others with discharge tubes and the discovery of cathode rays, which led to J. J. Thomson's critical experiments with this "atom of electricity," the particle of the cathode rays.[97] He demonstrated that the speed of the cathode rays was below the speed of light and measured their charge to mass (e/m) ratio, showing that it was 2,000 to 4,000 times greater than that found for the hydrogen atom. This discovery provided a mechanical foundation for electromagnetism (the theory of electrons) and was consistent with theories of chemical affinity and electrolysis. The first models of the atom emerged shortly thereafter, proposed by J. J. Thomson and Phillip Lenard.[98] Thomson's assumption that the entire mass of the atom was due to its electrons was soon shown to be untenable, and scattering experiments by Charles Barkla and others between 1904 and 1910 demonstrated that the number of electrons per atom was about half its atomic weight.

*B*y the time the nineteenth century neared its close, scientists had made considerable progress in bringing together the sometimes antagonistic theories of chemical and physical atomism. The kinetic theory of gases definitively removed the physical atom from speculation to reality, and the understanding of atomic and molecular spectra achieved by the 1860s and 1870s, which made possible the use of spectroscopy in chemical analysis, went far toward bridging the gap between the two manifestations of the microscopic structure of matter. The final resolution came only with a detailed theory of atomic structure, which had to wait for—and help to create—the quantum revolution. Therefore, we have put aside the role of atomic spectra in elucidating atomism until chapter 8, where we will examine the critical problems in late nineteenth-century physics that fostered this great revolution in microphysics.

But for most physicists, the nineteenth century ended, once and for all, the often fruitless, often speculative, but ultimately crucial debates about the microscopic structure of matter and provided convincing proof of the reality of atoms. Not *indivisible* atoms, to be sure: In the last third of the century, models of the internal structure of the atom were being seriously proposed. But the nineteenth century saw Dalton's work verified in both chemistry, which blossomed into a full-fledged science at the end of the eighteenth century, and physics, primarily through the kinetic theory and statistical mechanics that finally gave atomism the microscopic support it had long needed. Some would say that the final blow to the opponents of atomism was Einstein's 1905 paper on Brownian motion, which showed that it was due to the motion of molecules.[99] But as late as 1906 Boltzmann was moved to commit suicide in the face of growing antipathy to the atomic hypothesis, which had put his life's work in question.[100]

In short, it is not too much to say that the great revolution in twentieth-century physics—the quantum theory—owes its birth to atomism, not merely in the strict historical sense but because the nineteenth-century success of the corpuscular theory prepared the way for the discontinuities and quantization that lie at the heart of the quantum theory. The old Greek idea of atoms was never entirely lost, and Walter Nernst was correct in saying that the new atomic theory "by one effort of modern science arose like a Phoenix from the ashes of the old Greek philosophy."

CHAPTER 7

The Kinetic Theory of Gases and Statistical Mechanics

$\mathcal{A}$lthough atomism was far from being universally accepted, thermodynamics developed vigorously between 1824 and the 1850s because it is a macroscopic science whose conclusions are largely independent of the microscopic structure of matter. As we learned in chapter 6, atomic theory for the most part developed separately in nineteenth-century chemistry and physics, quickly becoming a new paradigm in chemistry (following the work of John Dalton and others) but occupying in physics a more equivocal place.[1] As late as 1862, William Thomson complained, "All that is valid of the unfortunately so-called 'atomic' theory of chemistry seems to be an assumption of . . . heterogeneousness in explaining the combination of substances."[2]

Unlike thermodynamics, however, the kinetic theory of gases rests squarely on the atomic theory of matter. As we have seen, Newton's influence was enormous, not only with respect to the atomic theory but in promoting the dynamical theory of heat as well.[3] The theory of gases he offered was a static one, whose main advantage was that it made calculations possible.[4]

Some properties of matter obviously depend on its microscopic structure—for example, specific heats, transport properties (conduction, diffusion, and so on), and even the equation of state of a gas—so that in the final analysis thermodynamics cannot entirely avoid confronting the issue of the small-scale structure of matter.[5] Initially, this required taking into account only the translational motion of atoms or molecules; but when this approach proved to be inadequate, internal degrees of freedom (rotation and vibration) had to be considered as well. Such complex internal structure could be incorporated into thermodynamics in the form of the empirical specific heats, thereby hiding the microscopic foundations of the theory; but ultimately a microscopic theory was required because the observed specific heats and their temperature dependence had no convincing macroscopic interpretation. Furthermore, the growing realization that a macroscopic quantity such as pressure was the result of molecular collisions with the walls of the vessel ultimately had to find expression in the form of a quantitative microscopic theory of molecular motion. In the 1860s Clausius and Maxwell finally accomplished this task in what we call the kinetic theory of gases.

Properties of Gases, Specific Heats: Early Kinetic Theory

During the 1770s and 1780s most of the permanent gases were identified, and much effort was given to measuring their properties, notably by Adair Crawford.[6] He was the first to measure the specific heats of gases and the first to distinguish specific heats at constant pressure and constant volume.[7] Crawford was an Irish physician educated at Glasgow and practicing in London. Influenced by William Irvine, he was also a friend of Joseph Priestley.[8] His work was taken up by Laplace, and, during the first third of the nineteenth century, by Delaroche and Berard, Dulong, and Regnault, among others. The first important generalization was the empirical law of Dulong and Petit (1819), which stated that the specific heats of solids, per gram (calories per gram per degree Celsius), multiplied by molecular weight (yielding what we call the molar specific heat, in calories per degree Celsius per mole) were the same for all substances. After Petit's death, Dulong showed that equal volumes of gases, under the same conditions, give up or absorb the same amount of heat when suddenly expanded or compressed and that the temperature changes are inversely proportional to the specific heats at constant volume, a result that Sadi Carnot had assumed in 1824.[9] Dulong also obtained the specific heats at constant volume and constant pressure, and it became clear that their ratio, $\gamma = c_p/c_v$, had a value of about 1.4 for all gases.[10] Victor Regnault showed that the specific heat of a gas was independent of pressure and temperature.[11] An understanding of this problem would require a dynamical, molecular theory of gases and the equipartition theorem, which would explain how the energy of a gas was shared among its degrees of freedom. Certainly the values of the specific heats could not be predicted without such a theory.

The true parents of the dynamical theory of gases based on atomic or molecular motions were Daniel Bernoulli and John Herapath, whose work was separated by a century.[12] In *Hydrodynamica*, Bernoulli was able to derive "Boyle's" law relating the pressure and volume of a gas that is compressed or rarified at constant temperature, and he also deduced from molecular motion that, "on account of the fact that through this velocity the number of impacts and their intensity will increase equally at the same time," the pressure must be proportional to the velocity squared (at constant volume).[13] With the further assumption that vis viva is proportional to temperature, it followed that pressure was proportional to temperature, hence the law of Charles and Gay-Lussac. Bernoulli understood that the temperature had to be defined in some absolute sense.[14]

In *Mathematical Physics* (1847), Herapath proposed that the properties of a gas resulted from the vis viva of its particles, which led to his becoming the first person to calculate the mean speed of atoms in a gas from a kinetic

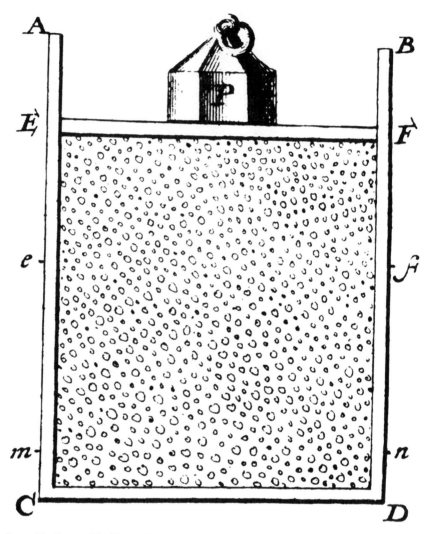

PLATE 18. Bernoulli's illustration of his kinetic theory of gases. *From Daniel Bernoulli, Hydrodynamica (1738).*

theory.[15] He found that the product pV was proportional to T^2, rather than T, because he took the temperature, T, to be proportional to the momentum rather than the vis viva. Although *Mathematical Physics* was published in 1848, Herapath had shown how to find the speed of a gas molecule eleven years earlier in a paper in his *Railway Magazine,* and there he claimed to have obtained the result four years before.[16] Neither Herapath's book, nor Joule's paper based on it and published in 1851, received much attention. Although Herapath's work was ignored by the British scientific community, Maxwell honored him as one of the founders of the kinetic theory in his 1859–60 paper.[17]

J. J. Waterston (1811–83) was also rebuffed when he tried to publish a mechanical theory of gases in 1845; indeed, his work was only rediscovered by Lord Rayleigh (who was then secretary of the Royal Society) in 1891 and finally published a year later. As a consequence, Waterston had little actual influence on the development of the kinetic theory. In Brush's words, this was "a classic example of the suppression of originality by an established scientific institution."[18] The abstract of Waterston's work published in the *Proceedings of the Royal Society* mentioned that the average kinetic energy of each kind of particle in a gas must be the same; and Brush suggests that this may have influenced August Karl Krönig, who in turn influenced Clausius.[19] Waterston's assertion was the first statement of something resembling the equipartition theorem, which describes how energy is shared among the degrees of freedom of the particles of a gas.[20] Later, Maxwell gave the generalization of this idea—namely, that energy is equally shared, on the average, among all degrees of freedom.[21] It appears that Waterston's influence was largely limited to Rankine, who heard of his theory as early as 1851 and cited it in the 1850s and 1860s but nonetheless thought it was inadequate.

Having embraced the dynamical theory of heat, Joule took the next step by trying to see how the motion of atoms was related, in a quantitative way, to the thermal properties of a substance. Using the theory and method of Herapath, Joule calculated the speed of hydrogen atoms in 1848; but his meager mathematical skills prevented further progress.[22]

Equipartition and the Kinetic Theory of Gases: Clausius and Maxwell

The modern theory of heat as an expression of molecular or atomic motions may be thought of as having begun with the 1856 paper by August Krönig (1822–1879) and the much more profound work of Rudolph Clausius, which was initiated with papers published in 1857–58. Clausius's first paper, "The Nature of the Motion Which We Call Heat" (1847), closely followed Krönig's "The Foundation of the Theory of Gases," which stimulated Clausius to publish his own independent speculations about the vis viva theory of heat.[23] In "On the Dynamical Theory of Gases" (1867), Maxwell declared Clausius's 1857 paper to be "a complete exposition of the molecular theory adopted in this [Maxwell's own] paper."[24] In Clausius's next paper, published in 1858 and translated in the *Philosophical Magazine* in 1859 (Maxwell read it in April) as "On the Mean Lengths of the Paths Described by the Separate Molecules of Gaseous Bodies," he introduced the idea of the mean free path of a molecule.[25] Maxwell's first words on the subject, "Illustrations of the Dynamical Theory of Gases," were delivered to the BAAS in September 1859 and published in 1860.[26]

The exchanges in print between Clausius and Maxwell during a fifteen-year period beginning in 1857 demonstrate the degree to which the kinetic theory was a direct result of this interaction, facilitated by rapid communication and publication of results and especially the quick appearance of English translations of Clausius's papers in the *Philosophical Magazine*. A few years earlier, a similar scientific "correspondence" between Clausius and Thomson had the same consequences for the rapid development of an understanding of the second law of thermodynamics.

In his first paper Clausius set about to explain the measured ratio of specific heats (c_p/c_v) for simple gases, which for air had been found to be 1.421. This clearly could not be accounted for by translational degrees of freedom alone, as in Krönig's model, for the following reason.[27] As Carnot and Clapeyron had shown, the difference between the specific heats at constant pressure and volume is constant. The result can be written $c_p - c_v = vR$, where R is now called the universal gas constant, and appears in the ideal gas law as $pV = vR(a + T)$, which is the way Clausius wrote it for a unit weight of a gas; v is the number of moles.[28] But from the equipartition theorem of Waterston and Maxwell, which states that energy of a gas is shared equally among the n degrees of freedom, the internal energy U of a gas is given by 1/2n v RT, or, per mole, $U = 1/2$ nRT. Because the specific heat at constant volume, c_v, is equal to $(\partial U/\partial T)_v$, it follows that $c_v = v$ nR/2. Combining the two results, $c_p - c_v = vR$ and $c_v = vnR/2$, yields, for $\gamma = c_p/c_v$, the result 1 + 2/n. If the number of degrees of freedom is 3 (that is, if we only consider translational vis viva), then $\gamma = 5/3 = 1.67$, compared to the measured value for air of 1.421. If, on the other hand, n were equal to 5, then $\gamma = 1.40$, very near the observed value.

Referring to the ratio γ, Clausius wrote: "The excess of this quotient over unity [that is, $\gamma - 1$] is therefore, for the different gases, inversely proportional to the specific heats of the same at constant volume, if these are referred to the unit of volume."[29] He noted further that this law had been verified by Dulong.[30] That is, $\gamma = c_p/c_v = 1 + v$ R/c_v, or $\gamma - 1 = vR/c_v$, in terms of specific heat per mole, $\gamma - 1 = R/c_v$. In "The Nature of Motion We Call Heat," Clausius expressed this in terms of the ratio of translational vis viva to the total, which he showed was K/H = 3/2 ($\gamma - 1$). This can also be written as 3/2 R/c_v = 3/vn. The measured value of $\gamma = 1.421$ implied that n was approximately 5, or that translational vis viva accounted for only 60 percent of the total. Clausius concluded by writing: "The *vis viva* of the translator motion does not alone represent the whole quantity of heat in the gas, and…the difference is greater, the greater the number of atoms of which the several molecules of the combination consist. We must conclude, therefore, that besides the translator motion of the molecules as such, the constituents of these molecules perform other motions, whose *vis viva* also forms a part of the contained quantity of heat."[31]

In this work Clausius made two crucial observations: first, that it was not necessary to assume that the collisions between the molecules and the walls of the container be elastic; second, that the velocities he obtained were only mean velocities, "which, for the totality of molecules give the same *vis viva* as would

PLATE 19. Rudolph Clausius (1822–88). *From Archiv für Kunst und Geschichte, Berlin.*

their actual velocities." He also noted that "it is possible that the actual velocities of the several molecules differ materially from their mean value."[32]

Just as Clausius's first paper on the kinetic theory was apparently stimulated by Krönig, his second paper on the subject, "On the Mean Lengths of Paths,"
was an answer to critics (Christoph Buys-Ballot, in particular) who protested
that, if particles in a gas traveled in rectilinear paths, given their high speeds,
gases would quickly mix, a fact that was not observed.[33] Clausius replied that
the fallacy lay in assuming that the particles traveled long distances unimpeded.
In fact, he argued, "those portions of a path of a molecule throughout which the
molecular forces are of influence in sensibly altering the motion of the molecule,
either in direction or velocity, must be of vanishing value compared with those
portions of the path throughout which such forces may be considered as inac-
tive."[34] In other words, intermolecular forces are of short range, or of a range
that is small compared to their mean separation. Clausius noted, however, that in
real gases the paths need only be small, not vanishingly small. By considering
that each molecule has a sphere of action, ρ, and by finding the probability that a
molecule would be detained in a layer of thickness Δx and area A and then multiplying by Δx and averaging over all values of thickness by integration, the mean
path length could be obtained. Rather than deal with the more difficult problem
of molecules' encountering each other, Clausius chose to consider one molecule's
encountering a large collection of stationary molecules, although he calculated
that the mean length in the case where all molecules move at the same velocity
would be only 3/4 the value obtained by the simpler method. Note the assumption that all molecules move at the same speed; Maxwell was the first to assume
that the speeds obeyed a probability distribution.

Clausius noted that the mean path λ (the average distance between collisions) was "in the same proportion to the radius of the sphere of action as the
entire space occupied by the gas, to that portion of the space which is actually
filled up by the spheres of action of the molecules."[35] Thus, $\lambda/\rho = V/(4/3\pi\rho^3)$,
or $\lambda = V/(4/3\pi\rho^3)$. Even after Maxwell presented his probability distribution for
molecular speeds in 1860, deriving it from a rather abstract argument (which
he later admitted "may appear precarious"), and further justified it in 1867,
Clausius declined to employ it.[36] An incidental result was a difference in the
numerical factors that appear in the expression for the mean path.

From Clausius's paper, Maxwell took the idea of a mean free path and
the suggestion that a statistical approach should be used to describe the molecular
motions in a gas.[37] He built thereon a description of the measurable properties
of a gas, including its transport properties, which depend on the random motions of its molecular constituents. Clausius immediately saw that Maxwell's
treatment of transport theory was inadequate and pointed this out in a paper published two years later (1862).[38] Maxwell's response, some five years later, was
"On the Dynamical Theory of Gases," perhaps the fundamental paper of the kinetic theory of gases and one that influenced Boltzmann and Gibbs, the founders

of statistical mechanics.[39] It carried kinetic theory far beyond the calculation of the thermodynamic properties of a gas to properties such as diffusion and viscosity and other transport properties.[40]

Maxwell's first paper on the subject, "Illustrations of the Dynamical Theory of Gases" (1860), to which we have earlier referred, had introduced the statistical description of a gas, which is at the heart of the kinetic theory. From the assumption that molecular motion in one direction, say the x-direction, does not affect motion in another perpendicular direction, he obtained an exponential distribution function giving the fraction of particles whose velocity lies in a certain range dv. This function allowed computation of the mean square velocity or of the average values of other properties of the gas. The result was that the probability that a molecule would have a speed in the range dv, centered on the value v, was proportional to $v^2 \exp(-v^2/a^2)$ dv.[41] The constant a^2 ultimately proved to be proportional to the absolute temperature. Maxwell then proceeded to investigate the viscosity or friction of a gas, the diffusion of one gas through another, and heat conduction.[42] He also arrived at the important conclusion that the average energies associated with each degree of freedom, translational and rotational, are the same. Having thus established the equipartition theorem, he could compare the observed ratio, γ, of the specific heats, which was about 1.4, with the theoretical prediction of 5/3, based on the assumption of point particles (only translational degrees of freedom), leading him to say that the result "seems decisive against the unqualified acceptation of the hypothesis that gases are such systems of hard elastic particles."[43] This fact troubled Maxwell from his first paper on the kinetic theory until the time of his death, prompting him to remark, "Here we are brought face to face with the greatest difficulty which the molecular theory has encountered."[44] It disturbed many other scientists as well, until quantum theory finally pointed the way to a solution; we have quoted Clausius on the subject already.

Maxwell's 1875 paper, "On the Dynamical Evidence of the Molecular Constitution of Bodies," stated the problem in terms of the total number of degrees of freedom, written as the number of translational plus rotational degrees of freedom, m, plus an additional unknown number, e, which contribute to the potential energy.[45] Thus, n = (m + e). For a point particle, n = 3 (m = 3, e = 0), while for a rigid body, which could rotate about three axes, n = 6 (3 translational and 3 rotational; m = 6, e = 0).[46] Maxwell did not speculate about the form of the potential energy that would determine e, but we now know that e = 2 for harmonic vibrations. Thus, for γ, Maxwell obtained the ratio (n + 2)/n = (m + e + 2)/(m + e).[47] For point particles, with n = 3, γ = 5/3, while for a rigid body γ = 4/3, a value nearly 10 percent below the measured values. Clearly, however, n = 5 would lead to γ = 1.4, very close to the experimental value of 1.408 used by Maxwell (which yields γ = 4.9). But why should n have the value 5 rather than 6 (as would a rigid body) or 8 (which would include vibrations)? As Maxwell said, "by establishing a necessary relation between the motions of translation and rotation of all particles not spherical, we proved that a system of such particles could not possibly satisfy the known relation between the two specific heats of all gases."[48] For a while, this discrepancy seemed, to him, to refute the kinetic theory and the theory of equipartition.

Boltzmann (and Bosanquet, independently) was able to offer a partial solution to the problem of specific heats in 1876 when he noted that, if an atom were a smooth sphere, its rotational degrees of freedom would not be excited by collisions or that, if a smooth molecule had a symmetry axis, rotations about this axis would not contribute to the kinetic energy.[49] Thus, diatomic molecules such as O_2 and N_2 would have only 5 translational and rotational degrees of freedom ($\gamma = 7/5$, $C_v = 5/2R$), consistent with observation.[50] Maxwell reacted coolly to this explanation, which seemed ad hoc to him. Lord Rayleigh, while defending the law of equipartition against critics such as Thomson, was nonetheless sorely troubled by the related issue: the failure of degrees of freedom corresponding to potential energy (that is, vibrations) to appear in the measured specific heats.[51] No resolution was possible until (1) the temperature dependence of specific heats was studied and the heat capacity found to be lower at elevated temperatures (implying that n increased with temperature) and (2) quantum theory could explain why certain degrees of freedom would be excited only at high temperatures. We revisit the question in chapter 8.

When Clausius criticized Maxwell's calculation of transport properties in 1862, the focus of his objections included Maxwell's introduction of what we now call the Maxwell or Maxwell-Boltzmann distribution, which amounted to the assumption that the velocities obey a normal distribution; in the end Clausius rejected the statistical approach altogether.[52] Earlier he had written: "In order to be able rightly to calculate these mean values, we must know the law which regulates the various velocities which occur....such a law was established by Maxwell, and it might perhaps be employed for calculating mean values. I prefer, however, not to discuss the subject here."[53] By the time Maxwell offered his answer to these criticisms, in "On the Dynamical Theory of Gases" (1867), he had largely adopted, and greatly improved, Clausius's approach to calculating transport properties. With respect to the deficiencies pointed out by Clausius, Maxwell wrote: "M. Clausius has recently published an investigation of the particular case of the conduction of heat through a gas which was very imperfectly treated by me in the paper referred to." Maxwell rederived his expressions for coefficients of viscosity, diffusion, and conductivity. In the process, he had in these two papers applied the statistical ideas of Laplace and Gauss to a physical problem for the first time. As a result, while Clausius was despairing that there was too little information to permit quantitative estimates, primarily because he did not have a way of determining molecular diameters, Maxwell was actually obtaining numerical values for the mean free path of molecules in air.

The origin of Maxwell's interest in and knowledge of statistical considerations has received considerable attention. Garber et al. suggest that Maxwell's introduction to statistical ideas began while he was a seventeen-year-old student at Edinburgh, when J. D. Forbes resurrected the Reverend John Michell's argument for the physical nature of binary stars, based on the observed frequency of close pairs and the statistical improbability that these were random associations.[54] Maxwell's encounter with Laplace's work came shortly thereafter, and he was likely influenced by John Herschel's review of a work by Quetelet on the laws of probability applied to the "moral and political sciences."[55] In "On the Dynamical Theory of Gases," Maxwell claimed to have proved that the exponential distribution function that he had obtained in the earlier paper had the property that, once established, it could not be changed by collisions, a result that is true only in an average sense.[56] Ultimately, Boltzmann gave a general proof.

After a hiatus during which he wrote *Treatise on Electricity and Magnetism* (1873), Maxwell returned to the kinetic theory of gases. In 1872 he wrote to William Thomson concerning the slight notice he (Maxwell) had given to Clausius in his *Theory of Heat* (1871): "Observe how my invincible ignorance of certain modes of thought have caused Clausius to disagree with me (in a digestive sense), so that I failed in my attempts to boil him down and he does not occupy the place in my book on heat to which his other virtues entitle him."[57]

By this time Ludwig Boltzmann had appeared on the scene, adopting Maxwell's statistical approach and carrying these ideas even further. One result was his famous collision equation, an integro-differential equation for the time rate of change of the function $f(x, t)$ that represented the fraction of gas molecules with a certain energy at time t, which he obtained by 1872. This argument leads inexorably to a statistical interpretation of the second law, a conclusion that Maxwell had already reached.[58] By this time Clausius was in ill health and had become inflexible in his approach to gas theory, and Maxwell had only seven years to live.[59] Thus, the torch was passed to Boltzmann. Maxwell, as it happens, was only slightly happier with Boltzmann's style than with Clausius's. He wrote to Tait: "By the study of Boltzmann, I have become unable to understand him. He could not understand me on account of my shortness and his length was and is an equal stumbling block to me. Hence I am about to join the glorious company of supplanters and to put the whole business in about 6 lines."[60]

When J. Willard Gibbs wrote Clausius's obituary in 1889, he emphasized the difference between the views of Maxwell and Boltzmann on the one hand and Clausius on the other: "In reading Clausius, we seem to be reading mechanics; in reading Maxwell, and in much of Boltzmann's most valuable work, we seem rather to be reading in the theory of probabilities. There is no doubt that the larger manner in which Maxwell and Boltzmann proposed the problems of molecular science enabled them in some cases to get a more satisfactory and complete answer, even for those questions which do not at first sight seem to require so broad a treatment."[61]

Ludwig Boltzmann

Ludwig Boltzmann (1844–1906) is usually thought of as the founder of statistical mechanics (a term coined by Gibbs), which provided the microscopic foundations for thermodynamics. Boltzmann was born in Vienna and studied physics at the university there, receiving his doctorate in 1866. At Vienna he was influenced by Joseph Stefan, with whom he worked until 1869, and by Josef Loschmidt. His first academic position was professor of mathematical physics at the University of Graz, and most of his career was spent in Austria. Nevertheless, he also spent time at Heidelberg and especially in Berlin, where he worked with Gustav Kirchhoff and Helmholtz. For three years beginning in 1873, Boltzmann was professor of mathematics at Vienna, after which he returned to Graz, where he stayed for nearly fifteen years as professor of experimental physics. He became professor of theoretical physics at Munich until Stefan's death in 1893, at which time he returned to Vienna, this time as professor of theoretical physics. For two years, 1900–1902, he was at the University of Leipzig, at Ostwald's invitation, and then returned to Vienna, where he remained until his death, even lecturing on the philosophy of science after Mach's retirement.[62] Although Boltzmann's first love was theoretical physics, he was experimentally and practically inclined, and there are many testimonies to his experimental skill. Indeed, much of his effort in the late 1870s was experimental. He was a renowned teacher whose prominent students included Lisa Meitner, Walter Nernst, Svante Arrhenius, and Paul Ehrenfest.

Boltzmann's first paper, written in 1868 when he was twenty-four years old, took up where Maxwell left off and generalized the latter's result to the case where an external field, such as gravity, is present.[63] The result, a factor exp $(-V/kT)$, when coupled with Maxwell's distribution function, which contained v^2, or, equivalently, the vis viva $1/2mv^2$, led to the famous Maxwell-Boltzmann distribution function $\exp(-E/kT)$, which has wide application in physics—namely, whenever a system of classical particles have established thermal equilibrium at temperature T due to collisions among themselves and/or with vessel walls.

For Boltzmann, it became a major goal to derive the second law of thermodynamics from the laws of mechanics, starting from Maxwell's statistical approach to the description of gases. In 1872 he presented his collision equation in a 126–page paper titled "Further Studies on the Thermal Equilibrium of Gas Molecules," providing the basis for his study of the second law of thermodynamics, which he had by then come to see as statistical in nature.[64] Boltzmann strove to understand how a gas approached equilibrium and how the equilibrium distribution was established as the result of collisions among the particles of the gas—that is, from the laws of mechanics. He finally showed that, if the distribution $f(x, t)$ was Maxwellian, then its time rate of change would vanish, and f would be constant. But if it were non-zero, collisions would make f approach the Maxwell distribution as time goes on.[65] Boltzmann then defined a quantity E (later called H and the theorem, the H-theorem), generally of the form

∫f ln f, with the negative of the entropy (within a constant), and thus had a proof of the second law. The fact that, for a Maxwell distribution, H is a constant provided a justification of that distribution function. In general, any non-Maxwellian distribution will evolve into a Maxwellian one over time due to collisions. Because the quantity -H has the property of the entropy function, Boltzmann had obtained a description of the approach to equilibrium, which is an irreversible process (in the statistical sense), but he did so from the time-reversible laws of collisions. In response to criticisms made by Wilhelm Ostwald, Boltzmann wrote:

> From the fact that the differential equations of mechanics are left unchanged by reversing the sign of time without changing anything else, Herr Ostwald concludes that the mechanical view of the world cannot explain why natural processes always run preferentially in a definite direction. But such a view appears to me to overlook that mechanical events are determined not only by differential equations, but also by initial conditions. In contrast to Herr Ostwald I have called it one of the most brilliant confirmations of the mechanical view of Nature that it provides an extraordinarily good picture of the dissipation of energy, as long as one assumes that the world began in an initial state satisfying certain conditions. I have called this state an improbable state.[66]

Although each microstate (which Boltzmann called a "complexion") is equally probable, a priori, the equilibrium configuration can be obtained in the greatest number of ways and thus is the most probable; the system evolves, through collisions, from a less probable to a more probable state. This problem of microscopic reversibility and macroscopic irreversibility was first raised by William Thomson in 1874 and by Loschmidt in 1876.[67] At the time Boltzmann said of Loschmidt's work that it "casts doubt on the possibility of a purely mechanical proof of the second law." If the microscopic laws are reversible, how can they lead to irreversibility of macroscopic processes? Although this question still causes heated discussions about "time's arrow," Boltzmann's answer, which is often given today, is that the irreversibility lies in the initial conditions, not in the microscopic laws of collision.[68]

Essentially the problem is that, although the laws of physics are invariant under the transformation t—> -t, the behavior of the macroscopic world is not. This fact provides an "arrow of time" because we can tell from watching a system of many particles evolve what the direction of time is. If a system is evolving toward equilibrium—that is, toward a Maxwell-Boltzmann distribution—or from a state of lower to higher entropy, we know that the direction of time is the "normal" direction; for by definition the direction +t is the direction in which the entropy increases in irreversible processes. This is true, as Boltzmann pointed out, only in a statistical sense, so that it is possible for the system to evolve from an equilibrium state to one very far from equilibrium—say, with all of the molecules in one corner of a vessel. But because the number of equilibrium or near-equilibrium states is enormous, the probability of this happening is very slight. Boltzmann addressed these issues in a series of papers, one of the most fundamental being "On the Relation between the Second Law of Thermodynamics and the Theory of Probability," published in 1877.[69]

In his 1868 paper Boltzmann had introduced another derivation of the distribution law, which represented the first application of a technique that is at the heart of statistical mechanics and is independent of the analysis of collisions. He used combinatorial analysis to distribute the total energy of a gas, divided into M small, discrete increments ε, such that n_1 systems had energy $M_1\varepsilon$, n_2 had energy $M_2\varepsilon$, and so on, where $M = \Sigma\ M_i$, and postulated that every such assignment of the energy $E = \Sigma M\varepsilon$ among the N particles, or microstate, had equal a priori probability. This eventually led, for a large number of particles, to the Maxwell-Boltzmann distribution. Although in principle the system of N particles could be in any one of the microstates, the most probable state, the one obtained in the largest number of ways (which, in equilibrium and for 10^{23} particles, would be by far the most likely), would be the state of the actual system (although fluctuations would occur around that most probable value). We can say that there are a huge number of microstates that yield the same thermodynamic or macrostate and that the actual equilibrium macrostate, for large N, will be the one that corresponds to the largest number of microstates.[70] Although discrete energy increments were introduced by Boltzmann only for convenience, his use of them almost certainly made it easier for Planck and Einstein to introduce the quantum of energy some thirty years later.[71]

In other work, Boltzmann generalized Maxwell's description of transport phenomena, arriving at a transport equation that bears his name (and often Maxwell's as well) and is widely applied in contemporary physics.[72] In his 1877 paper, Boltzmann derived the famous relation between the entropy, S, and the number of microstates, $S = k \log W$, in which W is the number of microstates, or ways of arriving at a particular energy for the system based on distributing that energy over "cells" in phase space; here k is the constant that, following Planck, we call Boltzmann's constant, which is simply the gas constant R divided by Avogadro's number. Fritz Hasenöhrl called this theorem (that entropy is proportional to the logarithm of the probability) "one of the most profound, most beautiful theorems of physics, indeed of all science."[73] Thus, Boltzmann had established that the entropy could be identified with the probability of a configuration, or the number of ways of arriving at it.

Recurrence Paradoxes

In 1896 Ernst Zermelo, a student of Max Planck's, raised seemingly formidable objections to Boltzmann's theory based on a recurrence theorem of Poincaré. Zermelo's objection was that all mechanical systems, including a system of a large number of gas molecules, are, from a mathematical standpoint, periodic and hence not irreversible, meaning that the claim that entropy must increase cannot be sustained.[74] While Boltzmann conceded the validity of

Poincaré's proof, he rejected Zermelo's conclusion that the mechanical viewpoint must be relinquished.[75] Instead, he reiterated that he had emphasized that "the second law of thermodynamics is from the molecular viewpoint only a statistical law" and went on to say that Zermelo's paper "shows that my writings have been misunderstood."[76] This failed to end the controversy, however; and Paul Ehrenfest, Boltzmann's former student at Vienna, and his wife Tatyana joined the argument after 1906, the year of Boltzmann's suicide.[77] Especially important was the article the Ehrenfests wrote for the *Encyklopädie der Mathematischen Wissenschaften* in 1911, which also included a discussion of Josiah Willard Gibbs's *Elementary Principles in Statistical Mechanics,* published in 1902.[78]

In 1898 a discouraged Boltzmann wrote: "It would be a loss to science if the kinetic theory were to fall into temporary oblivion because of the present, dominantly hostile mood. . . . I am conscious of how powerless the individual is against the currents of the times. But in order to contribute whatever is within my powers so that when the kinetic theory is taken up once again, not too much will have to be rediscovered."[79] The year before, he had written, with a nod to Galileo, that "I think I can still safely say of molecules: nevertheless they do move!"[80] Even Planck was, for a time, a follower of the positivist Mach and an opponent of Boltzmann. The event that can be said to have brought about the final triumph of atomism, and hence the kinetic theory and statistical mechanics based upon atomism, was the explanation of the phenomenon of Brownian motion: by Einstein in his *annus mirabulis* of 1905 and the next year by Marion von Smoluchowski, a student of Stefan, Exner and Boltzmann.[81]

Josiah Willard Gibbs

In the century following Franklin's pioneering work on electricity, perhaps Joseph Henry (1799–1878) was the only American to make any major contribution to physics, unless we count Benjamin Thompson, who did none of his work in the United States.[82] This was due in part to the isolation of American science but perhaps even more to the nation's general disregard of science for science's sake. Thus, while American engineering and technology thrived, its physics was mostly derivative.

Josiah Willard Gibbs (1839–1903), who received his doctorate from Yale in 1863, was the person to break that mold, even though he also was initially disposed toward applied science and developed a number of patentable devices; his dissertation was entitled "On the Form of the Teeth of Wheels in Spur Gearing."[83] In August 1866 Gibbs went to Europe, where he attended mathematical lectures by Chasles, Duhamel, and Liouville and became acquainted with the work of the "French" mathematical physicists Laplace, Lagrange, Poisson, Fresnel, and Cauchy.[84] The following year (1866–67) Gibbs was at the University of Berlin, where he attended courses by Magnus, Kundt, Weierstrass, Kronecker, Kummer, and others and apparently read the works of the great German and English scientists of the mid-nineteenth century, with the notable exception of

Clausius and Maxwell. At Heidelberg during 1867–68, Gibbs would have been able to hear Kirchhoff, Helmholtz, and Bunsen. These studies, rather than his engineering education at Yale, formed the basis for his scientific career.[85] Thus, America's first great theoretical physicist actually received much of his education in Germany and France. Two years after returning to America, Gibbs was appointed to a professorship at Yale without salary.[86] In the 1870s, stimulated now by the works of Clausius and Maxwell's *Treatise* of 1873, Gibbs began to move away from applied science toward theoretical physics.

Beginning in 1871–72, Gibbs turned his attention toward thermodynamics and during the next half-dozen years published several important papers in the field. By 1875 his work had attracted Maxwell's attention. While controversy raged over the meaning of *entropy,* in part because of Tait's mistake in interpreting it as unavailable energy and Maxwell's propagation of that error (see chapter 4), Gibbs was making entropy the heart of his thermodynamical treatment. In fact, Gibbs is probably due the credit for straightening out Maxwell's ideas about entropy. In 1878 Gibbs published a three-hundred-page monograph, "On the Equilibrium of Heterogeneous Substances," which was important in extending the realm of thermodynamics far beyond the relatively simple problems of gases to chemical, elastic, and electromagnetic systems.[87] He studied the nature of chemical equilibrium and first introduced the idea of a chemical potential. His work became known on the continent very slowly; and although Boltzmann was aware of it by 1883, it was not until 1892 that Ostwald's translation of Gibbs's works popularized his ideas in Germany.

During the following year, Gibbs, already in his fifties, began to turn his attention to what he called "statistical mechanics." Over the next decade he lectured regularly on "the a priori deduction of thermodynamics from the theory of probabilities." In 1902, at Yale's bicentennial, he published these ideas in *Elementary Principles in Statistical Mechanics Developed with Special Reference to the Rational Foundations of Thermodynamics.*[88] Here Gibbs formulated statistical mechanics in terms of ensembles of systems, which were collections of large numbers of copies of the system of interest, and introduced the microcanonical, canonical, and grand-canonical ensembles representing, respectively, isolated systems, systems in contact with a reservoir at temperature T (that is, closed isothermal systems with a specified mean energy), and systems with a given average energy and average number of particles (that is, open isothermal systems). Gibbs used the canonical ensemble as a means of describing a thermodynamical system and found that the quantity F = U - TS, now known as the Helmholtz free energy, is the fundamental or characteristic function in systems that can be described by the canonical ensemble—that is, when N, V, and T (number of particles, volume, and temperature) are the independent thermodynamic variables.[89] Gibbs's methods came to be the standard methods of statistical mechanics, and still are today. He died at the age of sixty-four, only a year after he published these ideas, having been a professor of physics at Yale for thirty-two years.

Albert Einstein

The extension of Boltzmann's ideas heavily involves the work of Gibbs but also that of Albert Einstein (1879–1955). Einstein first addressed the subject, which he called the kinetic-molecular theory of heat, in 1902. During the next two years, largely self-taught and mostly isolated from the mainstream of German science, he rederived results that had been obtained by Boltzmann, and especially by Gibbs, and rediscovered "all essential elements of statistical mechanics," as Max Born put it. If Gibbs's *Elementary Principles of Statistical Mechanics* had not appeared in 1902, Einstein might be remembered almost as much for these original contributions to statistical mechanics, especially for the 1904 paper on fluctuations and Boltzmann's constant, as for the papers on the light quantum (1905), Brownian motion (1905), special relativity (1905, 1907), general relativity (1915), and cosmology (1917).[90] Much of his work—in particular, that on the light quantum, blackbody radiation, thermodynamic fluctuations, critical opalescence, and Brownian motion—was built on this formulation of statistical mechanics.

*T*he kinetic theory and its progeny, statistical mechanics, arose primarily in the decade and a half between 1857 and 1872 at the hands of Clausius, Maxwell, and Boltzmann and reached mature development when Gibbs and later Einstein turned their attention to it. It provided, for the first time, a microscopic basis for thermodynamics; lent proud support to the atomic theory of matter; and played a crucial role in the creation of quantum theory, notably in the applications by Planck and Einstein. In the twentieth century it was fully mated with quantum theory to become one of the most powerful tools of modern theoretical physics. In the form of statistical mechanics, it is now widely applied to both classical and quantum systems; to the properties of condensed matter, magnetic materials, plasmas, and so on; as well as to gases.

CHAPTER 8

Fin de Siècle

$\mathcal{A}$s we turn our attention to the last two or three decades of the nineteenth century and prepare to bring down the curtain on this extraordinary era in the history of physics, we take note of a metaphor used by William Thomson (Lord Kelvin), who as much as anyone personifies the great successes of nineteenth-century physics. Commenting on the current and future state of physics, he spoke with concern of clouds on the horizon that dimmed the otherwise shining achievements of physics at the century's end. The story of this quarter-century is simultaneously a tale of the triumph of classical physics and of the first serious evidence of deep problems lurking in the phenomena. As a result, we will yield slightly to the temptation to extend this survey into the early years of the new century, on the grounds that to stop arbitrarily at the year 1900 would do violence to the historical coherence of the narrative. It is not, of course, our intent to give a thorough description of the way in which the remarkable discoveries of the first quarter of the twentieth century depended on the state of late nineteenth-century physics. But it seems unreasonable to turn away from the final scene in the drama of nineteenth-century physics, when the edifice of classical physics that had been built during that century began to crack, in some areas, under the weight of new unsolved problems and new discoveries.

We will, by now, have persuaded the reader that the study of nineteenth-century science needs no external or teleological justification and must be taken on its own terms. Nevertheless, the fact that it provides a springboard into the twentieth century is inescapable. One may choose to argue that the transformation of classical physics that occurred at the end of the century might have come at any other time, making fin de siècle arguments empty. (The end of the eighteenth century, for example, was far more a political than a scientific watershed.) But, in fact, a crucial transformation did occur in the two decades surrounding 1900. We shall not repeat those arguments, nor those that others have given, nor shall we even hint at the revolutionary changes that were occurring or were about to occur and that affected every area of human intellectual and artistic expression. Whatever one's view of cause and effect in this situation, whatever metaphor one uses to describe the way in which physics was carried along by this river of change, or to what extent one believes that scientific change was substantially responsible for the fin de siècle revolution on the larger scale, we can hardly fail to try to gain some insight into the way in which the nineteenth-century consensus was transformed.

Having already achieved some understanding of the state of physics by the 1870s, we shall turn our attention to those problems that arose in late nineteenth-century physics, which we now know pointed the way toward the revolutionary developments of the first quarter of the twentieth century and toward quantum theory in particular.[1] Specifically, we shall concentrate on (1) the problem of emission and absorption of radiation—in particular, blackbody radiation; (2) the specific heats of gases and solids; (3) atomic and molecular spectra; (4) the photoelectric effect; (5) the theory of the electron; (6) X rays and X-ray scattering; and (7) natural radioactivity.[2] Several of these issues have not previously been hinted at in this book, and although all are nineteenth-century discoveries (at least technically), they are also inherently nonclassical; their understanding would ultimately require a break with classical physics. Yet even they had to be addressed, at the outset, in classical terms, there being no alternative or competing paradigm. The problem of the aether, which might otherwise occur in this list, has been dealt with (briefly) in earlier chapters; the story of its demise is fascinating but long and runs far into the twentieth century.

The reader should be forewarned that the problems just enumerated were not chosen because physicists of 1890 or 1900 necessarily considered them to be most important (although, in fact, many were). Rather, with hindsight, we know that they held the key to the quantum revolution (and, to a lesser extent, relativity). There was, for example, an enormous interest in problems in continuum mechanics (fluids and elasticity), the theory of sound, irreversible and nonequilibrium thermodynamics, particle mechanics, optics, and so on.

Our thesis is an obvious one: that the ingredients of the quantum theory that came together in the years after 1900 grew naturally out of classical nineteenth-century physics. The techniques used to analyze the problems we will discuss were those with which a nineteenth-century physicist such as Thomson would have been comfortable. Indeed, Thomson, whose extraordinary scientific career had spanned more than sixty years from its beginning in the mid-1840s, and whose name stood, in the last quarter of the century, at the very top of the scientific world, epitomizes the triumphant and thus thoroughly conservative state of classical physics at that time.[3] To be sure, even he saw difficulties and unsolved problems as the century drew to a close. But as a man in his mid-seventies with an Edwardian outlook, he was unable to shed old nineteenth-century ways of looking at the world or welcome the new physics.[4] He came to represent a reactionary force in physics during the quarter-century after Maxwell's death.

There was a general crisis in British theoretical physics, and scientists watched with some alarm the growing ascendancy of German science and technology. The contributions of Maxwell's successor as Cavendish professor, Lord

Rayleigh (William Strutt), and his contemporary James Jeans were profound and numerous; and in experimental physics, the Cavendish Laboratory continued to lead the world under the direction of Rayleigh, J. J. Thomson, and Ernest Rutherford. But beginning in the 1870s, with the establishment of theoretical physics as a separate career in Germany, and with the rise of German industrial might, that nation increasingly took the lead in physics.

Heinrich Hertz played the greatest role in promoting the acceptance of Maxwell's theory: first, by carefully studying the foundations of Maxwell's equations; second, by demonstrating the existence of electromagnetic waves propagating with a finite speed. The young Hertz carried out these researches mostly at Kiel and Karlsrühe during the 1880s; but his life was cut tragically short in 1894, when he was thirty-six years old. Indeed, when one considers that Maxwell died in 1879, Kirchhoff in 1887, Hertz and Helmholtz in 1894, Boltzmann in 1906, and Kelvin in 1907, it is logical to suggest that the disappearance of these mighty authorities was a factor that facilitated the gestalt shift of which we are speaking.

Because the atomic hypothesis was generally accepted by 1890 (even though it was still controversial), one cannot plausibly argue that atomism per se was one of the decisive issues in the drift toward revolution. Nonetheless, the extraordinary dislocations that took place would have been impossible without it. The corpuscular nature of electricity, however, was far more controversial. Not until J. J. Thomson (1856–1940) showed that cathode rays had a fixed charge to mass ratio was it clear that they *had* to be taken to be corpuscular and that, by inference, electricity consisted of electrical atoms.[5] The chosen name, *electron,* was taken from a term that G. J. Stoney had applied to the unit of electrical charge, that on monovalent ions in electrolysis. The electron theory of the great Dutch theorist Henrik Antoon Lorentz, which predated Thomson's discovery, had as a fundamental assumption the molecular nature of electricity.

The aether postulate was as all-pervasive as the aether itself was supposed to be; to scientists as different as Thomson and Lorentz, no problem was as important as that of the nature of the aether, the medium in which electromagnetic waves were thought to propagate, and its relation to ordinary (ponderable) matter. We have seen in preceding chapters how important this hypothesis was in the work of almost all nineteenth-century physicists. By 1904 Lorentz's electron theory had ascribed the mass of an electron entirely to its electromagnetic field, concluded that the velocity of light was a limiting velocity for motion through the aether, and explained the Faraday and Zeeman effects. McCormmach argues that it was the clarity of Lorentz's electron theory, with its discrete particles and continuous fields, that drove Einstein to conclude that light was a discontinuous medium and to reformulate Lorentz's theory in the 1905 relativity paper. The reformulation was an attempt to reconcile mechanics and electromagnetism, a program that may have been rooted in Ostwald's energeticism.[6]

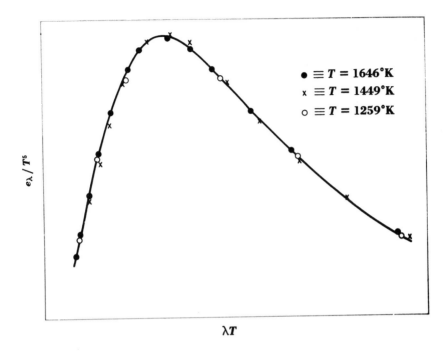

FIGURE 8–1. Experimental blackbody curves for different temperatures, with E_λ/T^5 plotted versus λT. The Wien displacement law is exhibited in the fact that the maxima occur for the same value of λT. *From F. K. Richtmyer, E. H. Kennard, and J. N. Cooper, Introduction to Modern Physics, 6th ed. (New York: McGraw-Hill, 1969), 121.*

Blackbody Radiation

The problem of blackbody radiation (that is, the radiation from an idealized perfect absorber and emitter of electromagnetic radiation) was first considered by Balfour Stewart and Gustav Kirchhoff in 1859–60.[7] Thus, the problem was not new in 1895 or 1900, but it had not been possible to measure accurately the radiation emitted by hot bodies over a range of wavelengths much before 1900.[8] From a quantitative point of view, the problem was one of trying to understand the blackbody spectrum—that is, the distribution of electromagnetic energy as a function of wavelength—when it is in equilibrium with the walls of a cavity containing the radiation (see figure 8–1). Most of the measurements involved source temperatures comparable to the sun's (in the visible part of the electromagnetic spectrum) and were often quite approximate. Stefan's law, often called the Stefan-Boltzmann law, which expressed the fact that the total radiant energy, integrated over all wavelengths or frequencies, is proportional to the fourth power of the temperature, stood as a challenge to theorists in the last quarter of the nineteenth century. While Stefan's result, obtained in 1879, was only an extrapolation from the results of experiments, Boltzmann was able to provide a derivation of it five years later.[9] The latter's work on thermal radiation was the most fundamental until Planck's and was based on his profound studies of the second law of thermodynamics and the approach to equilibrium of a gas. The crucial insight was to consider electromagnetic radiation in a cavity to behave like a perfect gas in equilibrium with the walls of its container. Extending earlier work, Wilhelm Wien (1864–1928), at the University of Berlin (who nearly became a farmer rather than a physicist), first showed that the distribution of energy $\rho(\lambda)$ with wavelength λ was proportional to λ^{-5} multiplied by a universal function of λT: that is, $\rho(\lambda) = \lambda^{-5}\phi\,(\lambda T)$.[10] Because ϕ is a function of the product $x = \lambda T$, it follows that the peak of the blackbody curve, which is when $d\phi/d\lambda = 0$, occurs at the value of x for which $d\phi/dx = 0$; the wavelength of the maximum in the blackbody spectrum times the absolute temperature is a constant. If T is doubled, the wavelength λ at which the maximum in the blackbody curve occurs is halved. This is the so-called Wien displacement law. The next step for Wien was to determine the function $\phi(\lambda T)$. At the time there seemed no way to arrive at this function except through experiment.

W. A. Michelson first proposed an exponential form for $\rho(\lambda)$.[11] German spectroscopist Frederick Paschen (1865–1947), assuming that $\rho(\lambda)$ had the form $\lambda^{-a}e^{-b/\lambda T}$, found experimentally that the value of a was in the range 5 to 6, generally consistent with the Wien law.[12] Later Wien attempted to show that the only value of a that could reconcile the Paschen law and the Stefan-Boltzmann law was 5; in 1896 he proposed what became known as the Wien distribution law: $\rho(\lambda, T) = a\lambda^{-5}\exp(-b/\lambda T)$.[13]

Until the 1880s, most of the experimental work involved the visible part of the spectrum; but in the 1880s American S. P. Langley began studying the emission and absorption of longer wavelength infrared radiation from planetary surfaces. Throughout the last decade of the century, attempts were made to verify the Wien law, especially in the infrared part of the spectrum, by Paschen himself and by Otto Lummer and Ernst Pringsheim, Ferdinand Kurlbaum, and Heinrich Rubens. Although the experiments initially confirmed the Wien formula, by 1899 the first departures were detected by Lummer and Pringsheim.[14] By September 1900, further infrared experiments convinced them that the Wien formula could not be valid.[15]

In the meantime, Rayleigh, who was Cavendish professor at Cambridge, pointed out on theoretical grounds that for high temperatures the Wien law predicted that the energy density of radiation would cease to depend on temperature, an implausible result.[16] In January 1900 Rayleigh wrote that what was wanted was "some escape from the simplicity of the general conclusion relating to partition of kinetic energy."[17] In a subsequent paper (June 1900), he complained: "Speculation upon this subject is hampered by the difficulties which attend the Maxwell-Boltzmann doctrine of the partition of energy. According to this doctrine every mode of vibration should be alike favored; and although for some reason not yet explained the doctrine fails in general, it seems possible that it may apply to the graver modes."[18]

Rayleigh was calling attention to a crucial failure at small values of λ— that is, at high frequency. The energy in the frequency range $d\nu$ was proportional to ν^2, a result that diverged as the frequency increased without limit: the so-called ultraviolet catastrophe.[19] Rayleigh took this to signal a failure in the law of equipartition and so multiplied the $\lambda^{-4}T$ factor for the $\rho(\lambda)d\lambda$ he got from equipartition by the Wien factor $e^{-a/\lambda T}$. This derivation of the radiation formula (in the June paper) was based on a model of standing electromagnetic waves in a cubical cavity, which used the equipartition theorem to obtain the average energy per mode.[20] He found that for large T, the radiation distribution should be proportional to T, as required. The full formula, with the exponential factor included, was equivalent to

$$\rho(\lambda, T) = k \, \lambda^{-4}T \, \exp(-a/\lambda T) \approx k\lambda^{-4} \, T \, [1 - a/\lambda T + (a/\lambda T)^2 + \ldots] \approx k\lambda^{-4}T$$

so that $\rho(\lambda, T)$ would indeed be proportional to T at high temperatures. The factor $e^{-a/\lambda T}$ addressed the temperature dependence and the ultraviolet catastrophe, albeit in an ad hoc fashion, and also agreed with the Wien formula. Yet what was going on was clearly a last ditch effort to patch up a theory that was beginning to come apart at the seams. Rayleigh was reluctant to question the law of equipartition in this June 1900 paper but made his point more clearly five years later.

It was at this time that Max Planck (1858–1947), who had been studying the problem during the previous five years, began to converge on an answer to the problem. For a decade Planck had been engrossed in studies of irreversible thermodynamics and the second law and had been studying the form that the entropy should have if the cavity radiation were considered to be in equilibrium with a set of oscillators or resonators.[21] Planck might be characterized as the gatekeeper of the twentieth century: the critical transitional figure, who unexpectedly and without an especially revolutionary turn of mind, opened the door to modern physics through his work on the problem of blackbody or cavity radiation. Planck was born only twenty-seven years after Maxwell but outlived him by sixty-eight years. His scientific model was Rudolph Clausius, whose *Mechanical Theory of Heat* he had studied on his own. Planck's path to the blackbody law grew directly out of that work as he devoted the first decade and a half of his career to explicating the second law of thermodynamics. His view of the second law was a conservative one in that he regarded it as absolutely true, rejecting, at first, Boltzmann's statistical description of it. As late as 1909 Planck wrote that a universe in which heat could return to a hot body "would not be the same as our universe."[22] Nor was he initially committed to the atomic or molecular viewpoint; rather, he was loathe to adopt a method of understanding the second law, and hence thermodynamics, that depended on this hypothesis. He thought that "obstacles, at present unsurmountable," stood in the way of the molecular theory and the mechanical interpretation of thermodynamics.[23] In

1882, when he was twenty-four, he wrote: "Despite the great success that the atomic theory has so far enjoyed, ultimately it will have to be abandoned in favor of the assumption of continuous matter."[24] Moreover, "anyone who studies the works of the two scientists who have probably penetrated most deeply in the analysis of molecular motions, Maxwell and Boltzmann, will not be able to resist the impression that the admirable display of physical ingenuity and mathematical cleverness shown in overcoming these problems is not in suitable proportion to the fruitfulness of the results achieved."[25]

Planck was educated primarily at Munich but spent a year at the University of Berlin in 1877–78, where he attended lectures by Kirchhoff and Helmholtz, although he apparently felt that he learned little from these two giants of nineteenth-century physics.[26] He received his Ph.D. from Munich in July 1879; but six years elapsed before he obtained an extraordinary professorship at the University of Kiel, for very few posts in the new field of theoretical physics were available in Germany. Four years later Planck succeeded Kirchhoff at Berlin, and in 1892 he became a full member of the Berlin Academy of Sciences (nominated by Helmholtz) and ordinary professor of theoretical physics, a position he held for thirty-four years.

During the 1890s Planck wrote on the role of entropy in thermodynamics and especially in irreversible thermodynamics—in particular, chemical thermodynamics and physical chemistry. The latter fields brought him into contact with scientists such as J. H. Van't Hoff who were pioneers in molecular physics. His involvement in the controversy between his student Zermelo and Boltzmann over the mechanical formulation of the entropy principle (see chapter 7) brought the problem of reconciling thermodynamics and mechanics to the front of his mind. While he was still reluctant to acknowledge the compatibility of entropy and atomism, Planck thought deeply about the problem, especially after 1897. The ultimate result, by no means entirely at Planck's hands, was the beginning of the quantum revolution.

By 1895 Planck had begun to relate his studies in irreversible thermodynamics to the growing interest in Maxwell's electrodynamics as elaborated by Helmholtz, Hertz, and Heaviside. This turned his attention to the problem of thermal radiation, in which he had been interested since his days as a student.[27] His interest coincided with the formulation by Paschen, Wien, and others of Wien's blackbody distribution law, which seemed to fit the observed data. Initially Planck set about to derive the Wien law and place it on a solid theoretical foundation, eventually using Boltzmann's statistical approach as the basis for his attack.[28] He published five papers on the subject between 1897 and 1899. But by 1899–1900, as we have seen, the experiments of Lummer and Pringsheim had begun to show a departure from the Wien law, and Rayleigh had begun to question the law of equipartition of energy.[29] It is not known whether Planck knew of this result of Rayleigh's in 1900 or 1901; Martin Klein believes he may have, but Planck makes no mention of the paper in his work in this period.[30] It does seem clear that the crisis was not a major motivating factor for him.

Planck was forty-three in 1900 and had devoted most of his scientific career to understanding the second law.[31] Both he and Boltzmann had studied the way in which a gas approaches equilibrium and, specifically, the irreversible approach to equilibrium of radiation in a cavity. But in 1900 Planck was unable to see the second law as being statistical in nature. Klein argues that it was Planck's distrust of statistical mechanics that allowed him to ignore the best classical result of the time: Rayleigh's. Indeed, Planck had earlier attempted to describe the irreversible approach to equilibrium based on the mistaken notion that the irreversibility arose from the asymmetry between emission and absorption of a spherical wave by an oscillating dipole.[32] Boltzmann, however, quickly disabused Planck of this notion, pointing out the time symmetry of the laws of electromagnetism.

Planck had at his disposal the proof in 1894 by Wilhelm Wien that spectral distribution function for blackbody radiation was of the form $\rho(v, T) = v^3 \phi (v/T)$, where v is the frequency.[33] He also understood the more specific, if more speculative, proposal that $\rho(v, T) = a\, v^3 \exp(-b\, v/T)$.[34] Then in 1900 the low frequency (long wavelength) measurements by Lummer and Pringsheim and by Rubens and Kurlbaum at Berlin, over a wide range of temperatures, showed disagreement with the Wien law.[35]

In 1899 Planck had already obtained a result for the relationship between the distribution law and the average energy $E_v (T)$ of an oscillator of frequency v: $\rho(v, T) = (8\pi v^2/c^3)\, E_v(T)$. So when he learned of the latest experimental results in the autumn of 1900, definitively showing the inadequacy of the Wien law, he was able to obtain in only a few days the correct mathematical form—which now bears his name—as the simplest generalization of Wien's law. On 19 October 1900, he presented it to the German Physical Society at the same meeting at which the data were presented, in the following form: $\rho(v, T) = A\, v^3/[\exp(bv/T)-1]$. This leads to the Wien law in the limit of high frequency or low temperature and is proportional to T, as required, at low frequencies (long wavelength). In both cases (that is, in the Wien law and the new one) Planck had been led to the blackbody law by considering the required form of the entropy S but had not yet shown why the entropy had to have that particular form.[36]

After arriving at a correct form for the blackbody distribution that fit the experiments of Rubens and Kurlbaum and had the appropriate long-wavelength limit, Planck set about, between October and December, to give a more fundamental derivation of the distribution law—rather, of the average energy per oscillator, which classically just comes from the equipartition theorem. His approach was to find the relation between the energy and entropy of an oscillator; once he had $E_v(T)$, the average energy per oscillator, the entropy could be found. The model he used was of the equilibrium of a system of radiation and damped, charged oscillators in a cavity.[37] If one assumes that the oscillator energies are given by a Maxwell-Boltzmann distribution, or simply uses the equipartition theorem, one obtains the Wien law.

It was nearly two months before Planck could offer a more substantial derivation of his blackbody formula, which he did on 14 December in a paper delivered to the German Physical Society titled "On the Theory of the Energy Distribution Law in the Normal Spectrum."[38] During those two months Planck had worked on the problem of providing a theoretical basis for the blackbody formula. He wrote: "On the very first day when I formulated this law, I began to devote myself to the task of investing it with a true physical meaning. This quest automatically led me to study the interaction of entropy and probability— in other words, to pursue the line of thought inaugurated by Boltzmann."[39] The approach Planck took was first to find the required form for the entropy S in terms of the energy that gave the Wien law and also satisfied the second law in the form $\partial S/\partial E = 1/T$.[40] Then he found the form for the entropy that gave his own, evidently correct, blackbody formula. Finally, he set out to justify, through a probabilistic argument, to derive that form for the entropy.[41]

The story of how Planck obtained, and then justified, what we now call the Planck blackbody formula is a fascinating one, especially because it is usually considered to represent the birth of the quantum revolution. Important ingredients in this episode include the way he used Boltzmann's statistical arguments, or misused them, in obtaining a formula for the entropy of a collection of oscillators; the way he viewed the energy increment ε, which in Boltzmann's theory would be allowed to go to zero in a limiting process; his knowledge of the equipartition theorem or at least its implications; and so on. What Planck actually intended by the finite energy elements he introduced into his derivation of the entropy is still a matter of heated controversy; the view of Thomas Kuhn, which we have followed (and believe to be sound), is by no means universally accepted.

There were actually two derivations offered by Planck. In the first, presented on 14 December 1900 to the German Physical Society, the probability W was simply written down.[42] The second was published in the *Annalen* a few weeks later (January 1901).[43] Many have regarded both derivations as unintelligible, although Kuhn seems to argue that each paper makes sense in view of the other, claiming, in particular, that December's incomplete proof was allowed to stay that way because Planck had the second derivation in mind.[44]

In 1905 James Jeans (1877–1946) wrote that "the function W (obtained by Planck through a combinatorial argument]) as at present defined, seems to me to have no meaning."[45] Jeans, of course, argued that Planck's method had no meaning unless the limit $\varepsilon \longrightarrow 0$ (or $h \longrightarrow 0$) was taken.[46] Acknowledging that Planck's result agreed with experiment, he concluded that this did not alter his "belief that the value $h = 0$ is the only value which it is possible to take." His view was "that the supposition that the energy of the ether is in equilibrium with that of matter is utterly erroneous in the case of ether vibrations of short wavelength under experimental conditions."[47]

Why did Planck not take the limit $\varepsilon \longrightarrow 0$ as Boltzmann would have done?[48] Klein argues that this is "undoubtedly related to Planck's apparent unawareness of the equipartition theorem and all it implied."[49] Kuhn claims that for Planck the energy increment or cell size ε, which was proportional to the frequency ν, was unrelated to the oscillator energies because the latter moved from cell to cell as their energies changed; thus, Planck did not have in mind anything like quantization.[50] Kuhn's case is meticulously argued and seems persuasive.

The essential question remains: How did Planck view the energy element ε, and how did he rationalize this concept of an atom or quantum of energy proportional to frequency, with the proportionality constant being h, now called "Planck's constant"?[51] Kuhn argues that nothing in Planck's writings of 1900–1901 indicate that he was thinking of quantization of the energy of the oscillators. From 1902 until 1906, he was silent on the question of radiation theory; and when he published *Lectures on the Theory of Thermal Radiation* (1906), he again made no explicit mention of a discontinuity.[52] Kuhn's argument is that Einstein and Ehrenfest saw that Planck's procedure required an energy discontinuity, while Planck himself came to that view only in 1908: "[He] did not publicly acknowledge the need for discontinuity until 1909, and there is no evidence that he had recognized it until the year before."[53] Was Planck, as Kuhn believes, responsible for the emergence of a belief he did not himself hold?

Both Lorentz and Ehrenfest (in 1903 and 1905, respectively) commented on Planck's introduction of finite units of energy, or minuscule energy particles; Kuhn takes these to be "misunderstandings" of Planck's ideas.[54] Indeed, Kuhn quotes Lorentz in a letter to Wien in 1908 that "according to Planck's theory resonators receive or give up energy to the ether in an entirely continuous manner (without there being any talk of a finite energy quantum)."[55] In 1909, Einstein, who had by then already introduced the light quantum and had shown the need for discontinuous exchange of energy between the resonators and the field, wrote: "Delighted as every physicist must be that Planck in so fortunate a manner disregarded the need, it would be out of place to forget that Planck's radiation formula is incompatible with the theoretical foundations which provide his point of departure."[56]

Specific Heats and the Quantum

The oldest of the problems that beset classical physics in the latter part of the nineteenth century was that of specific heats, particularly the specific heats of gases. We have considered this subject from the classical point of view in chapters 4 and 7, and in the latter chapter we saw how it troubled the founders of the kinetic theory, Maxwell and Clausius. In 1875, three years before his untimely death, Maxwell observed of the problem of specific heats that "here we are brought face to face with the greatest difficulty which the molecular theory has encountered."[57] If one accepted Boltzmann's argument that rotation about the symmetry axis of a diatomic molecule (chapter 7) did not contribute to the specific heat (and Maxwell did not), one could account for the 5 degrees of freedom observed in diatomic molecules by considering 3 translations and

2 rotations. But Rayleigh (and others) worried about the absence of vibrational degrees of freedom: Molecules should not only rotate but vibrate as well. Harmonic vibrations would raise the number of degrees of freedom to 7; hence, $c_v = 7/2R$, or $\gamma = 9/7 = 1.29$, compared to the observed value of γ, which was near 1.4.

There was independent evidence of internal degrees of freedom lurking in the spectra of molecules, such as HCl. But unlike the spectra of atoms, which could be fruitfully studied using visible light and whose interpretation, after the innovations of Niels Bohr, was at least potentially straightforward, molecular absorption spectra were found primarily in the infrared and their significance realized only at the end of World War II. Theoretical understanding came still later.

The problem resulted from the fact that the specific heats had been measured largely near room temperature so that the critical temperature dependence was only hinted at, and there was no theory that would explain such a temperature dependence. Kelvin and Rayleigh, to name just two of the physicists puzzled by the intractability of the problem, saw this failure of classical physics as due to the inadequacy of the equipartition theorem. Although the problem of specific heats substantially predated that of blackbody radiation, the solution to the former problem grew directly out of the latter. Kuhn rightly points out that the problem of specific heats was a more suitable vehicle for introducing revolutionary ideas into physics and chemistry than that of blackbody radiation. This problem was intertwined with the everyday experience of many physicists and chemists and involved an area of physics generally thought to be describable in straightforward mechanical terms. Perhaps it is also important that the problem was longstanding, lurking within the dynamical theory of gases since Maxwell's time.

Figure 8–2 shows the specific heat of diatomic hydrogen as a function of temperature over the range 10K to 5,000K.[58] The apparent constancy of the specific heat found from measurements near room temperature (300K) is seen to be accidental. At low temperature $c_v = 3/2\ R = 2.98$ cal/mole • deg, corresponding to translational motion (3 degrees of freedom) only. From about 250K to 750K, $C_v = 5/2R$, as 2 rotational degrees of freedom are excited; and at very high temperatures, above 4,000K, c_v makes a transition to 7/2R, indicating the excitation of 2 vibrational degrees of freedom. The full quantum mechanical solution required recognizing that the vibrational levels are quantized so that they are excited only when the thermal energy is high enough to produce a reasonable population of these vibrational states. Although this problem of the specific heats of gases was actually addressed only after the problem of solids was broached, it was nonetheless more pressing and more difficult. The problem of specific heats of solids was in a sense equally troublesome, and the solution came only after 1907, when Einstein demonstrated one of the earliest applications of the quantum idea. That partial solution was to a problem not generally appreciated to exist.[59]

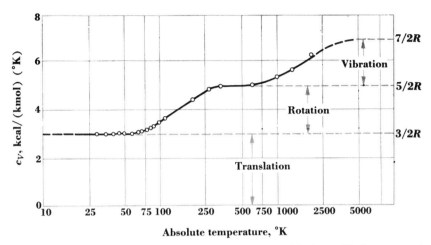

FIGURE 8–2. Observed temperature dependence of the specific heat of hydrogen gas at constant volume (c_v). *From F. K. Richtmyer, E. H. Kennard, and J. N. Cooper,* Introduction to Modern Physics, *6th ed. (New York: McGraw-Hill, 1969), 545.*

Atomic Spectra

The problem of atomic spectra originated in Wollaston's first observation of solar absorption lines in 1802. Unfortunately, this discovery was made in the context of studies of color, and the discrete dark lines attracted little interest.[60] When Joseph von Fraunhofer (1787–1826) rediscovered the solar absorption spectrum, he immediately made the connection between the dark lines he observed and the emission lines he had seen in flame spectra—notably, the D-line, which was later found to be due to sodium and which Fraunhofer showed was actually a pair of lines. Although Fraunhofer ultimately catalogued 574 lines in the solar spectrum, there was no explanation available for their origin; and he seems to have had the good sense not to pursue that problem, which we know from the perspective of the twentieth century required a fully developed atomic theory.[61] Nonetheless, John Herschel correctly suggested that "it is no impossible supposition that the deficient rays in the light of the sun and stars may be absorbed in passing through their atmospheres."[62]

In the 1820s and 1830s, Fox Talbot and Herschel noted the emission spectra associated with certain compounds. Talbot began to isolate the emission lines due to certain elements, including sulphur, strontium, and lithium. The general state of spectroscopy, however, was confused by the presence of large amounts of impurities in the samples being subjected to analysis; the D-line was found in the spectra of a variety of salts and metals, including, of course, sodium chloride.

In the 1830s Charles Wheatstone and others studied electric-spark spectra and found that the metals used as the electrodes could be identified by their emission spectra, although they were still observed to share common lines.

Anders Jonah Ångström discovered gaseous emission spectra in the early 1850s, providing additional impetus in the program to develop chemical analysis through spectroscopy. It was common in the 1830s to assume that the emission lines were somehow due to vibrations of matter; for example, this was Ångström's view. Similarly, about 1850 George Gabriel Stokes had suggested that the coincidence between the D-line in the solar spectrum and the emission line of sodium could be explained "by supposing that a certain vibration capable of existing among the ultimate molecules of certain ponderable bodies, and having a certain periodic time belonging to it, might either be excited when the body was in a state of combustion, and thereby give rise to a bright line, or be excited by luminous vibrations of the same period, and thereby give rise to a dark line by absorption."[63] Kelvin recalled in 1902 that in 1859, only two or three days before he received a letter from Helmholtz informing him of Kirchhoff's discovery of metals in the solar atmosphere, he had been talking to his students at Glasgow about the problem. He told them that he had previously learned from Stokes that the D-lines proved the presence of sodium in the sun and that, as he told Helmholtz, he had been "giving this doctrine regularly in my lectures for several years."[64] This conjecture by Stokes had actually been verified experimentally by Léon Foucault in 1849.

It was Kirchhoff, however, who, prompted perhaps by the work of William Swan of Edinburgh and, through him, by the earlier studies of Foucault, definitively established the equivalence of the solar D absorption line and the ubiquitous yellow emission line by viewing with a spectroscope sunlight that had passed through a salt flame. In so doing, he was using a method developed with Bunsen. By the time he published "On the Sun's Spectrum" (1859), he had determined that whether one observed a dark or a bright line in this experimental situation depended on the temperatures of the two sources (in this case, the sun and the salt flame).[65] This unexpected result—that the flame could absorb more radiation than it emits—was discovered by Balfour Stewart at about the same time.[66] Bunsen and Kirchhoff published a series of papers in the early 1860s on the use of spectroscopy in chemical analysis.[67]

In the 1860s Robert Clifton of Manchester and Johnstone Stoney of Queens College in Galway, Ireland, attempted to explain the observed spectra on the basis of molecular vibrations. Stoney used an argument that we would now call adiabatic to show that, because the time scales associated with the rectilinear motion of a molecule (that is, the time between collisions) is very large compared to that associated with the molecular vibrations, the vibrations will be essentially undisturbed by the collisions. Concluding that the time between collisions would be about 14×10^{-11} seconds, he noted that "this fragment of time, tiny as it is, is nevertheless more than 50,000 times the vastly shorter period that suffices for a double vibration of red light, and more than 100,000 times the duration of a double vibration of the extreme violet ray."[68] Julius Plücker and J. W. Hittorf's discovery that the spectra of gases depend on their temperature and Adolph Wullner's discovery of the dependence on pressure were important; but their acceptance was again greatly complicated by the presence of impurities, including those introduced by the mercury vacuum pump itself. Indeed, it was 1877 before the correctness of the observations was established.[69]

The attempts to interpret atomic and molecular spectra were founded on both laboratory observations and observations of stellar spectra. Other phenomena exhibited the properties of atoms and molecules as well. Avogadro's law and Avogadro's number, the definition of the mole, and the law of mass ratios and the concept of valency, all of which assume the existence of atoms or molecules, were established by or about 1860. The periodic system of the elements dates from the end of the succeeding decade. None of this, of course, in any way threatened classical physics. Even Pieter Zeeman's (1865–1943) discovery of the splitting of spectral lines in an external magnetic field in 1897 and Lorentz's explanation in 1897 required only the assumption of the corpuscle of electricity, the electron, which we credit to J. J. Thomson but for which Lorentz may deserve equal credit. J. J. Thomson also found, in 1903, that the number of electrons per atom was roughly the same as the atomic weight; it quickly became apparent that the correct number was about one-half that.

It was assumed that spectral lines originated either from the interaction between atoms and the aether or from the internal structure of atoms (or molecules) themselves. In the latter case, it became important to try to understand that structure. Early models that assumed a pointlike positive nucleus included J. B. Perrin's and Hantaro Nagaoka's.[70] William Thomson in 1902 and J. J. Thomson a year later each proposed a model that assumed that the positive charge was uniformly distributed; these models are now often called "plum-pudding" models.[71] J. J Thomson, whom we often forget was a Cambridge mathematical physicist before succeeding Rayleigh as Cavendish professor, offered an elaborate ring model of electronic structure, which, however, shed no light on atomic spectra. The model advocated by Perrin in 1901 foreshadowed the Rutherford-Bohr atom in its analogy with the planetary system.[72] Frederick Paschen and Ernst Back, in a study of the Zeeman effect in alkali metal atoms, discovered the "anomalous Zeeman effect," which was only explained after the introduction of electron spin by two students of Ehrenfest, George Uhlenbeck and Samuel Goudsmit, and also by R. Krönig, in 1925.[73]

The experimental clues to the distribution of positive charge in the atom came first from Lenard's early experiments with cathode rays, in which it was found that electrons passed readily through a thin foil, suggesting that atoms were either far apart or somewhat transparent to cathode rays.[74] More clues followed from Hans Geiger (1882–1945) and Ernest Marsden's (1886–1976) experiments in Ernest Rutherford's (1871–1937) laboratory and from Rutherford and his collaborators shortly before World War I.[75] The inescapable conclusion, based on Rutherford's calculations of the scattering of alpha particles by matter around 1910, was that the positive charge of the atom was concentrated in a pointlike center.[76] Because the sign of the charge of the alpha particle was not known, the sign of the nuclear charge could not be determined either. It was, however, the same as that of the alpha particle and was assumed to be positive. By 1912 scientists knew that the charge on the nucleus and the atomic number were the same.[77]

Early in 1912 the young Niels Bohr (1885–1962) went to the University of Manchester to work with Rutherford, having found J. J. Thomson's Cavendish Laboratory not to his liking. Late that year, influenced by Rutherford's nuclear atom and the quantum ideas of Planck, Einstein, and Poincaré, Bohr had grasped the basic elements of what we call the Bohr theory of the atom.

Cathode Rays, the Photoelectric Effect, and the Light Quantum

Lenard and Hertz had considered cathode rays to be vibrations in the aether, a view based on experiments by Hertz that had shown that they could not be electrostatically deflected. But the experiment was defective because of Hertz's poor vacuum, and in 1897 J. J. Thomson showed that cathode rays were indeed deflected by electric fields in a series of experiments that demonstrated that they had a fixed charge to mass ratio, which he found to be 770 times that of the hydrogen ion.[78] When considered in the context of 1897, Lenard's view after he repeated Thomson's experiments, that cathode rays were "pieces of the ether," was not really so different from Thomson's. Although we have by now dispatched the aether, we can hardly claim today to know any better than they did in 1897 what an electron "is."

In 1898–99 J. J. Thomson was able to measure the charge on the electron directly, using the early version of the cloud chamber developed by C.T.R. Wilson and J. S. Townsend at the Cavendish for observing the ionization produced by X rays.[79] Thomson's book *Conduction of Electricity Through Gases* is a good source of information on these experiments and the understanding of the ionization of gases in 1903. R. A. Millikan's oil-drop experiment, based on a suggestion of Felix Ehrenraft and first performed in 1913, would allow a more precise determination of the electron charge.

In 1907, Johannes Stark (1874–1957) and Wien separately explained the production of X rays when cathode rays were decelerated by a metal target and employed the Planck-Einstein expression $E = h\nu$ for the energy of the radiation.[80] Nonetheless, the light quantum hypothesis was generally treated with skepticism until the discovery of the Compton effect in 1922. Moreover, as late as 1924, with the "modern" quantum theory in the process of being born, Bohr, along with H. A. Kramers and John Slater, continued to reject the photon in a paper in the *Philosophical Magazine.*[81]

The history of the photoelectric effect can be said to date from the experiments of Hertz in 1887 in connection with his discovery of the reality of electromagnetic waves. Lenard first encountered the photoelectric effect in 1889 and pursued the matter as assistant to Hertz at Bonn two years later. It is not surprising that there was great interest in this phenomenon of the transformation of light into electricity; indeed, it has had enormous practical application.

During the following decade, Lenard's experimental work was devoted to un-covering the nature of cathode rays and understanding their interaction with thin metal foils. His results depended crucially on being able to achieve and main-tain a high vacuum. Lenard's development of well-sealed discharge tubes with aluminum windows led directly to Röntgen's discovery of X rays, which he made with a tube of Lenard's design. Lenard and Röntgen shared the Baumgartner Prize as well as the Royal Society's Rumford Medal in 1896.[82] Lenard narrowly missed sharing the first Nobel Prize with Röntgen in 1901, for by this time his claim to the discovery of X rays was being discounted. Nevertheless, he was the fifth recipient, in 1905, for his work on the photoelectric effect "and for his services to theoretical physics."

Lenard's critical investigations into the photoelectric effect took place be-tween 1898 and 1902, by which time he had discovered that the effect took place in a vacuum as well as in air and that the photocurrent originated in the cath-ode. He also found the well-known results that the photocurrent is proportional to the intensity, that there was a cutoff potential that stopped the photocurrent, that the kinetic energy of the cathode rays (or electrons) were not affected by intensity, and that the maximum electron velocity depended on the type of light (frequency) used. His culminating paper was "On the Photoelectric Effect" (1902), which was the twenty-three-year-old Einstein's principle source of in-formation about the phenomenon.[83]

Einstein's seminal "On a Heuristic Point of View about the Creation and Conversion of Light" was not really a paper about the photoelectric effect.[84] The subject is raised primarily in the final section, "On the Production of Cathode Rays by Illumination of Solids."[85] Assuming for the moment the validity of Wien's law, Einstein concluded that monochromatic radiation behaves "in a ther-modynamic sense, as if it consisted of mutually independent energy quanta of magnitude R/Nv."[86] This introduction of the light quantum represents the birth of the photon.

In 1909 Einstein explored the question now described as wave-particle duality at a meeting in Salzburg and in a paper published in *Physikalische Zeitschrift*.[87] The central part of the argument involved statistical energy fluc-tuations in a gas of light quanta. This duality gained support only after 1922, with the discovery of the Compton effect and of electron diffraction by C. J. Davisson and L. H. Germer, G. P. Thomson, and E. Rupp. When O. Lummer and E. Gehrcke showed in 1902 that split beams of light with a path difference of about a meter (greater than 10^6 wavelengths) could interfere, they obtained a result that was difficult to reconcile with any corpuscular picture of light. Einstein fleshed out the picture of the light quantum with papers in 1916–17, in which a momentum $p = E/c$ was associated with the photon. Thus emerged a picture of light as possessing both corpuscular and wave properties, the final resolution of the controversy that had raged since the time of Newton, with first one side and then the other gaining ascendancy.

In 1916, Einstein wrote "On the Quantum Theory of Radiation," in which he gave a complete theory of the emission and absorption of radiation, including what we now know as stimulated emission.[88] He also repeated a derivation of the Planck distribution that he had given the year before.[89]

X Rays and Radioactivity

In 1895 Wilhelm Conrad Röntgen, professor of physics at Würzburg since 1888, discovered X rays produced by cathode rays striking the glass wall of a discharge tube, which now is one of the most familiar stories in the history of science.[90] Shortly before the discovery, Helmholtz had suggested (in 1893) that short-wavelength light might travel through solid matter; and a consensus quickly grew that X rays were some kind of electromagnetic phenomenon. In a letter to William Thomson a year after the discovery, J. J. Thomson said that he doubted "whether the cathode rays have anything much to do with Röntgen's rays. . . . I am inclined to think that they are given out by the gas itself under the influence of the electrical discharge."[91]

In *The Tiger and the Shark,* Bruce Wheaton describes the attempt to understand X rays and their acceptance as electromagnetic.[92] He concludes that the most popular theory between 1898 and 1912 was that X rays were some kind of impulse propagating *through* the electromagnetic field, although some people (including Mach) simply saw X rays as extremely high-frequency electromagnetic waves.[93] Röntgen himself thought that his rays might be the long-sought longitudinal aether waves, a view that Boltzmann, and especially William Thomson, supported. Although electromagnetic interpretations were demanded by the fact that no deflection of X rays by electric or magnetic fields could be observed, Oliver Lodge thought that they might be gravitational waves.[94] X rays were first diffracted in 1899, and Charles Barkla polarized them in 1904 at Liverpool. The interference of X rays scattered by a crystal was first observed by Max von Laue in Munich in 1912, essentially deciding the question of their electromagnetic origin. The problem of the identity of the continuous X rays (*Bremmstrahlung*) produced in an X-ray tube and the characteristic X rays emitted by irradiated matter, not to mention the confusion of X rays and γ rays, slowed the process of understanding X rays. The conflicting theories of Stark, Arnold Sommerfeld, and others failed to resolve the many questions raised by the experiments.

By 1912 William Henry Bragg (1862–1942) had come to favor a dualistic view of light that would allow a simultaneous corpuscular and wave description, in part because of the seeming tendency of X rays to travel in well-defined channels in a crystal.[95] Bragg's son, William Lawrence Bragg (1890–1971), interpreted von Laue's theory of X-ray diffraction as the analog of light diffraction by a grating.[96] By 1913 he had discovered X-ray emission lines; and the first applications of X-ray spectroscopy soon followed, especially from Henry Moseley and C. G. Darwin at Manchester.[97]

In 1896 Henri Becquerel (1852–1908) discovered radioactivity in uranium, raising a whole new set of questions about the atom and, in a sense, founding the study of nuclear physics, although the nuclear atom was still more than a decade away. William Thomson told Becquerel: "Your discovery of the electric conductivity produced in air by uranium is one of the most wonderful discoveries of the century. . . . I now consider it the most wonderful discovery which has ever been made in respect to properties of matter."[98] Becquerel had been stimulated by the discovery of X rays to investigate the salts that were known to fluoresce under them and inadvertently found that uranium salt had darkened a photographic plate even though it had been kept in darkness.[99] Marie Curie (1867–1934) first introduced the term *radioactive* to describe these substances that emitted Becquerel rays.

Rutherford began working on uranium a year later (in 1897), first as a means of producing ionization in gases and then for its own sake. He soon showed that the radiation from uranium was of two kinds: one that was rapidly absorbed by a few thin layers of aluminum foil, another that was much more penetrating. He named these alpha radiation and beta radiation, respectively, in a paper completed in September 1898.[100] Rutherford also first described the detection of γ rays from thorium in this work (Paul Villard discovered them independently in 1900) and named them in his book *Radioactivity* in 1904.[101] The ß rays were soon found to be deflected by magnetic fields (and Marie and Pierre Curie [1859–1906] found them to be negatively charged), leading to their identification with cathode rays, while the α rays showed no evidence of deflection and were thought at first to be electromagnetic.

As early as 1899, Rutherford was considering the troubling problem of where the energy came from that allowed uranium or radium to radiate for tens of millions of years and what light that might shed on the structure of the atom. By 1903 the release of heat by radium was measured by Pierre Curie and A. Laborde. In 1900 Rutherford discovered the radioactive gas radon, which he called "emanation" (although a full understanding of what was happening, including the source of the induced, or excited, quickly decaying radioactivity, came later), and noted that its radioactivity declined rapidly while thorium's remained constant.[102] The Curies noted the same effect in the case of the radium and polonium that Marie had extracted from pitchblende in 1898, as did A. Debierne for actinium, which he had discovered in 1899.[103] Rutherford and Frederick Soddy (1877–1956) liquified thorium emanation, and William Ramsay and Joseph Gray showed that it was an inert gas with atomic weight 222: radon. During the next decade, several new radioactive elements were found, including radium D, radiothorium, and ionium. In 1902, Rutherford and Soddy developed the theory of radioactive transformations and introduced the idea of a decay constant and radioactive half-life.[104] Rutherford's 1901–2 collaboration with the brilliant young chemist Soddy in Montreal (at McGill University) was enormously productive if somewhat testy at times; Soddy's knowledge of chemical separation techniques was especially valuable. Rutherford and Soddy discovered transmutation by radioactive decay during their work together, and Soddy went on to show that the charge to mass (e/m) ratio for alpha particles was one-half that of the hydrogen ion. In 1913 he discovered isotopes.[105]

The Special Theory of Relativity

Quantum theory and the special theory of relativity constitute the two great transformations in physics in the twentieth century. Together they represent a conceptual revolution that may be unprecedented in the history of science. We have already discussed the place of the quantum discontinuity in opening the way toward the solution of several pressing problems in classical physics at the end of the nineteenth century; the further evaluation of quantum theory, which matured around 1925, is beyond the scope of this work. But what of relativity? In what sense is it also a consequence of nineteenth-century physics or, more particularly, the product of unresolved problems in classical physics?

In the 1880s and 1890s, there arose a set of issues that were almost purely theoretical-philosophical and thus contrast sharply with the other problems we have discussed in this chapter, which were largely driven by experiment. At the center was the tension between a purely mechanical picture of the world, which necessitated complex mechanical models of the aether, and a purely electromagnetic picture, in which even the particles might emerge from the field; in a sense, the core problem was the aether. The basic raw material was the electron theory of Lorentz, the Lorentz force, which was built on Maxwell's equations, and the assumption of the aether. One of the troubling points was how to transform Maxwell's equations in moving coordinate systems. These issues would lead to the theories of Lorentz in 1904 and Poincaré in 1905–6, which not only set the stage for Einstein's revolutionary 1905 paper but very nearly anticipated him.

There was, on the experimental side, the famous Michelson-Morely experiment, whose null result slowly made it apparent that the speed of the Earth through the aether could not be measured.[106] Einstein gave conflicting answers when asked about the role that this experiment played in his thinking and claimed that knowledge of stellar aberration and the Fitzeau experiment were sufficient for him. But one knows better, by now, than to rely only on the memory of the innovator in evaluating what influenced him or her. Furthermore, Einstein was by no means the sole creator of what we now call special relativity. To pursue this question in depth will not be possible, but we will sketch briefly the main issues and developments that culminated in Einstein's revolutionary step in 1905.

Some, notably Edmund Whittaker, have tried to deny Einstein any important role in the evolution of special relativity, a position that is simply untenable.[107] At the same time, however, the view that Einstein created relativity out of whole cloth is no less wrong. Not only were there significant precursors to Einstein's "On the Electrodynamics of Moving Bodies" (published in *Annalen der Physik* in 1905), but most of Einstein's results had been previously presented by the Dutch physicist Lorentz or the French mathematician Poincaré.[108] The true roots of the considerations that led to special relativity are older still.

The special theory is an important exception to the picture we have drawn of experimental results that failed to yield to the best assaults of classical physicists (specific heats, for example). On the contrary, relativity theory arose almost entirely from attempts to cast Maxwell's theory of electromagnetic waves in the aether interacting with charged particles (electrons) into a form that properly treated moving bodies and explained the null result of the Michelson-Morely experiment. The rather ad hoc Lorentz-Fitzgerald contraction hypothesis could explain the inability to measure the Earth's speed through the aether, but the central concerns of the theoretical programs of Lorentz and Poincaré (and Max Abraham) were much more fundamental yet thoroughly tied to the aether.

The critical period was 1904–5, which saw the great papers of Lorentz, Einstein, and Poincaré. Einstein, who was twenty-six in 1905 and a clerk in the Swiss patent office in Bern, knew of Lorentz's earlier work and had read Mach's *Science of Mechanics* (which rejected Newton's ideas of absolute space and time), Abraham's *Theory of Electricity,* and Poincaré's *Science and Hypothesis.*[109] What he knew of the Michelson-Morely experiment, its importance for him, remains a matter of controversy.[110]

Einstein's paper, "On the Electrodynamics of Moving Bodies," begins with two famous axioms or postulates, from which he develops the transformation equations from one uniformly moving coordinate system to another, which we now call the Lorentz transformation. These transformations were first derived by Lorentz as a means of explaining the Michelson-Morely experiment. Although there are no citations to any previous work on the problem and no references to experiment, Lorentz, until at least 1909, regarded Einstein's work as equivalent to his. Only in 1911, following a fundamental paper by Ehrenfest, did it become apparent just how different the approaches were and how revolutionary was Einstein's.

Neither of the two postulates, the *relativity principle* and the *light principle* (the constancy or the speed of light), was entirely new; furthermore, both can be seen as generalizations or abstractions from empirical evidence.[111] Newton, in corollary 5 of the *Principia,* clearly stated the relativity principle, and some would trace it to Galileo.[112] For more information about the other forms of the light principle, in the context of pre-Einstein aether theories, the reader is referred to the literature.[113]

The story of the elaboration of Einstein's theory by Minkowski and others, and of its somewhat grudging reception, is told elsewhere.[114] But it is worth emphasizing again that relativity theory, at least the special theory, grew out of problems that were inherent in classical physics of the late nineteenth century, mostly of a theoretical nature.

As the nineteenth century drew to a close, many saw classical physics as "complete" or very nearly so. In this chapter we have tried to show how misguided this prediction was (and, by inference, similar predictions, including some being made near the end of the twentieth century). At almost the same moment that William Thomson was declaring the triumph of classical physics (although he, too, had some misgivings, as we have seen), a new paradigm was being born. The essential theoretical innovation that defined the quantum revolution was, obviously, the introduction of the quantum discontinuity. It now seems clear that this step was first taken by Einstein rather than Planck and took place between 1905 and 1907. Einstein introduced the quantum into discussions of light, the specific heats of solids, and blackbody radiation; thus, its power became evident in spite of much skepticism about how one or another of these problems ought to be solved. In 1913, Bohr credited Einstein with first pointing out "the general importance of Planck's theory for the discussion of the behavior of atomic systems."[115] The Franck-Hertz experiment and the Bohr theory of the hydrogen atom, both dating from about 1912–13, virtually settled the issue of the quantum per se.[116] (Recall, however, that the idea of the light quantum was controversial up to 1922, and even after.) Thus, the final picture is very complex, involving experimental results and discoveries that apparently had no explanation in classical terms (blackbody radiation, specific heats, atomic spectra, X rays, radioactivity) along with theoretical innovation that both explained and motivated further experiments. The solution in each of these cases proved ultimately to follow from the introduction of the quantum discontinuity by Einstein and by the decisive figure of Max Planck, who, groping toward an answer to the radiation problem, only reluctantly accepted the revolutionary implications of his own solution. It is proof again that scientific discovery is an unpredictable, almost chaotic, process that defies any rules or prescriptions.

CHAPTER 9

Epilogue

$$\text{---}\rightleftharpoons\text{---}$$

$\mathcal{A}$s we stand at the end not only of another century but also a millennium, it is difficult to avoid looking one more time at the nineteenth century. The great issues of nineteenth-century natural philosophy that we have described can be viewed in several ways. Above all, the nineteenth century was the period in which physics as we know it came to exist; it congealed from a collection of loosely related sciences that were not always seen as part of one whole. More specifically, one can argue that, if the eighteenth century was concerned especially with mechanics of particles and rigid bodies, the nineteenth was on the whole much more interested in what we now call continuum mechanics and with other continuous properties of matter (or what seemed to be continuous properties), especially heat and, after Faraday, electromagnetism. Alternatively, it can be seen as a time when particle and continuum theories collided directly as natural philosophers weighed the theories of the aether and other imponderable fluids and attempted to reconcile them with atomistic descriptions of matter. Although atomism was still controversial when Einstein wrote his 1905 paper on Brownian motion, in part because of the eloquent opposition of positivists Ernst Mach and Pierre Duhem, the issue had been settled for many people by John Dalton almost precisely a century before.[1] If not then, it surely had been settled by the 1860s, when Clausius and Maxwell developed the kinetic theory of gases. Furthermore, the successes of the kinetic theory, in the hands of Clausius, Maxwell, and Boltzmann, made the efficacy of the atomic theory evident.

At the same time that the concept of heat as an imponderable fluid was being rejected in favor of its interpretation as dynamic or motional, Faraday was introducing the electromagnetic field. The role of the aether in either or both of these phenomena represented a problem not fully resolved until the 1920s; arguably, it is still unresolved.[2] But the discovery that much of the phenomena of nature could be modeled by a very small set of partial differential equations and that the same partial differential equations arose in areas of continuum physics that seemed superficially unrelated was a profound one that was not lost on important nineteenth-century figures such as Joseph Fourier, Claude Louise Marie

Henri Navier, George Stokes, and William Thomson. It was especially important that this description was almost entirely independent of the controversy over the detailed structure of matter.

To a considerable extent, the first two-thirds of the nineteenth century was preoccupied with the development of the theories of electromagnetism, optics, heat, elasticity, and fluid mechanics. Perhaps the greatest issue of nineteenth-century mechanics was that of energy conservation, which turned out to be inseparable from the theory of heat and even electromagnetism.[3] As it happened, this idea arose quite early in the strictly mechanical context but was not generalized and framed as a conservation law of general applicability until about 1850. It took even longer to recognize the energy principle as a global concept of great power in describing the behavior of systems. This problem arose in mechanics in the previous century, when the issue was conservation of vis viva and momentum, an outgrowth of collision theory, and found its general solution about midcentury in the work of Mayer, Joule, Helmholtz, and others, especially in the context of the theory of heat.

Much of classical electromagnetic theory as we know it today can be found in Maxwell's *Treatise;* and the unification of the two phenomena, electricity and magnetism, along with the "invention" of the field by Faraday and Maxwell, stand as the crowning achievements of that science in the nineteenth century. The modern theory of heat began to develop after about 1820, with the writings of Fourier and Carnot and the slow demise of the caloric theory in the face of a mechanical theory of heat.[4] The next half-century led to an understanding of the interconvertibility of heat and work, the first statements of the conservation of energy, and the beginnings of the kinetic theory of gases, all but the last due in considerable measure to the "last great amateur" of science, James Prescott Joule.[5] The first law of thermodynamics, which expresses the conservation of energy in thermodynamic processes, dates from the late 1840s; and although the second law is sometimes credited to Sadi Carnot in 1824, it was clearly formulated by Clausius and Thomson in 1850–51. This list of classical fields of physics, which includes heat, electricity and magnetism, and, by implication, several areas of continuum mechanics (elasticity, hydrodynamics, and acoustics) is not exhaustive; and we do not claim to have done justice to even these.

The problem of the aether formed a backdrop—indeed, a substrate—for much of the elaboration of each these areas of classical physics, but it was a problem was not destined to be resolved in classical terms. As we have tried to show, the aether was part and parcel of almost any discussion of physical phenomena but especially of heat and electromagnetism. The influence of the aether persisted into the twentieth century as the last surviving manifestation of the idea that imponderable, "subtle" fluids characterized the internal structure and the interactions of matter.

Science and Technology at the End of the Century

The transition from classical physics to the quantum theory at the end of the nineteenth century was the most important and most wrenching dislocation in physics since the seventeenth century.[6] The beginnings of the quantum theory, which between 1870 and World War I grew from a situation of mounting tension between classical theory and increasingly precise experiment and observation, forms an interesting contrast with relativity theory, which developed alongside it during the period before World War I. Unlike the latter, quantum theory is an example of a situation in which experiment and observation largely determined the course that theoretical developments would take. These include measurements of the blackbody spectrum (deviations from the Wein distribution law) and the failure of classical theory to describe its high-frequency behavior; the multitude of problems raised by observations of atomic and molecular spectra, including the Zeeman and anomalous Zeeman effects; the Paschen-Back effect; the photoelectric effect; the discovery of X rays and natural radioactivity; and the problem of specific heats of solids and gases.

All too often studies of scientific change give scant attention to the roles of measurement, observation, and experimentation and to the instruments that make those measurements possible. The history of physics, when seen in this way, is purely intellectual, a history of the interplay of scientific ideas that ignores the critical role of instrumentation and technology. Yet every physical scientist knows the crucial role of experimentation in scientific discovery. There are times in any scientific field when theory may dominate over experiment or the reverse, and there are areas of scientific exploration in which one or the other of these approaches may be most fruitful at any given time. Ultimately, however, in all the physical sciences, theoretical and empirical approaches are complementary and interdependent, the one essential to the other. Theory without experimental support or falsification is little more than speculation, and unguided experimentation more often than not generates useless data.

Who among our readers knows the name of an outstanding experimental physicist of the late twentieth century?[7] Among the important figures of nineteenth-century science, Michael Faraday sometimes seems to stand almost alone as a scientist revered for his experimental rather than his theoretical discoveries. After Faraday, the name of Ernest Rutherford may spring to mind.[8] The critical experiments of Joule are given their due, although we often pay more attention to the conclusions he drew from them concerning conservation of energy and the nature of heat than to the experiments themselves and how those conclusions were drawn. The giants of nineteenth-century physics, Thomson, Maxwell, Helmholtz, Clausius, Boltzmann, and Kirchhoff, are all renowned for their contributions to what we now know as theoretical physics; yet all but one were quite at home in the laboratory. In the twentieth century, however, intense specialization has largely meant the passing of the all-around physicist who is both a talented experimenter and an able theorist, especially in frontier areas of modern physics. (There do remain areas of applied physics where that is not necessarily the case, and it is manifestly not true in some of the other sciences.) From the seventeenth century into at least the first third of the nineteenth, the practice of natural philosophy was assumed to entail experimental investigation; but as technological advances made increasingly complex scientific instrumentation available to the experimentalist, and the mathematical tools of the theorist became increasingly sophisticated, the divide between them grew. While this was inevitable, the imbalance, if there is one, lies in the greater laurels given to the explainer of the phenomena than to their discoverer.

The *role* of experiment in scientific discovery is an issue that has been thoroughly dealt with by Karl Popper, Ian Hacking, and others.[9] Yet it is important to note specifically that a major element in scientific progress in the nineteenth century was the development of new and more accurate techniques of measurement and the development of new technology, such as vacuum techniques, which made it possible to study phenomena not previously known to exist. Any list of techniques and technology that drove or facilitated scientific discovery would include timing devices; electrical technology such as lighting, telegraphy, and power generation; vacuum technology; machine tools; the internal combustion engine; photography; cryogenics; and industrial chemistry. Other techniques that opened up new areas and new phenomena for study include the use of calorimetry techniques for measuring specific heats of solids and gases; the need and ability to measure temperatures more accurately than was possible with the mercury thermometer, which led to the development of the resistance thermometer; developments in optics for microscopy and for spectroscopy outside the visible part of the spectrum; the invention of Gauss meters for measuring magnetic fields; manometers and flow meters for studying fluid flow; strain gauges for studying elasticity; sensitive analytical and chemical separation techniques; and ultimately, although mostly after 1900, the development of X-ray sources, radio-frequency generators, regulated power supplies, ultraviolet and infrared spectroscopy, and so on.

Remember, however, that during most of the nineteenth century very simple equipment sufficed. This was in considerable measure because the phenomena being explored were those of the macroscopic world, where effects are large—that is, in electricity and magnetism, hydrodynamics, elasticity, acoustics, and so on. As it happens, even the atom and the molecule absorb and emit a significant amount of energy in the optical part of the electromagnetic spectrum, where a variety of light sources were available, and which, late in the nineteenth century, became increasingly monochromatic.

But experimental physics crossed a threshold in the late nineteenth century from a science in which rather simple measurements sufficed to the one we know today, which stretches modern technology to its limits. What is indisputable is that the sophisticated instrumentation we use today, which evolved from that used in the first decade or so of this century and makes possible new discoveries in physics, is the immediate by-product of the research of the previous generation. To use a later example, the transistor evolved out of studies of the physics of solids, became the heart of all electronics devices, was miniaturized, and now is the basis for the enormous computational power on which almost every experiment depends. But the important inventions of the nineteenth century—the voltaic pile, the electric motor, electromagnetic induction and the dynamo, the mercury vacuum pump, the internal combustion engine, and so on—that made twentieth-century physics possible seem rather small in number considering the period of eighty years or more during which they arose. And while the tools available to Kirchhoff in the 1860s and 1870s were more sophisticated than those Faraday had at his disposal in 1831 or 1845, there is a much greater difference between those available in 1910 compared with 1870 than there is between 1820 and 1870.

Science and Culture

The sciences, both as professions and as a body of knowledge, are as much a product of the times in which they are being pursued as they are a shaper of those times. But the insight that the social and intellectual ferment of the years before World War I included a sort of scientific radicalism does not show how Planck or Einstein or Bohr was influenced by this rejection of the past or how they or any of the other innovators in physics during this period received the precious gift of originality, creativity, or skepticism that they inherited in common with contemporaneous painters, sculptors, composers, and writers. Each had a unique genetic and cultural heritage that may have been crucially important. Many had extraordinary mentors. Yet it is abundantly clear from the intellectual history of humanity that original and creative individuals in conservative and backward-looking times may simply work within and strive to perfect an accepted paradigm rather than break new ground. Thus, while we may not establish direct connections between avant-garde movements in music and painting and innovations in physics, it seems improbable that an Einstein brought up in an intellectual world dominated by the ferment of German and Swiss café life and by fin-de-siècle Vienna was not profoundly affected by his times.[10] It is surely no coincidence that the two great twentieth-century revolutions in physics were an integral part of a much wider revolution in human thought and culture, which included all intellectual effort. To a considerable extent this was a Jewish intellectual movement and was molded by all those factors that affected Jewish culture and Jewish-European intellectual life in the late nineteenth century.

In the end, we are to remember the nineteenth century for what it accomplished in physics rather than for what it led to. Although our world view is undeniably Newtonian in the sense of being rational, mechanical, and causal, the nineteenth century brought to completion what Newton (and the polemicists) had only begun: the final establishment of the mechanical world view. There is, of course, that sense in which the mechanical view of matter was fundamentally compromised; we might say that Michael Faraday, when he delivered his

"Thoughts on Ray Vibrations," destroyed the mechanical world view for ever. But that is largely a semantical argument. Certainly, if we use *mechanical* in a naïve and restricted sense, we can say that the world was no longer mechanical. And perhaps nothing has undermined the mechanical world view, in this sense, more than quantum theory—quantum "mechanics." Yet in the end, the mechanical view has triumphed, and our modern search for the Higgs boson or attempts to understand the first instants of creation are only an elaboration of it.

NOTES

The following journal abbreviations appear throughout the notes:

Ann. d. Chem.	*Annalen der Chemie*
Ann. Chim. Phys.	*Annales de Chimie et de Physique*
Ann. d. Phys.	*Annalen der Physik*
Ann. Phil.	*Annals of Philosophy*
Ann. Sci.	*Annals of Science*
Arch. Hist. Exact Sci.	*Archive for the History of Exact Sciences*
BAAS Report	*Report of the British Association for the Advancement of Science*
Br. J. Hist. Sci.	*British Journal of the History of Science*
Camb. Math. J. or Camb. and Dub. Math. J.	*Cambridge or Cambridge and Dublin Mathematical Journal*
Edin. Phil. J.	*Edinburgh Philosophical Journal*
Hist. Sci.	*History of Science*
Hist. Stud. Phys. Sci.	*Historical Studies in the Physical Sciences*
J. Chem. Soc.	*Journal of the Chemical Society*
L. Phil. J.	*London Philosophical Journal*
Mem. Proc. Manchester Lit. Phil. Soc.	*Memoirs and Proceedings, Manchester Literary and Philosophical Society*
Phil. Mag.	*Philosophical Magazine*
Phil. Trans. Roy. Soc.	*Philosophical Transactions of the Royal Society of London*
Phys. ZS.	*Phyiskalische Zeitschrift*
Pogg. Ann.	*Poggendorff's Annalen (an early series of Annalen der Physik)*
Proc. Am. Acad. Arts Sci.	*Proceedings of the American Academy of Arts and Sciences*
Proc. Roy. Soc.	*Proceedings of the Royal Society of London*
Proc. Roy. Soc. Edin.	*Proceedings of the Royal Society of Edinburgh*
Stud. Hist. Phil. Sci.	*Studies in the History and Philosophy of Science*
Trans. Camb. Phil. Soc.	*Transactions of the Cambridge Philosophical Society*

Trans. Roy. Soc. Edin.	*Transactions of the Royal Society of Edinburgh*
Verh. d. D. Phys. Ges.	*Verhandlungen der Deutschen Physikalischen Gesellschaft*
Wiener Ber.	*Sitzungberichte der Akademie der Wissenschaften in Wien*

Preface

1. See William Berkson, *Fields of Force: The Development of a World View from Faraday to Einstein* (New York: Wiley, 1974), ix. This is true, I think, in spite of Peter Harman's *Energy, Force, and Matter* (Cambridge: Cambridge University Press, 1982). This little book is, however, highly recommended, especially the concluding "Bibliographic Essay," despite the fact that it is current only through 1980. C. Hackfoort expressed this same idea recently when he asked, "Has it ever happened to you, that you go to fetch a book from the shelves and find that it isn't there? Do you have the same experience of wanting to read a book and discovering that it has obviously not been written yet?" ("The Missing Syntheses in the Historiography of Science" (*Hist. Stud. Phys. Sci.* 29 (1991): 207–16).

2. Thomas Kuhn, "Revisiting Planck," *Hist. Stud. Phys. Sci.* 14 (1984): 231–52. On the other hand, Kuhn himself does not follow his own advice in his important work on Planck, and fortunately so. Thus, Kuhn should be heeded but not followed blindly.

3. Lewis Wolpert's statement that "scientists can be very proud to be naive realists" marks him as a clear exception (*The Unnatural Nature of Science* [New York: Faber and Faber, 1992], 117). He is not, however, immune to the charge of scientism.

4. Ernst Mayr, "When Is Historiography Whiggish?," *Journal of the History of Ideas* 51 (1990): 301–9. See E. Harrison, "Whigs, Prigs, and Historians of Science," *Nature* 329 (1987): 213–14, for a stronger criticism of the anti-Whig detractors of history written by scientists. See also David Hull, "In Defense of Presentism," *History and Theory* 18 (1979): 1–15.

5. See Stephen Brush, "Scientists As Historians," *Osiris* 10 (1994): 215–31.

6. Note especially the Kelvin-Stokes correspondence housed at Cambridge, recently published by David Wilson: *The Correspondence between Sir George Gabriel Stokes and Sir William Thomson, Baron Kelvin of Largs,* 2 vols. (Cambridge: Cambridge University Press, 1990). The first volume of P. M. Harman's edition of *The Scientific Letters and Papers of James Clerk Maxwell* has also been published (Cambridge: Cambridge University Press, 1990).

7. *Histoire général des sciences,* 2d ed. (Paris: Universitaires de France, 1966).

8. Crosbie Smith and Norton Wise, *Energy and Empire: A Biographical Study of Lord Kelvin* (Cambridge: Cambridge University Press, 1989).

9. See S. F. Cannon, *Science As Culture: The Early Victorian Period* (New York: Dawson and Science History, 1978), 236.

10. See David Wilson, *Rutherford: Simple Genius* (Cambridge: MIT Press, 1983), 281. For a recent work, see Jed Buchwald's paper in the Kuhn Festschrift (May 1990). For some of Kuhn's writing on the subject, see "Mathematical Versus Experimental Traditions in the Development of the Physical Sciences," and "The Function of Measurement in Modern Physical Science," in *Essential Tension* (Chicago: University of Chicago Press, 1977).

   Among my reasons for embarking on this book, one of the most important has been the imbalance between the emphasis given to theory compared with that given to experiment in most writings on the history of science, A typical view—if a rather extreme statement of it—is Henry Guerlac's, quoted here. It has been obvious to me for some years that there is a tendency to overvalue theory—or, more properly, to devalue experiment—and it seems nowhere as clear as in the developments in physics that led to quantum theory at the end of the nineteenth century. I have been both pleased and shocked to discover in recent years that this view was anticipated in the writings of philosophers of science such as Ian Hacking. The advantage, as far as this book is concerned, is that we may assume that the point is, at the very least, under discussion and already explicated to some degree, making it unnecessary to tackle in general terms, which was my original intention. The philosophical question can be pushed into the background, and one may expect that the importance of experimentation will arise organically and naturally from the historical development.

The facile answer (that, of course, experimentation is important, and no one denies it, but . . .) is inadequate. See, for example, Ian Hacking, *Representing and Intervening* (Cambridge: Cambridge University Press, 1983). J. L. Heilbron has written of "a pervasive bias in the recent historiography of science: the tendency to make general theory, or world view, or deep principle, the driving force in the growth of scientific ideas" (*Elements of Early Modern Physics* [Berkeley: University of California Press, 1982], ix). See also Allan Franklin, *The Neglect of Experiment* (Cambridge: Cambridge University Press, 1986), and David Gooding, Trevor Pinch, and Simon Schaffer, eds., *The Uses of Experiment* (Cambridge: Cambridge University Press, 1989).

11. It is revealing that an early reviewer of this work assumed that my support of experimental physics meant that I was an experimentalist, which is not at all the case.
12. In the twentieth century, Enrico Fermi also comes to mind.
13. I pursue this question throughout the book, but it is worth mentioning here Jed Buchwald's lecture on the subject at the 1990 Kuhn Festschrift.
14. Max Born, *Experiment and Theory in Physics* (Cambridge: Cambridge University Press, 1944). Born was then Tait Professor of Natural Philosophy at the University of Edinburgh, having fled Nazi Germany.
15. In his 1983 AAAS Sarton Lecture.
16. Philosophers of science seem sometimes to forget that microphysics is not a good model for all of science and that even in many branches of physics the objects of study are "objectively" real and palpable.
17. These are the smoking guns of Leon Lederman's irreverent and reductionist *The God Particle* (Boston: Houghton Mifflin, 1993).
18. See also *Annals of Science, Journal of the History of Ideas, Centaurus,* and others.

CHAPTER 1   *Prologue*

1. In that war Helmholtz served as a military doctor, and his countryman Clausius was wounded.
2. A. W. Levi, *Philosophy and the Modern World* (Bloomington: University of Indiana, 1959). As is well known, Darwin himself became an unbeliever.
3. F.W.H. Myers, *Essays Modern* (London: Macmillan, 1897), 268–69.
4. Lionel Stevenson, *Darwin among the Poets* (New York: Russell and Russell, 1963).

5. A useful source for the history of elasticity and strength of materials is Stephen Timoshenko, *History of Strength of Materials* (New York: McGraw-Hill, 1953; reprint, New York: Dover, 1983).

6. R. H. Silliman described physics in 1800 as "immature [and] undisciplined . . . with indefinite limits and little cohesiveness among its various concerns." See his "Fresnel and the Emergence of Physics As a Discipline," *Hist. Stud. Phys. Sci.* 4 (1974): 137–62.

7. Mary Sommerville noted this fact in *On the Connexion of the Physical Sciences* (London: Murray, 1834): "The progress of natural science, especially within the last five years, has been remarkable for a tendency to . . . unite detached branches. . . . there exists such a body of union, that proficiency cannot be gained in one branch without a knowledge of the others."

8. Note, however, the importance of physical astronomy (as it was called) in the Cambridge Tripos at midcentury.

9. Even mechanics, which at first might seem like an exception to this statement, is not. Although its most important and fundamental principles were formulated in the half-century after Newton's death by d'Alembert, Maupertuis, and Euler, it became fully mature in the hands of Lagrange and Laplace at the close of the eighteenth century and has continued to evolve to the present.

10. On American science, see Daniel J. Kevles, *The Physicists* (New York: Knopf, 1978). The growth of American physics, in both numbers of physicists and quality, was almost meteoric after about 1890. The careers of Henry, Rowland, Michelson, Gibbs, and a few others, along with the establishment of the *Physical Review* and the American Physical Society, highlight the two decades after 1885. The awarding of the Nobel Prize to Michelson in 1907 was a critical event in the maturing of physics research in the United States. Kevles mentions that in the early 1890s there were only about two hundred practicing physicists in the country; in 1994 there were 8,200 physics faculty members, and the number of practicing physicists was many times greater.

11. See Robert D. Purrington, "Romantic Science: The Role of Samuel Taylor Coleridge" (Paper presented at the annual meeting of the Society for Literature and Science, Portland, Oreg., 1991).

12. For a discussion of some of these issues, see Rudolph Carnap, *Philosophical Foundations of Physics* (New York: Basic Books, 1966).

13. John Herschel, *Preliminary Discourse on the Study of Natural Philosophy* (Chicago: University of Chicago Press, 1987), 219 [facsimile of 1830 edition].

CHAPTER 2   *Nineteenth-Century Science in Context*

1. See Norton Wise, "The Maxwell Literature and British Dynamical Theory," *Hist. Stud. Phys. Sci.* 13 (1982): 175–205.

2. S. F. Cannon, *Science As Culture: The Early Victorian Period* (New York: Dawson and Science History, 1978); Robert Fox, "Science, the University, and the State," in *Professions and the French State, 1700–1900*, ed. G. L. Geison (Philadelphia: University of Pennsylvania Press, 1984), 70.

3. The first meeting of the Royal Society was held on 28 November 1660.

4. By the end of the nineteenth century, the academy was primarily an honorific body.

5. See Maurice Crosland, *The Society of Arcueil* (Cambridge: Harvard University Press,

1967); Geison, ed., *Professions and the French State;* Maurice Crosland, ed., *The Emergence of Science in Western Europe* (New York: Science History Publishing, 1976), especially "Scientific Careers in Eighteenth Century France" and "The Development of a Professional Career in Science in France"; N. Dhombres and J. Dhombres, *Naissance d'un nouveau pouvoir: Sciences et savants en France, 1793–1824* (Paris: Bibliothèque Historique Payot, 1989); H. W. Paul, *From Knowledge to Power: The Rise of the Science Empire in France, 1860–1939* (Cambridge: Cambridge University Press, 1985); Mary Jo Nye, *Science in the Provinces: Scientific Communities and Provincial Leadership in France, 1860–1939* (Berkeley: University of California Press, 1986); C. C. Gillispie, *Science and Polity at the End of the Old Regime* (Princeton, N.J.: Princeton University Press, 1980); Maurice Crosland, *Science under Control: The French Academy of Sciences, 1795–1914* (Cambridge: Cambridge University Press, 1991).

6. Crosland, *The Society of Arcueil.*
7. See Fox, "Science, the University, and the State."
8. Gillispie, *Science and Polity in France at the End of the Old Regime,* chap. 2, p. 74.
9. Paul, *From Knowledge to Power,* chap. 4.
10. Fox, "Science, the University, and the State," 106. Note that this position—namely, that the decline in vigor and originality in French science, especially in theoretical physics, was due to increased bureaucratization—also has critics.
11. Ibid., 71. This point is not universally accepted, however; and even if it seems evident when one looks at the contribution of French scientists to the pace of scientific discovery in the last half of the nineteenth century, increased bureaucratization of science was not necessarily the only or even the primary cause. I thank Stephen Brush for comments concerning this point.
12. Fox, "Science, the University, and the State."
13. Nye, *Science in the Provinces,* 15.
14. Cannon, *Science As Culture.*
15. At best one-third of the Fellows of the Society were scientists in its early years, and a century and a half later the situation was similar. According to Peter Alter in *The Reluctant Patron: Science and the State in Britain, 1850–1920* (trans. Angela Davies [Oxford: Berg, 1987]), ten Anglican bishops were counted among the Royal Society's members in 1830. The founding of the BAAS was a reaction to this amateurism (see, among others, ibid., chap. 6), although the movement to found it predates Herschel's contested election.
16. Some of these clergymen were important figures in the scientific reform movement—for example, George Peacock, later Dean of Ely Cathedral. On skeptics, see Cannon, *Science As Culture,* 3, 12.
17. L. Stewart, *The Rise of Public Science* (Cambridge: Cambridge University Press, 1992). Stewart's book is largely devoted to the period before the Industrial Revolution.
18. See Cannon, *Science As Culture.*
19. Gillispie, *Science and Polity in France,* 78; D. S. Cardwell, *The Organization of Science in England* (London: Heinemann, 1972), 246.
20. One of the important vehicles was the Analytical Society.
21. See Cannon, *Science As Culture.*
22. Professor William Hamilton (not William Rowan Hamilton, the Irish mathemati-

cian) was influenced by both the Kantian dynamical and the Scottish commonsense traditions; he later became Sir William. In all probability, Maxwell obtained the position he briefly held at Aberdeen's Marischal College through Herschel's influence.

23. Hippolyte Taine, *Notes on England,* trans. E. Hyams (Fairlawn, N.J.: Essential Books), 118; originally published as *Notes sur l'Angleterre* (Paris, 1872).

24. See Alter, *The Reluctant Patron.*

25. One could mention also the laboratories of the London Institution, established in 1819, and those of University and King's colleges, established in the 1820s.

26. See, for example, J. L. Heilbron and T. S. Kuhn, "The Genesis of the Bohr Atom," *Hist. Stud. Phys. Sci.* 1 (1969): 211–90. One is tempted to ask the ahistorical question, What if? Actually, Bohr soon left Cambridge for Manchester and Ernest Rutherford's laboratory.

27. Robert Schofield, *The Lunar Society of Birmingham* (Oxford: Oxford University Press, 1963).

28. See, for example, George E. Davie, *The Democratic Intellect: Scotland and Her Universities in the Nineteenth Century* (Edinburgh: Edinburgh University Press, 1961).

29. The wranglers were the top students on the Cambridge Mathematical Tripos examinations, the "first wrangler" being the student who scored highest.

30. See J.R.R. Christie, "The Rise and Fall of Scottish Science," in *The Emergence of Science in Western Europe,* ed. Crosland, including the chapter's numerous references. See also L. L. Laudan, "Thomas Reid and the Newtonian Turn of British Methodological Thought," in *The Methodological Heritage of Newton,* ed. R. E. Butts and J. W. Davis (Toronto: University of Toronto Press, 1970).

31. The Tripos received its name because of the three-legged stools used by the examinees. For more about the exam, see David B. Wilson, "The Educational Matrix: Physics Education at Early-Victorian Cambridge, Edinburgh, and Glasgow Universities," in *Wranglers and Physicists: Studies on Cambridge Mathematical Physics in the Nineteenth Century,* ed. P. M. Harman (Manchester: Manchester University Press, 1985), and C.W.F. Everitt, "Maxwell's Scientific Creativity," in *Springs of Scientific Creativity,* ed. Rutherford Aris, H. Ted Davis, and Roger H. Steuwer (New York: Atheneum, 1983).

32. Everitt, "Maxwell's Scientific Creativity," 102

33. Ibid., 105

34. Christie, "The Rise and Fall of Scottish Science," 114. See also J.R.R. Christie and Sally Shuttleworth, *Nature Transfigured: Science and Literature, 1700–1900* (Manchester: Manchester University Press, 1989).

35. Christie, "The Rise and Fall of Scottish Science."

36. When Davy joined the Royal Institution, Thomas Young was Professor of Natural Philosophy and editor of the journal. His lectures became the fashion, especially for young ladies, who, it is said, were as interested in his eyes as in the content of his lectures. Davy wrote, "The audience is assembled by the influence of fashion merely; and fashion and chemistry form a very incongruous union" (quoted in Anne Treneer, *The Mercurial Chemist* [London: Methuen, 1963], 86, 87).

Rumford, made count of the Holy Roman Empire in 1792, was born in Massachusetts and educated at Harvard College. He escaped to England after having chosen to support the Crown in the American Revolution. His famous experiments were carried out in Munich. He never returned to London after leaving in 1802 but married Lavoisier's widow (see chapter 4).

37. See Treneer, *The Mercurial Chemist*, 16.
38. The term *scientist* was apparently first used by Whewell in 1840, as was *physicist*, reflecting his interest in etymology. Both terms gained currency in the 1870s.
39. University College, the predecessor of the University of London, was founded by Jeremy Bentham and John Stuart Mill in 1826 specifically to serve non-Anglicans.
40. The annual proceeds from the Royal Society's endowments were £1,400 in 1900. The society also administered funds allocated by Parliament, but individual grants rarely exceeded £200. See Alter, *The Reluctant Patron*, 20–21.
41. Early German contributions to natural history and chemistry are another matter.
42. Although part of Hegel's 1817 *Encyclopedia of Philosophical Works*, it was not really available until his collected works were published in 1842.
43. D. M. Knight, "German Science in the Romantic Period," in *The Emergence of Science in Western Europe*, ed. Crosland.
44. When Weber applied for the position, his references were Gauss, Humboldt, Oersted, and Berzelius. He was dismissed in 1838 for refusing to take a loyalty oath and did not hold a position for six years. See C. Jungnickel and R. McCormmach, *The Intellectual Mastery of Nature: Theoretical Physics from Ohm to Einstein*, 2 vols. (Chicago: Chicago University Press, 1986).
45. See David Cahan, "The Institutional Revolution in German Physics, 1865–1914," *Hist. Stud. Phys. Sci.* 15 (1985): 1–65.
46. The earliest were at Bonn in 1825 and Schweigger's at Halle in the late 1820s. Neumann's at Königsberg was approved in 1828 but did not begin until six years later and then survived for only four years. See Jungnickel and McCormmach, *The Intellectual Mastery of Nature*.
47. See Cahan, "The Institutional Revolution in German Physics," 6–11, and R. S. Turner, "The Growth of Professorial Research in Prussia, 1818–1848: Causes and Context," *Hist. Stud. Phys. Sci.* 3 (1971): 137–82.
48. A *cabinet* was an experimental laboratory with its collection of instruments. See Cahan, "The Institutional Revolution in German Physics," 8.
49. See Jungnickel and McCormmach, *Intellectual Mastery of Nature*, for details on the hierarchy of German academic appointments, which ran, in order of decreasing prestige, from ordinary professor (*ordinarius*) to *extraordinarien* to *privatdozenten*. See also Cahan, "The Institutional Revolution in German Physics," and K. M. Olesko, *Physics As a Calling: Discipline and Practice in the Königsberg Seminar for Physics* (Ithaca, N.Y.: Cornell University Press, 1991).
50. Olesko, *Physics As a Calling;* Jungnickel and McCormmach, *Intellectual Mastery of Nature*, chap. 15.
51. The Privatdozent is an entry-level professorial designation that may not actually represent employment. There is no U.S. or English equivalent.
52. See Alter, *The Reluctant Patron*, 13.
53. If we were to add Dalton and Priestley, we would soon recognize the role played by dissenting English scientists with traditions totally distinct from those of the gentry.
54. The *Proceedings of the Royal Society* started by publishing abstracts of the *Philosophical Transactions* and then became a separate journal.
55. R. M. Gascoigne, *A Chronology of the History of Science, 1450–1900* (New York: Garland, 1987).

56. For journals in chemistry, see R. M. Gascoigne, *A Historical Catalogue of Scientific Periodicals, 1665–1900* (New York: Garland, 1985).

57. Jungnickel and McCormmach, *Intellectual Mastery of Nature.*

58. Trevor Levere, *Affinity and Matter: Elements of Chemical Philosophy, 1800–1865* (Oxford: Oxford University Press, 1971); John Hendry, *James Clerk Maxwell and the Theory of the Electromagnetic Field* (Bristol, England: Hilger, 1986).

59. At the same time, be aware of the strong reaction, often negative, that Hendry's book has received—specifically, the skepticism about the usefulness of this division. In particular, see reviews by P. M. Harman (*Ann. Sci.* 44 [1987]: 651–52), Bruce Hunt (*Isis* 79 [1988]: 735–36), Paul Theerman (*Brit. J. Hist. Sci.* 20 [1987]: 365–66), and Cyril Domb (*Nature* 326 [1987]: 26). According to Harman, Hendry "persistently confuses philosophical categories about ontologies with programmes of explanation in mathematical physics" and "construes philosophical categories as modes of scientific argument." Hunt acknowledges that the categories may be useful but says that they "eventually buckle under the weight Hendry puts on them."

60. By *positivist* I simply mean anti- or nonspeculative: the refusal to posit hypothetical structures that are not accessible to measurement. Thus, Duhem and Ernst Mach rejected atoms when there was no direct, incontrovertible, empirical evidence of their existence.

61. P. Laplace, *Complete Works,* 14 vols. (Paris: Gauthier-Villars, 1878–1912), 12:295.

62. J. Fourier, *The Analytical Theory of Heat,* trans. A. Freeman (Cambridge: Cambridge University Press, 1878), 1–2.

63. Pierre Duhem, *The Aim and Structure of Physical Theory,* trans. P. P. Wiener (Princeton, N.J.: Princeton University Press, 1954), 51. The book was originally published in 1914.

64. Ibid., 70.

65. Robert Kargon and Peter Achinstein, eds., *Kelvin's Baltimore Lectures and Modern Theoretical Physics* (Cambridge: MIT Press, 1987).

66. Duhem, *The Aim and Structure of Physical Theory;* quoted from H. Poincaré, *Electricité et optique,* vol. 1 (Sceaux, France: Gabay, 1990).

67. Duhem, *The Aim and Structure of Physical Theory,* 99.

68. Ibid., 99–100; from Hermann Helmholtz's preface to Heinrich Hertz, *Die Principien der Mechanik* (Leipzig: Barth, 1894), published in English as *The Principle of Mechanics Presented in a New Form,* trans. D. E. Jones and J. T. Walley (London: Macmillan, 1899).

69. Robert Fox, "The Rise and Fall of Laplacian Physics," *Hist. Stud. Phys. Sci.* 4 (1974): 89–136.

70. Ibid., 110.

71. See Crosbie Smith and Norton Wise, *Energy and Empire* (Cambridge: Cambridge University Press, 1989), 162.

72. Ibid., 131. But Ampère is not easily pigeonholed, as we shall see.

73. See Jungnickel and McCormmach, *Intellectual Mastery of Nature,* vol. 1, *The Torch of Mathematics, 1800–1870,* 25. The quotation is from G. W. Muncke, *System der Atomistischen Physik Nach den Neuesten Erfahrunger und Versuchen* (Hannover, 1809).

74. G. F. Parrot, *Grundriss der Theoretischen Physik zum Gebrauche fur Vorlesungen* (Leipzig: Meinshaufen, 1809–15). Parrot, who was a professor at Dorpat, made

many attempts to reproduce Coulomb's famous experiment. He is quoted in Jungnickel and McCormmach, *Intellectual Mastery of Nature,* 1:26.

75. This comparison cannot be carried too far because some of those scientists whom we would classify as dynamicists were certainly not positivists: Although they rejected hypothetical entities such as atoms, they indulged in wide-ranging speculations that lacked any empirical support and used terms totally devoid of any operational definition.

76. Lagrange was not French by birth; he was born in Turin, where he signed himself as Lodovico or Luigi Lagrange. Although he worked at the Berlin Academy from 1766 until 1787, he spent the remainder of his life in Paris, regularly participating in the prize competitions of the Paris Academy. All of Lagrange's important work was published in French (including the *Mémoires* of the Berlin Academy).

77. This illustrates Harman's point that Hendry sometimes confuses philosophical predispositions and scientific methodology.

78. James Clerk Maxwell, *A Treatise on Electricity and Magnetism,* 3d ed. (Oxford: Clarendon, 1892), 199–200.

79. Wise, "The Maxwell Literature and British Dynamical Theory."

80. Ibid.

81. Hendry, *James Clerk Maxwell.*

82. On Whewell, see R. Yeo, *Defining Science: William Whewell, Nature, Knowledge, and Public Debate in Early Victorian Britain* (Cambridge: Cambridge University Press, 1993).

83. In this, Hamilton was no doubt influenced by Thomas Reid, Chair of Moral Philosophy at Edinburgh a half-century before him. See Hendry, *James Clerk Maxwell,* 25–27

84. The second edition of the book (1847) was reprinted in 1967 (London: Frank Cass). See M. Fisch and S. Schaffer, eds., *William Whewell: A Composite Portrait* (Oxford: Oxford University Press, 1991), and Yeo, *Defining Science.*

85. See Hendry, *James Clerk Maxwell.*

86. See ibid.; Joseph Agassi, *Faraday As a Natural Philosopher* (Chicago: University of Chicago Press, 1971); and the discussion in this chapter of dynamical and mechanical philosophies.

87. Y. Elkana, *The Discovery of the Conservation of Energy* (Cambridge: Harvard University Press, 1974).

88. E. Whittaker, *A History of the Theories of Aether and Electricity* (London: Nelson, 1951–53; reprinted in one volume, New York: Dover, 1989). On theories of aether specifically, see Whittaker; K. F. Schaffner, ed., *Nineteenth-Century Aether Theories* (Oxford: Pergamon, 1972); and L. S. Swenson, *The Ethereal Aether: A History of the Michelson-Morely Aether Drift Experiments, 1880–1930* (Austin: University of Texas Press, 1972).

89. See excepts from Lorenz Oken, *Elements of Physiophilosophy* (London, 1847) in *The History of Science in Western Civilization,* vol. 3, *Modern Science, 1700–1900,* ed. L. P. Williams and H. J. Steffens (Washington: University Press of America, 1978).

90. L. Königsberger, *Hermann von Helmholtz* (Oxford: Clarendon, 1906), 103–6.

91. Ibid., 159. On Kant and Helmholtz, see S. P. Fullinwinder, "Hermann von Helmholtz: The Problem of Kantian Influence," *Stud. Hist. Phil. Sci.* 22 (1990):

57–90. On transcendental idealism, see R. Harré, *The Anticipation of Nature* (London: Hutchinson, 1965), 104, and the final chapter of Arthur Lovejoy, *The Great Chain of Being* (New York: Harper and Row, 1960).

92. F.W.J. von Schelling, *Ideas for a Philosophy of Nature*, 2d ed. (1803), trans. E. E. Harris and P. Heath (Cambridge: Cambridge University Press, 1988), especially the introduction by Robert Stern. On Kant, see, for example, P. M. Harman, *Metaphysics and Natural Philosophy* (Brighton, England: Harvester, 1982).

93. Schelling, *Ideas for a Philosophy of Nature*, 156.

94. Knight, "German Science in the Romantic Period," 164.

95. Schelling, *Ideas for a Philosophy of Nature*, 97.

96. See Edith Selow's entry on Schelling in the *Dictionary of Scientific Biography*.

97. Knight, "German Science in the Romantic Period," 164–65. Note his comments on the preoccupation with Naturphilosophie as "whig history."

98. Obituary, *Ann. d. Phys.* 76 (1824): 468–69; see Jungnickel and McCormmach, *Intellectual Mastery of Nature*, 1:28.

99. See Thomas Nickles, "From Natural Philosophy to Metaphilosophy of Science," in *Kelvin's Baltimore Lectures*, ed. Kargon and Achinstein.

100. B. S. Gower, "Speculation in Physics: The History and Practice of *Naturphilosophie*," *Stud. Hist. Phil. Sci.* 3 (1973): 301–56.

101. See especially Daniel M. Siegel, *Innovation in Maxwell's Electromagnetic Theory: Molecular Vortices, Displacement Current, and Light* (Cambridge: Cambridge University Press, 1992).

102. Hendry, *James Clerk Maxwell*, 47.

103. My apologies to Gödel.

104. Lewis Campbell and William Garnett, *The Life of James Clerk Maxwell* (London: Macmillan, 1882), 348.

105. James Forbes, "On the Refraction and Polarization of Heat," *Phil. Mag.* 6 (1835): 134–42, 205–14, 366–71; Hendry, *James Clerk Maxwell*, 48.

106. Campbell and Garnett, *The Life of James Clerk Maxwell*, 235–244, especially 236. The Apostle's Club was an organization of Cambridge undergraduates.

107. Hendry, *James Clerk Maxwell*, 50.

108. For an important discussion of the use of models in formulating theory, see Duhem, *The Aim and Structure of Physical Theory*.

109. See the proceedings of the 1990 Kuhn *Festschrift*.

110. There are innumerable sources to which one might direct a reader interested in this issue, including Mary B. Hesse, "Positivism and the Logic of Scientific Discoveries," in *The Legacy of Logical Positivism*, ed. Peter Achinstein and Stephen F. Baker (Baltimore: Johns Hopkins University Press, 1969); Edgar Zilsel, "The Problems of Empiricism," in *The International Encyclopedia of Unified Science*, vol. 2, no. 8 (Chicago: University of Chicago Press, 1941); and much of the work of Carl Popper, including *The Logic of Scientific Discovery* (New York: Basic Books, 1959).

111. In Einstein's case this statement is valid as applied to his analysis of the photoelectric effect.

112. How such a statement might be viewed several centuries from now is another matter. Our division of classical and modern physics is entirely relative both to our own time and to physics as we know it at the end of the twentieth century.

CHAPTER 3   *Electromagnetism*

1. See E. Whittaker, *A History of the Theories of Aether and Electricity*, 2 vols. (1951–53; reprint, New York: Dover, 1989), 81. Kant believed that electricity and magnetism were *both* due to common action of fundamental forces. The Leyden (or Leiden) jar is a condenser or capacitor, a device for storing electric charge.

2. I. B. Cohen, *Franklin and Newton* (Philadelphia: American Philosophical Society, 1956), 251, 254–55. The influence of the Newtonian Stephen Hales is also clear, as is Hales's influence on Franklin. J. L. Heilbron is an excellent source of information on physics, especially electrical science, in the seventeenth and eighteenth centuries; see his *Electricity in the 17th and 18th Centuries: A Study of Early Modern Physics* (Berkeley: University of California Press, 1979), and *Elements of Early Modern Physics* (Berkeley: University of California Press, 1982).

3. Cohen, *Franklin and Newton*, 269–79. See Whittaker's chapter "Electric and Magnetic Science prior to the Introduction of the Potentials," in his *A History of the Theories of Aether and Electricity*, vol. 1.

4. Joseph Priestly, *The History and Present State of Electricity, with Original Experiments* (London, 1767). There was, of course, a powerful prejudice in favor of the inverse-square force.

5. It was published as *System of Mechanical Philosophy* and edited by David Brewster. The result that Robison obtained was $1/r^{2.06}$, and he assumed that it was actually $1/r^2$.

6. See Heaviside's strong comments on Cavendish in Oliver Heaviside, *Electromagnetic Theory* (London, 1893), 1:3–4. Cavendish did give the Royal Society an elaboration of Aepinus's mathematical theory but no account of his own experiments.

7. On Coulomb, see C. S. Gilbert, *Coulomb and the Evolution of Physics and Engineering in Eighteenth Century France* (Princeton, N.J.: Princeton University Press,1971).

8. Michell was a remarkable Englishman who contributed not only to electricity and magnetism but to astronomy as well. He performed an early calculation showing that, if an object were sufficiently small but massive, at a certain radius the escape velocity would exceed the velocity of light. Thus, he foreshadowed the twentieth-century idea of the black hole. See *Memoirs of John Michell* (Cambridge: Cambridge University Press, 1918). Newton had inserted at least three comments on magnetism into the *Principia*. See Sir Isaac Newton, *Mathematical Principles of Natural Philosophy,* trans. Andrew Motte (1729), intro. by I. B. Cohen (London: Dawson, 1968), 1:37 and 2:79, 225 (notably), and 313. For details, see Whittaker, *A History of the Theories of Aether and Electricity,* 1:52–57.

9. Whittaker, *A History of the Theories of Aether and Electricity,* 1:57. See also J. L. Heilbron's article on Wilcke in *Dictionary of Scientific Biography.*

10. Alessandro Volta, "On the Electricity Excited by the Mere Contact of Conducting Substances of Different Kinds," *Phil. Trans. Roy. Soc.* 90 (1800): 403–43 (or in the abridged transactions, 18 [1809]: 744–46); *Phil. Mag.* 7 (1800): 288–311.

11. William Nicholson and Anthony Carlisle, "Account of the New Electrical or Galvanic Apparatus of Sig. Alex. Volta, and Experiments Performed with the Same," *Nicholson's Journal* 4 (1800): 179–87. Ritter obtained the same result shortly thereafter, and Wollaston and others soon showed that frictional and galvanic electricity served equally well and were apparently identical.

12. See Whittaker, *A History of the Theories of Aether and Electricity,* 1:73–78.
13. See L. Pearce Williams, *Michael Faraday: A Biography* (London: Chapman and Hall, 1965), 17.
14. According to the contact theory, contact between the metal layers of the pile (in this case, copper and zinc) allowed the negative and positive electric fluids to accumulate on either side of the junction. For the opposing view, see William Hyde Wollaston, "Experiments on the Chemical Production and Agency of Electricity," *Phil. Trans.* 91 (1801): 427.
15. Sturgeon demonstrated the discovery to members of the Society of Arts on 23 May 1825. It was obvious that a current had magnetic effects; if Oersted's work had not made that clear, then the experiments of Biot, Savart, and Ampère did. E. Bauer credits Arago with inventing the electromagnet in 1820, when he observed that iron filings clung to a current-carrying wire. See René Taton, ed., *A General History of the Sciences,* trans. A. J. Pomerans (London: Thames and Hudson, 1963–66); vol. 3 published as *Science in the Nineteenth Century* (1965), 191. Schwigger's invention, the "multiplier," was essentially a galvanometer.
16. Simeon Denis Poisson, "Mémoire sur la distribution de la l'électricité à la surface des corps conducteurs," *Mémoires de l'Institut* 1 (1811): 1–92; 2 (1811): 163–274. The memoir was published in several parts in various places, 1811–13. See *The Royal Society Catalog of Scientific Papers, 1800–1863* (London, 1867–72). Quoted in Whittaker, *A History of the Theories of Aether and Electricity,* 1:60.
17. V would be written, in modern notation, $V = \int \dfrac{dM}{x - \bar{x}'}$.
18. Pierre Simon, Marquis de Laplace, "Théorie du mouvement et de la figure elliptique des planets" (1784); Poisson, "Mémoire." Note that I am using modern notation for clarity.
19. Poisson, "Mémoire," 247. For explication, see Whittaker, *A History of the Theories of Aether and Electricity,* 1:62–65.
20. Green refers specifically to two of Poisson's papers in the *Mémoires de l'Académie des Sciences* (1821 and 1822). George Green, "An Essay on the Application of Mathematical Analysis to the Theories of Electricity and Magnetism" (1828), reprinted in *The Mathematical Papers of the Late George Green* (London, 1871). For details, see D. M. Cannell, *George Green: Mathematician and Physicist, 1793–1841* (London: Athlone, 1993).
21. In *Dictionary of Scientific Biography,* P. J. Wallis ends his description of Green's work with the following statement: "Through Thomson, Maxwell, and others, the general mathematical theory of potential developed by an obscure, self-taught, miller's son would lead to the mathematical theories of electricity underlying twentieth-century industry." Cannell reveals many heretofore unknown details of Green's life in her new biography, *George Green.* She includes the fact that, as father of seven children born out of wedlock in Nottingham, he had poor prospects for entry into the upright social and scientific circles of Cambridge and the Church.
22. For the role of James MacCullagh of Ireland, see J. Larmour, *Mathematical and Physical Papers* (Cambridge: Cambridge University Press, 1929), 1:392ff. Although Leonardo da Vinci and Grimaldi both took notice of what we call diffraction, Robert Hooke and Christian Huygens were the first to champion the wave theory of light.
23. See, for example, Leonhard Euler, *Opuscula varii argumenti* (Berlin, 1746). See

also R. W. Home, "Leonhard Euler's 'Anti-Newtonian' Theory of Light," *Ann. Sci.* 45 (1988): 522; and A. I. Sabra, *Theories of Light from Descartes to Newton* (Cambridge: Cambridge University Press, 1981). For a different view of the problem of the nature of light, see Jed Z. Buchwald, "Kinds and the Wave Theory of Light," *Stud. Hist. Phil. Sci.* 23 (1992): 39–74.

24. See David Wilson, "The Reception of the Wave Theory of Light by Cambridge Physicists (1820–1850): A Case Study in the Nineteenth-Century Mechanical Philosophy" (Ph.D. diss., Johns Hopkins University, 1968); and G. N. Cantor, "The Reception of the Wave Theory of Light in Britain," *Hist. Stud. Phys. Sci.* 6 (1975): 109–32. Among the best works on the wave theory of light are Whittaker's classic *A History of the Theories of Aether and Electricity* and especially Jed Buchwald, *The Rise of the Wave Theory of Light: Optical Theory and Experiment in the Early Nineteenth Century* (Chicago: University of Chicago Press, 1989).

25. Huygens discovered the polarization of light while investigating the phenomenon of double refraction, or birefringence.

26. Young played a crucial role in the decipherment of the Rosetta Stone and thus Egyptian hieroglyphics.

27. It is interesting to note that the discovery of the polarization of X rays in the early twentieth century was taken as evidence of their wave character.

28. See L. Pearce Williams and Henry John Steffens, *The History of Science in Western Civilization,* vol. 3, *Modern Science, 1700–1900.* Fresnel was not the loser in these encounters; in fact, he carried the day in the conflict between the wave theory of light and the Newtonian corpuscular theory, supported by Biot. See L. Pearce Williams, "Faraday and Ampère: A Critical Dialogue," in *Faraday Rediscovered,* ed. David Gooding and F.A.J.L. James (Basingstoke, England: Macmillan, 1985). See also Whittaker, *A History of the Theories of Aether and Electricity,* 1:108; and M. Born and E. Wolf, *Principles of Optics,* 6th ed. (Oxford: Pergamon, 1980), 375. For more on theories of the aether, see K. F. Schaffner, ed., *Nineteenth-Century Aether Theories* (Oxford: Pergamon, 1972); and L. S. Swenson, *The Ethereal Aether: A History of the Michelson-Morely Aether Drift Experiments, 1880–1930* (Austin: University of Texas Press, 1972).

29. A good deal of controversy surrounds this remarkable and counterintuitive discovery. Because apparently Poisson and others saw it as an erroneous prediction of the wave theory, its verification was electrifying but not considered definitive; both Poisson and Biot resisted the wave theory for at least another decade. This turn of events should not surprise us: The reception of a scientific theory is not a thoroughly rational process, nor are longheld commitments easily given up, even in the face of empirical evidence.

The bright spot was the only novel prediction of Fresnel's theory; his detailed description of the single-slit diffraction pattern was new but not considered decisive either. Moreover, the novel prediction was not decisive in awarding the prize to Fresnel because he had only one opponent. John Worrall offers this incident in support of his (and Zahar's) theory of empirical support for scientific theories. Whether that view is correct is another matter because Worrall's formalistic notion of empirical verification seems not to give sufficient weight to nonquantifiable factors. History shows that items of empirical evidence, apparently equal in their formal status as predictions of a theory (either before or after the fact), may have a very different impact on the acceptance of a theory. See John Worrall, "Fresnel,

Poisson and the White Spot: The Role of Successful Prediction in the Acceptance of Scientific Theories," in *The Uses of Experiment,* ed. David Gooding, Trevor Pinch, and Simon Schaffer (Cambridge: Cambridge University Press, 1989), 135–157. I thank Stephen Brush for bringing Worrall's paper to my attention.

30. See James Clerk Maxwell, *The Scientific Papers of James Clerk Maxwell,* ed. W. D. Niven (Cambridge: Cambridge University Press, 1890). The matter was not, of course, really settled. There was continued skepticism on the part of continental scientists, especially Germans, into the last decade of the nineteenth century and beyond. The writings of Heaviside and Hertz, particularly Hertz's generation of Maxwell's electromagnetic waves, finally decided the issue for most physicists.

Ampère, Leslie, and W. R. Grove were among those who held the view that light and electricity or magnetism were one. John Leslie (1766–1832) was professor of mathematics at Edinburgh after 1805; his important work is *Experimental Inquiry into the Nature and Propagation of Heat* (London, 1804). He discovered the flux law for radiant energy—that is, the dependence on the projected area, or the cosine of the angle of deviation from line of sight. See Henry John Steffens, *James Prescott Joule and the Concept of Energy* (Folkestone, England: Dawson, 1979), 84; and D. S. Cardwell, *From Watt to Clausius: The Rise of Thermodynamics in the Early Industrial Age* (Ithaca, N.Y.: Cornell University Press, 1971), 107–20.

31. Oersted's thesis topic was Kant's *Metaphysical Principles of Natural Science.*

32. Quoted in John Hendry, *James Clerk Maxwell and the Theory of the Electromagnetic Field* (Bristol, England: Hilger, 1986), 57. See H. C. Oersted, *Recherches sur l'identité des forces chimiques et électrique* (Paris, 1813); and *H. C. Oersted: Scientific Papers,* ed. and trans. Kirstine Meyer (Copenhagen: Royal Danish Society of Sciences, 1920).

33. H. C. Oersted in David Brewster's *Edinburgh Encyclopaedia,* quoted in Williams, *Michael Faraday,* 139.

34. A translation of the written version of this brief paper appears in Oersted, *Scientific Papers.* It originally appeared in *Skrifter* 2:214, but Faraday read it in *Ann. Phil.* 16 (1820): 273–76 as "Experiments on the Effect of a Current of Electricity on the Magnetic Needle." The experiment is described in elementary physics texts. If the needle is represented by a magnetic moment μ, and the magnetic field due to the current-carrying wire is denoted by B, then the torque, which causes a rotation of the needle, is given by μ x B, which acts at right angles to both μ and B.

35. Bern Dibner, *Oersted and the Discovery of Electromagnetism* (Norwalk, Conn.: Burndy Library, 1961), 27.

36. Oersted, *Scientific Papers.*

37. See Williams, *Michael Faraday,* 138–39.

38. Arago was a member of the Society of Arcueil and a close friend of Humboldt, Gay-Lussac, and Petit. His greatest interest was the undulatory theory of light. He received the Copley Medal of the Royal Society in 1825 for discovering an effect caused by magnetic induction: the damping effect of a piece of metal held near an oscillating compass needle. August de la Rive was a friend of Ampère and Arago and one of the luminaries of the Paris Academy, but he also corresponded extensively with Faraday, as did August's brother Gaspard. Neither brother contributed very much to the scientific knowledge of their age, although August was heavily involved in the controversy over the contact theory of the voltaic pile.

39. From L. De Launay, ed., *Correspondence du Grand Ampère,* 3 vols. (Paris, 1936–43), quoted in Williams, *Michael Faraday,* 142–43.

40. Jean-Baptiste Biot and Felix Savart, "Note sur le magnetisme de la pile de Volta," *Ann. Chim. Phys.* 15 (1820): 222–23. An excerpt appears in W. F. Magie, *A Source Book in Physics* (Cambridge: Harvard University Press, 1935), 441–42. The law know known as the Biot-Savart law is usually represented as giving the magnetic field as i **dl** x n/r^2 , where n is a vector of unit magnitude in the direction of r and idl is an infinitesimal current element. This has the magnitude dl sin θ/r^2, where θ is the angle between dl and r. Biot (1774–1862) experimented with magnetism with Gay-Lussac and Humboldt and carried out investigations in optics, sound, geodesy, and electricity.

41. See Hendry, *James Clerk Maxwell,* 54, 61; Christine Blondel, *Ampère et la création de l'électrodynamique* (Paris: Bibliothèque Nationale, 1982); and Williams, *Michael Faraday,* 141–42.

42. The definitive work on Ampère to this point is Blondel, *Ampère.*

43. From J. Tyndall, "Faraday's Experiments," in *Free Evening Lectures* (London: Chapman and Hall, 1876), 118–33, one of a set of lectures delivered at the South Kensington Museum.

44. L. P. Williams, "Andre Marie Ampère," in *Dictionary of Scientific Biography.*

45. Pierre Duhem, *The Evolution of Mechanics,* trans. Michael Cole (Germantown, Md.: Sijthoff and Noordhoff, 1980), 25.

46. André-Marie Ampère, "De l'action exercée sur un courant électrique par un autre courant, le globe terrestre ou un aimant," *Ann. Chim. Phys.* 15 (1820): 59–76. All of Ampère's papers on electrodynamics are available in *Mémoires sur l'électrodynamique,* 2 vols. (Paris, 1885–87).

47. ∇ x H = j and $\nabla \cdot$ B = 0. See any textbook on electromagnetic theory—for example, J. D. Jackson, *Classical Electrodynamics,* 2d ed. (New York: Wiley, 1975).

48. The line connecting the elements is the line of intersection of the planes because it is the only line common to the two planes. See Blondel, *Ampère.*

49. In other words, they acted along the line between the elements.

50. Ohm (1787–1854) was a secondary-school teacher in Bonn, gaining a university position only at the end of his life. His work, "The Galvanic Circuit Investigated Mathematically" (Berlin, 1827), was virtually ignored for more than a decade.

51. A. Ampère, "Mémoire sur le théorie mathématique des phénomènes électro-dynamiques," *Mem. de l'Acad.* 6 (1825): 175.

52. See Williams, *Michael Faraday,* and Williams, "Ampère's Electrodynamic Molecular Model," *Contemporary Physics* 4 (1962): 113–23.

53. He reached this conclusion in the *Ann. Chim. Phys.* paper (1820); see note 46. See also Williams, *Michael Faraday,* 144.

54. William Wollaston, friend of Humphrey Davy, also held this view—that magnetism was due to circulating currents.

55. See Williams, *Michael Faraday,* for a discussion of Ampère, the aether, and the interaction of parallel currents.

56. Ampère and De la Rive had seen the effect as early as 1822, but Ampère dismissed it without much thought. Following Faraday's discovery, Ampère in 1832 claimed it as his own. See Williams, "Faraday and Ampère."

57. His case is one of those in which these labels are of little use.

58. Letter to Roux, 11 March 1814, quoted in Hendry, *James Clerk Maxwell,* 62.

59. R.A.R. Tricker, *Early Electrodynamics* (Oxford: Pergamon, 1965), 156.

60. Letter from Faraday to Ampère, 17 November 1815. Quoted in Williams, *Michael Faraday,* 143.

61. William Whewell, *Philosophy of the Inductive Sciences,* 2d ed. (1847; reprinted, London: Cass, 1967). See Hendry, *James Clerk Maxwell,* 71.

62. James Clerk Maxwell, *A Treatise on Electricity and Magnetism* (Oxford: Clarendon, 1873; reprint of 3d ed. (1891), New York: Dover), 175–76.

63. Although Ampère was greatly influenced by Diderot and the *Encylopédie,* he was also devoutly religious, especially when faced by the misery of his later life. See Williams, "Faraday and Ampère.

64. See Williams, *Michael Faraday,* 140–42. Prechtl's theory pictured a conductor as consisting of many small magnets, with one pole on the axis of the conductor and the other on the surface.

65. Faraday was born in London on 22 September 1791, shortly after his family moved to the city. His father was in poor health, and the family's situation was rather difficult. In 1801 they were on public relief. For details of Faraday's early life, see H. Bence Jones, *The Life and Letters of Faraday,* 2 vols. (Philadelphia: Lippincott, 1870). See also the preface to Frank A.J.L. James, ed., *The Correspondence of Michael Faraday,* vol. 1, *1811–Dec. 1831* (London: Institute of Electrical Engineers, 1991). In addition to Williams's biography, see Geoffrey Cantor, David Gooding, and Frank A.J.L. James, *Faraday* (London: Macmillan, 1991); and John Meurig Thomas, *Michael Faraday and the Royal Institution: The Genius of Man and Place* (Bristol, England: Hilger, 1991).

66. "Electricity," *Encyclopaedia Britannica,* 3d ed. (1794), 6:418–45. According to Williams, Faraday was also strongly influenced by Isaac Watts's *The Improvement of the Mind,* which he obtained from the book shop (Williams, *Michael Faraday,* 12–13). See also Bence Jones, *The Life and Letters of Faraday,* 11.

67. *Encyclopaedia Britannica* (1794), 6:419. Concerning Franklin's discovery of the identity of electricity and lightning, the article's author wrote: "The suspicion of Dr. Franklin's was verified in 1752, and the discovery is perhaps the only one in the whole science that hath not been the result of accident."

68. Joseph Agassi, *Faraday As Natural Philosopher* (Chicago: University of Chicago Press, 1971).

69. Nevertheless, it is true that Faraday's earliest discoveries, including the rotation of a wire about a magnet, were made by varying the conditions of the experiment as much as possible (Michael Faraday, *Experimental Researches in Electricity,* 3 vols. [London, 1839–55], 2:147–58; Williams, *Michael Faraday*, 156). They were not driven by any theory, for no theory yet existed in his mind. See also Hendry, *James Clerk Maxwell,* 73–74.

70. See the discussion later in this chapter. In fact, Thomson embraced what he *thought* Faraday's views were. See Hendry, *James Clerk Maxwell,* and Jed Z. Buchwald, "William Thomson and the Mathematization of Faraday's Electrostatics," *Hist. Stud. Phys. Sci.* 8 (1977): 101–36.

71. James Clerk Maxwell, "On Faraday's Lines of Force," *Trans. Camb. Phil. Soc.* 10 (1856), part 1. Reprinted in Niven, ed., *The Scientific Papers of James Clerk Maxwell.*

72. The story of how Faraday came to be hired by Davy is fascinating but beyond the scope of this discussion. See, for example, the preface to volume 1 of James, ed., *The Correspondence of Michael Faraday.*

73. The simple experiment consisted of a magnet, held vertically in a bath of mercury with a wire, connected to a circuit that completed through the mercury bath. When the current in the circuit was turned on, the wire rotated around the magnet. Faraday published this result in *Experimental Researches in Electricity,* 2:147, and the *Quarterly Journal of Science* (October 1821). See Fig. 3–3.

74. See David Knight's description of the affair in "Davy and Faraday: Fathers and Sons," in Gooding and James, eds., *Faraday Rediscovered.*

75. In a letter to Professor C. Hansteen, now in the Royal Institution, dated 16 December 1857, Faraday spoke of his solitary work habits and his reluctance to take on students: "I have never had any student or pupil under me . . . but have always prepared and made my experiments with my own hands, working and thinking at the same time. I do not think I could work in company, or think aloud, or explain my thoughts at the time . . . all this being the consequence of the *solitary* & *isolated* system of investigation."

76. Williams, *Michael Faraday,* 104. The Sandemanians interpreted the Bible literally and believed in direct revelation. They had no clergy or church hierarchy. On the possible influence of Faraday's religious beliefs on his scientific career, see Geoffrey Cantor, *Michael Faraday: Sandemanian and Scientist: A Study of Science and Religion in the Nineteenth Century* (London: Macmillan, 1991), and L. Pearce Williams's review of the book in *Isis* 85 (1994): 120–24. See also Cantor's "Reading the Book of Nature: The Relation between Faraday's Religion and His Science," in Gooding and James, eds., *Faraday Rediscovered.*

77. Faraday, *Experimental Researches in Electricity,* and *Faraday's Diary, Being the Various Philosophical Notes of Experimental Investigations Made by Michael Faraday* (1820–62), 8 vols., ed. Thomas Martin (London: Bett, 1932–36).

78. See Sophie Forgan, "Faraday—From Servant to Savant: The Institutional Context," in Gooding and James, eds., *Faraday Rediscovered.* On Faraday's long relationship with the Royal Institution, see Thomas, *Michael Faraday and the Royal Institution.* A number of other valuable books about Faraday were published in or about 1991, the bicentenary of his birth.

79. His sample of benzene, still extant at the Royal Institution, was recently analyzed by chemists at British Petroleum and found to be 97 percent pure. Frank James has noted that Faraday's glass research was an unfortunate diversion of energy from his studies in electricity. See Frank A.J.L. James, "Faraday's Work on Optical Glass," *Physics Education* 26 (1991): 296–300.

80. Williams argues that R. Phillips's request for such a paper for the *Annals of Philosophy* was the impetus for Faraday's interest in electromagnetism (*Michael Faraday,* 153). "The Historical Sketch" was published anonymously in the journal in September and October 1821 and February 1822. The effect had been discovered during the previous month, in September 1821. William Wollaston, who had speculated about electromagnetic rotation, accused Faraday of appropriating his idea. See, for example, the letters exchanged between the two in Bence Jones, *The Life and Letters of Faraday.*

81. "Historical Sketch of Electromagnetism"; see Williams, "Faraday and Ampère." Putting aside whether or not Faraday was a positivist, current scholarship shows that he saw charges as a product of the field (see my discussion later in the chapter) rather than the source of it; at the very least they were secondary—what Daniel M. Siegel calls "epiphenomena" of the field (*Innovation in Maxwell's Electromag-*

netic Theory [Cambridge: Cambridge University Press, 1991]). After 1821 Faraday questioned the existence of an electrical fluid—that is, any current in a wire. He wrote: "There are many arguments in favor of the materiality of electricity, and but few against it, but still it is only a supposition; and it will be as well to remember, while perusing the subject of electro-magnetism, that we have no proof of the materiality of electricity, or of the existence of any current through the wire" ("Historical Sketch of Electromagnetism," *Ann. Phil.* 2 (1821): 195; see also Williams, *Michael Faraday*, 154).

82. The fascinating correspondence between Ampère and Faraday during this period can be found in James, ed., *The Correspondence of Michael Faraday.*
83. Williams, "Faraday and Ampère," 86–91. One of the articles was by John Herapath. Williams also discusses Faraday's "Historical Sketch of Electromagnetism."
84. Williams, *Michael Faraday*, 173.
85. Obviously this was not framed in terms of fields because the concept had not yet emerged in this context.
86. Faraday, *Experimental Researches in Electricity*, 1:2. It was read to the Royal Society on 24 November 1831 and published in *Philosophical Transactions.*
87. What Ampère and De la Rive actually observed were transient effects due to currents' being induced in a copper ring as current was set up in a nearby coil of wire. Although the experiment, which included a large horseshoe magnet, was designed to determine if one current could induce another, the small effect was dismissed. As Ampère later wrote to De la Rive, "unfortunately neither of us thought of analyzing the phenomenon." See, for example, E. Bauer, "Electricity and Magnetism," in *Science in the Nineteenth Century*, ed. Taton, 190–91.
88. Ryan D. Tweney, "Faraday's Discovery of Induction: A Cognitive Approach," in Gooding and James, eds., *Faraday Rediscovered*, 204–6.
89. In *Maxwell on the Electromagnetic Field: A Guided Study* (New Brunswick, N.J.: Rutgers University Press, 1996), Thomas K. Simpson observes that this elementary error, which any first-year physics student could correct, led Faraday to a discovery that constitutes one of the most important conceptual revolutions in physics.
90. Electricity was emerging from the field rather than acting as a source of it.
91. Nevertheless, Doran points out that he rejected an essential property of the Boscovichean atom: action-at-a-distance; see Barbara Giusti Doran, "Origins and Consolidation of Field Theory in Nineteenth-Century Britain: From the Mechanical to the Electromagnetic View of Nature," *Hist. Stud. Phys. Sci.* 6 (1975): 134–260. Faraday, *Experimental Researches in Electricity*, 2:291.
92. This remarkable passage appears in *Experimental Researches in Chemistry and Physics* (London, 1821–57), 371.
93. See Faraday, *Experimental Researches in Electricity*, 3:3–17. In this experiment, carried out in the fall of 1845, Faraday passed light through a rectangular prism of glass and viewed the light through crossed polarizing filters, which extinguished the light. When an intense magnetic field was applied (in the proper direction), the light reappeared, demonstrating the rotation of the plane of polarization.
94. William Thomson to Faraday, 6 August 1845. See Silvanus P. Thompson, *The Life of William Thomson, Baron Kelvin of Largs*, 2 vols. (London: Macmillan, 1910), and Williams, *Michael Faraday*, 383. As I have noted, Faraday's correspondence is being published by James as *The Correspondence of Michael Faraday.* See also L. Pearce Williams, ed., *Michael Faraday: Selected Correspondence* (Cambridge:

Cambridge University Press, 1971–). In November 1845 Faraday found that a rectangular glass prism would align itself *across* an intense magnetic field, unlike a piece of ferromagnetic material, which would align itself along the magnetic field lines. Previously Faraday had found no effect at all. See Faraday, *Experimental Researches in Electricity*, vol. 3. He was aided in these experiments by the use of a much stronger electromagnet.

95. See Williams, *Michael Faraday*, 409.

96. Ibid., 393.

97. The crucial experiment is described in Faraday, *Experimental Researches in Electricity*, vol. 3. Anyone interested in Faraday should also examine his diary. Faraday first used the term *diamagnetic* to describe the magneto-optic effect and then adopted the terms that Whewell suggested in a letter in December 1845; Whewell also suggested *paramagnetism*. These letters will be included in James's second volume of Faraday correspondence, presently being compiled. In the extensive Whewell-Faraday correspondence (largely in the Wren Library, Trinity College, Cambridge), readers continually find Faraday asking Whewell for a word to describe a new phenomenon he has observed. On 12 November 1844 Whewell wrote: "I am glad that you are working, and come to a point where you want new words; for new words with you infer new things" (letter in the Whewell Collection, Wren Library).

98. Faraday, "Thoughts on Ray Vibrations," in *Experimental Researches in Chemistry and Physics*, 370; *Phil. Mag.* 28 (1846). Reprinted in *Experimental Researches in Electricity*, 3:447–52.

99. See Williams, *Michael Faraday*, 376–79, and Doran, "Origins," especially 162ff, including 165 and 169n. See also Faraday, "Thoughts on Ray Vibrations." In *Experimental Researches in Electricity* (2:293), Faraday says: "The view now stated of the constitution of matter would seem to involve necessarily the conclusion that matter fills all space, or, at least, all space to which gravitation extends (including the sun and its system); for gravitation is a property of matter dependent on a certain force, and it is this force which constitutes the matter. In that view matter is not merely mutually penetrable, but each atoms extends, so to say, throughout the whole of the solar system, yet always retaining its own centre of force."

100. Faraday, *Experimental Researches in Electricity*, 3: 450–51; quoted in Doran, "Origins," 169–70.

101. A good source is P. M. Harman, *Energy, Force and Matter: The Conceptual Development of Nineteenth-Century Physics* (Cambridge: Cambridge University Press, 1982).

102. Nancy Nersessian, *Faraday to Einstein: Constructing Meaning in Science* (Dordrecht, Holland: Nijhoff, 1984), chap. 3.

103. Those who believed in action-at-a-distance (Newton and Ampère, for example) saw this action as mediated by some intervening medium—that is, the aether. See Siegel, *Innovation in Maxwell's Electromagnetic Theory*, 7.

104. Nersessian, *Faraday to Einstein;* Joseph Agassi, *Faraday As a Natural Philosopher;* David Gooding, "Final Steps to the Field Theory: Faraday's Study of Magnetic Phenomena," *Hist. Stud. Phys. Sci.* 11 (1981): 231–75; W. Berkson, *Fields of Force: The Development of a World View from Faraday to Einstein* (New York: Wiley, 1974).

105. Nancy Nersessian, "Faraday's Field Concept," in Gooding and James, eds., *Faraday Rediscovered*, 182; Faraday, *Experimental Researches in Electricity*, 1:74–75.

106. See Gooding, "Final Steps to the Field Theory."

107. See Nersessian, "Faraday's Field Concept," 182–86. The question was whether the lines of force were paths (in the aether?) or the vehicles of transmission.

108. For example, see Michael Faraday, "On the Conservation of Force," *Phil. Mag.* 17 (1859): 166. His views on the subject, however, go back much further.

109. Michael Faraday, "On the Physical Character of the Lines of Force," *London and Edinburgh Philosophical Magazine* 5 (1852): 401; reprinted in *Experimental Researches in Electricity,* 3:407. Four years later Maxwell gave his approval to this idea in "On Faraday's Lines of Force" (discussed later in the chapter). Faraday expressed his gratitude in a letter to Maxwell dated 25 March 1857, saying that "it . . . gives me much encouragement to think on." He added: "I was at first almost frightened when I saw such mathematical force made to bear upon the subject and then relieved to see that the subject stood it so well" (letter in the Maxwell Collection, Cambridge University Library).

110. This was especially the case in Germany. See, for example, the electrodynamics of Wilhelm Weber in Jungnickel and McCormmach, *Intellectual Mastery of Nature,* vol. 1, chap. 6.

111. As Faraday wrote to Whewell, "I wish most sincerely that some mathematician would think it worth his while to do for the facts [Faraday's experimental discoveries] that which I am not able to do for them" (letter dated 16 September 1835, in the Whewell Collection, Wren Library, Trinity College, Cambridge).

112. In the 1840s Faraday suffered from severe memory loss and weakness, which his letters record. He wrote to John Barlow on 18 October 1844: "My mind and memory became quite bewildered" (letter in the Royal Institution, London). In a letter to Thomson dated 23 November 1883, John Tyndall quoted from a memorandum he received from Faraday near the time of Faraday's last lecture on 20 June 1862: "He wrote of 'loss of memory and physical endurance of the brain,' and an 'inability to draw upon the mind for the treasures of knowledge it has previously received.' He talked of 'dimness, forgetfulness of one's former self-standard in respect of right, dignity, and self-respect,' and of 'strong duty of doing justice to others, yet inability to do so.' " Tyndall said, "This is very pathetic" (letter in the Cambridge University Library).

113. Niven, ed., *The Scientific Papers of James Clerk Maxwell,* 358, 360.

114. According to Crosbie Smith and Norton Wise, "J. P. Nichol, with his charismatic personality and popular lecture style, made Fresnel and Fourier the heroes of truth and beauty in the analytical art" (*Energy and Empire* [Cambridge: Cambridge University Press, 1989], 167). Nichol, however, had apparently not read the mathematical parts of Fourier's work.

115. Ole Knudsen, "Mathematics and Physical Reality in William Thomson's Electromagnetic Theory," in *Wranglers and Physicists,* ed. P. M. Harman (Manchester: Manchester University Press, 1985).

116. Smith and Wise, *Energy and Empire,* 149.

117. See Buchwald, "William Thomson and the Mathematization of Faraday's Electrostatics," and Robert Murphy, *Elementary Principles of the Theories of Electricity, Heat, and Molecular Action* (Cambridge, 1832).

118. Smith and Wise, *Energy and Empire,* chap. 2.

119. In "William Thomson and the Mathematization of Faraday's Electrostatics,"

Buchwald argues that Thomson was not aware of the dichotomy until he read Green's "Essay" in 1845.

120. Ibid., 121.

121. Ibid., 135.

122. Pierre Duhem, *The Aim and Structure of Physical Theory,* trans. P. P. Wiener (Princeton, N.J.: Princeton University Press, 1954).

123. On "mixed mathematics," see Siegel, *Innovation in Maxwell's Electromagnetic Theory,* chap. 1. Note Thomson's idea of a physical method of proof, discussed by Smith and Wise in *Energy and Empire,* 211.

124. See Smith and Wise, *Energy and Empire,* chap. 6. Laplace said that "Fourier's equations are right, but the true bases of them are to be found in the doctrine of the action of molecules *ad distans*" (*Energy and Empire,* 166). See also G. G. Stokes, "On the Theories of the Internal Friction of Fluids in Motion, and of the Equilibrium and Motion of Elastic Solids," *Trans. Camb. Phil. Soc.* 8 (1849): 287–319.

125. Gregory died in 1844, and R. L. Ellis filled in until 1845. Within two years, Thomson was becoming disenchanted with the *Journal,* which was dominated by purely mathematical papers. See Smith and Wise, *Energy and Empire,* 188.

126. Ibid., chap. 6.

127. William Thomson, "On the Uniform Motion of Heat in Homogeneous Solid Bodies, and Its Connexion with the Mathematical Theory of Electricity," *Camb. Math. J.* 3 (1843): 71–84.

128. H. I. Sharlin describes the paper as "one of those diverging points in the progress of scientific theory whose scope cannot be appreciated until theory has moved some distance away" (*Lord Kelvin: The Dynamic Victorian* [University Park: Pennsylvania State University Press, 1979], 23). See also Hendry, *James Clerk Maxwell,* 93, and Buchwald, "William Thomson and the Mathematization of Faraday's Electrostatics."

129. James Clerk Maxwell, review, *Nature* 7 (1873): 218–22.

130. Thomson, "On the Uniform Motion of Heat." He had already read a translation of Carl Friedrich Gauss's "General Theorems Relating to Attractive and Repulsive Force" in *Taylor's Scientific Memoirs* (1842); see Buchwald, "William Thomson and the Mathematization of Faraday's Electrostatics." Gauss's paper was originally published in *Resultate aus den Beobachtungen des Magnetischen Vereins im Jahre 1839,* ed. Carl Friedrich Gauss and Wilhelm Weber (Leipzig, 1840).

131. William Thomson, "Demonstration of a Fundamental Proposition in the Mechanical Theory of Electricity," *Camb. Math. J.* 4 (1845): 223–26. George Green, "Essay on the Application of Mathematical Analysis to the Theories of Electricity and Magnetism" (Nottingham, 1828). Thomson noted how he acquired a copy of Green's paper: "This I found in a reference to his memoirs, in Murphy's first memoir on definite integrals. Ever since I have been trying to see Green's memoir, but could not hear of it from anybody till to-day, when I have got a copy from Mr. Hopkins. Jan. 25, 1845" (footnote to William Thomson, "Propositions in the Theory of Attraction," *Camb. and Dub. Math. J.* [1842–43]).

See Cannell, *George Green,* for a discussion of how Thomson eventually found Green's work. The Tripos of Thomson's time was so devoid of material on electromagnetism that, despite the fact that Hopkins tutored young William in his rooms almost every day for two years, he never mentioned Green's work. Thomson incorporated Green's ideas in the paper "Note sur les lois élémentaires de l'électricité statique" in Liouville's *Journal de Mathématiques Pures et Appliquées* 10 (1845): 209–21. The extended English version was "On the Elementary Laws of Static Electricity." Thomson had Green's work published in *Journal für die Reine und Angewandte Mathematik* 39 (1850): 73–79; 44 (1852): 356–74; 47 (1854): 161–221; and it was translated into German in *Ostwald's Klassiker der Exakten Wissenschaften* 61.

132. William Thomson, "On the Mathematical Theory of Electricity in Equilibrium," *Camb. and Dub. Math. J.* 1 (1845): 75–95; 3 (1848): 131–48, 266–74; 5 (1850): 1–9. These are collected in *Reprint of Papers on Electrostatics and Magnetism* (London, 1872).

133. Snow Harris, "On Some Elementary Laws of Electricity," *Phil. Trans. Roy. Soc.* 124 (1834): 213–45. The succeeding article is a famous one on dynamics by William Rowan Hamilton.

134. He may have first heard them in 1841 during a conversation with David Thomson, a cousin of Faraday's. See Buchwald, "William Thomson and the Mathematization of Faraday's Electrostatics," 105.

135. William Thomson, "On the Elementary Laws of Static Electricity," in *Reprint of Papers on Electricity and Magnetism,* 15–37.

136. In 1845 Thomson wrote: "Mr. Faraday's researches on electrostatical induction . . . were undertaken with a view to test an idea which he had long possessed, that the forces of attraction and repulsion exercised by free electricity, are not the resultant of actions exercised at a distance, but are propagated by means of molecular action among the contiguous particles of the insulating medium surrounding the electrified bodies. . . . By this idea he has been led to some very remarkable views upon induction, or, in fact, upon electrical action in general. As it is impossible that the phenomena observed by Faraday can be incompatible with the results of experiment which constitute Coulomb's theory, it is to be expected that the difference of his ideas from those of Coulomb must arise solely from a different method of stating, and interpreting physically, the same laws: and farther, it may, I think, be shown that either method of viewing the subject, when carried sufficiently far, may be made the foundation of a mathematical theory which would lead to the elementary principles of the other as consequences" (William Thomson, "On the Mathematical Theory of Electricity in Equilibrium," *Camb. and Dub. Math. J.* 1 [1846]: 75–95).

137. Smith and Wise, *Energy and Empire,* 215–36.

138. See David Wilson, ed., *The Correspondence of Sir George Gabriel Stokes and Sir William Thomson, Baron Kelvin of Largs* (Cambridge: Cambridge University Press, 1990). Wilson's book gives readers access to this great resource of nineteenth-century British science, most of which resides in the manuscript collection at Cambridge University Library.

139. Sharlin, *Lord Kelvin,* 70.

140. Nor had Coulomb appeared to offer his two-fluid model as real: "I have no other intention than to present with as few elements as possible the results of calculation and experiment, and not to indicate the true causes of electricity." Quoted by Thomson (in French) in *Reprint of Papers on Electricity and Magnetism,* p. 15. From C. A. Coulomb, *Histoire de l'Académie* (Paris, 1788), 673. English translation by W. Reed.

141. In spite of absorbing Fourier's aversion to hypothetical entities, Thomson came to epitomize the "British" mode of building theory upon the secure foundation of mechanical models.

142. Thomson, *Diary,* 31 October 1846.

143. William Thomson, "On a Mechanical Representation of Electric, Magnetic, and Galvanic Forces," *Mathematical and Physical Papers* (London, 1872), 1:76–80.

144. Sharlin, *Lord Kelvin,* 86

145. It is a travesty to relegate Stokes's deep and lasting contributions to a footnote, but, alas, such must be the case. For more information, consult George Gabriel Stokes, *Mathematical and Physical Papers* (New York: Johnson Reprint Corporation, 1966), which reprints the second edition (1880–1905) and includes a preface by Clifford Truesdell.

146. Letter from Stokes to Thomson, 10 April 1847, Kelvin Collection, Cambridge University Library; reprinted in Wilson, *The Correspondence of Sir George Gabriel Stokes and Sir William Thomson,* 24.

147. Letter from Thomson to Stokes, 30 March 1847, Stokes Collection, K18, Cambridge University Library; reprinted in Wilson, *The Correspondence of Sir George Gabriel Stokes and Sir William Thomson,* 7.

148. Sharlin, *Lord Kelvin.*

149. See David Wilson, "The Educational Matrix: Physics Education at Early Victorian Cambridge," in *Wranglers and Physicists,* ed. Harman, 41.

150. Knudsen, "Mathematics and Physical Reality in William Thomson's Electromagnetic Theory," 155.

151. Letter from Thomson to Faraday, 11 June 1847; quoted in Norton Wise, "The Flow Analogy to Electricity and Magnetism," part 1: "William Thomson's Reformulation of Action at a Distance," *Arch. Hist. Exact Sci.* 25 (1981): 19–70. See also Hendry, *James Clerk Maxwell,* 103.

152. See Smith and Wise, *Energy and Empire,* 242–43.

153. Sharlin, *Lord Kelvin,* 91–92.

154. William Thomson, "An Account of Carnot's Theory of the Motive Power of Heat, with Numerical Results Deduced from Regnault's Experiments on Steam," *Trans. Roy. Soc. Edin.* 16 (1849): 541–74.

155. The first volume was revised in 1879, the second published in 1883. The original project was never completed.

156. Sharlin, *Lord Kelvin,* 165.

157. Fitzgerald died in 1901 at the age of forty-nine, the same year that Tait died. The reference to "mere analogy" was in a letter to Thomson dated 25 November 1898.

158. See Stephen Brush, "Note on the History of the Fitzgerald-Lorentz Contraction," *Isis* 58 (1967): 230–32. Fitzgerald's paper, "The Ether and the Earth's Atmosphere," was published in *Science* 13 (1899): 390.

159. Thompson, *The Life of William Thomson,* 2:984.

160. Nevertheless, Maxwell was not overwhelmed with honors in his lifetime. In Ivan Tolstoy's words, he was "modest, reticent, and detached" and indifferent to fame (*James Clerk Maxwell: A Biography* [Edinburgh: Canongate, 1981], 7). The best current biography of Maxwell is C.W.F. Everitt, *James Clerk Maxwell: Physicist and Natural Philosopher* (New York: Scribner's, 1975).
    Maxwell's father's name was originally Clerk. When he married into the Max-

well family, he assumed the Maxwell name. Properly, therefore, we should speak of James Clerk-Maxwell, as he was often indexed in publications of the time (such as *Nature*).

161. Their letters are wonderfully relaxed and humorous and provide a glimpse into Maxwell's broad and generous character. Much of the correspondence is in the Maxwell Archives at Cambridge University Library. See also P. M. Harman, ed., *The Scientific Letters and Papers of James Clerk Maxwell*, 3 vols. (Cambridge: Cambridge University Press, 1990–).

162. According to his obituary, he moved to Trinity because "the Peterhouse men were all classics or pure mathematicians" and he had made few friends. Although Routh was a competitor at Peterhouse, Maxwell did not have a designing character, meaning that the obituary's explanation is probably sufficient. Nevertheless, the opportunity to become a Fellow of Trinity may have been greater than it was at Peterhouse, where there were already several distinguished Fellows in mathematics.

163. Among his interests was writing poetry, some of which was published in *Blackwood's Magazine*. For more on the influence of Scottish commonsense philosophy, see Richard Olson, *Scottish Philosophy and British Physics, 1750–1880* (Princeton, N.J.: Princeton University Press, 1975), and George E. Davie, *The Democratic Intellect: Scotland and Her Universities in the Nineteenth Century* (Edinburgh: Edinburgh University Press, 1961). On Hamilton, Forbes, and Kelland, see Hendry, *James Clerk Maxwell*, 46–48, 111–115; Crosbie Smith, "'Mechanical Philosophy' and the Emergence of Physics in Britain, 1800–1850," *Ann. Sci.* 33 (1976): 3–29; Smith and Wise, *Energy and Empire;* and H. W. Becher, "William Whewell and Cambridge Mathematics," *Hist. Stud. Phys. Sci.* 11 (1980): 1–48. According to Martin Goldman, Forbes was interested in what we now call geophysics; invented the seismometer; and made the first ascent of Sgurr nan Gillean, one of the Cuillins on the Isle of Skye (*The Demon in the Aether* [Edinburgh: Harris, 1983]).

164. Hendry, *James Clerk Maxwell*, 111.

165. This was almost an apprenticeship or, as Siegel calls it, an "epistolary tutelage" (*Innovation in Maxwell's Electromagnetic Theory*, 13). In the preface to the first edition of *Treatise on Electricity and Magnetism* (1873), Maxwell wrote that he owed to Thomson "most of what I have learned on the subject." Some of their correspondence is in the Cambridge University Library, some at Edinburgh University. See Harman, ed., *The Scientific Letters and Papers of James Clerk Maxwell*.

166. From the preface to the first edition of Maxwell's *Treatise on Electricity and Magnetism*. For his part, Faraday never stopped hoping that the results of Thomson and Maxwell could be put into nonmathematical terms: "When a mathematician engaged in investigating physical actions and results has arrived at his conclusions, may they not be expressed in common language as fully, clearly, and definitely as in mathematical formulae?" (Letter from Faraday to Maxwell, 13 November 1857, in the Maxwell Collection, Cambridge University Library).

167. He read the paper in December 1855 and February 1856 and published it in *Trans. Camb. Phil. Soc.* 10 (1856): part 1, 27–83. It was reprinted in *Scientific Papers*, 1:155–229. While he was working on the paper, he was also doing revolutionary work on color vision. On 20 February 1854 he wrote to Thomson inquiring about how to "attack Electricity." He asked, "Suppose a man to have a popular knowledge of electrical show experiments and a little antipathy to Murphy's Elec-

tricity, how ought he to proceed in reading & working so as to get a little insight into the subject wh may be of use in further reading? If he wished to read Ampère Faraday &c how should they be arranged, and at what stage & in what order might he read your articles in the Cambridge Journal?" (letter from Maxwell to Thomson, 20 February 1854; quoted in Harman, ed., *Scientific Letters and Papers,* 1:237).

See also the letters to Thomson dated 13 November 1854 and 15 May 1855 (in ibid., 1:254–63, 305–13). In a letter dated 13 September 1855, Maxwell discusses what he is about to present in a few months to the Cambridge Philosophical Society: "As there can be no doubt that you have the mathematical part of the theory in your desk all that you have to do is to explain your results with reference to electricity. I think that if you were to do so publicly it would introduce a new set of electrical notions into circulation & save much useless speculation." He went on to say: "I do not know the Game-laws & patent-laws of science. Perhaps the Association may do something to fix them but I certainly intend to poach among your electrical images" (in ibid., 1:319–24).

168. Maxwell, "On Faraday's Lines of Force," in *Scientific Papers,* 1:157–58.

169. Maxwell noted: "This analogy between the formulae of heat and attraction was, I believe, first pointed out by Professor William Thomson in the *Camb. Math. Journal,* Vol. III" (see ibid., 1:157).

170. For mathematical details, see Siegel, *Innovation in Maxwell's Electromagnetic Theory,* chap. 2, or Hendry, *James Clerk Maxwell,* 133f.

171. Maxwell, "On Faraday's Lines of Force," in *Scientific Papers,* 1:197.

172. Ibid., 1:202. We would write, in differential form, $\nabla \times H = j$, where j is the current density, the equivalent of Maxwell's quantity of electricity.

173. Ibid., 1:207.

174. W. Weber, "Electro-Dynamic Measurements," translated in *Taylor's Scientific Memoirs,* vol. 5. Maxwell specifically refers to Thomson's "On a Mechanical Representation of Electric, Magnetic, and Galvanic Forces," *Camb. and Dub. Math. J.* 2 (1847): 61–64, and "Mathematical Theory of Magnetism," *Phil. Trans. Roy. Soc.* 141 (1851): 212–85.

175. William Thomson, "On Electrical Images," *BAAS Report* 17 (1847)" 6–7. See also his "On the Mathematical Theory of Electricity in Equilibrium, III" *Camb. and Dub. Math. J.* 3 (1848): 141–48.

176. Maxwell, "On Faraday's Lines of Force," in *Scientific Papers,* 1:187–88.

177. Maxwell's obituary in *Nature* on 13 November 1879, written by William Garnett, said: "He had not the power of making himself clearly understood by those who listened but casually to his pithy sentences, and consequently was not a so-called popular lecturer; nor was he a most successful teacher of careless students."

178. Glenlair was the family estate in Galloway near Knockvenie. The house burned in 1929 and is now in ruins but is still being cared for.

179. James Clerk Maxwell, "On Physical Lines of Force," *Phil. Mag.* 21 (1861): 161–75, 281–91, 338–48; 22 (1862): 12–24, 85–95. Reprinted in *Scientific Papers,* 1:451–513.

180. The first part of Maxwell's "A Dynamical Theory of the Electromagnetic Field" was read on 8 December 1864 and published in *Phil. Trans. Roy. Soc.* 155 (1865): 459–512. "On the Dynamical Theory of Gases" appeared in *Phil. Trans. Roy. Soc. London* 157 (1867): 49–88 and is reprinted in *Scientific Papers,* 2:26–78.

181. This issue, and the related one of Maxwell's progress toward the "cleaner" description in the *Treatise,* is detailed in Siegel, *Innovation in Maxwell's Electromagnetic Theory.*

182. Thomson thought Helmholtz had proved the stability of vortex motion (contrary to what Stokes believed). The source of Thomson's interest in vortex motion is thought to have been his brother James and Rankine. Rankine framed his discussion of the interconvertibility of heat and work in terms of vortex atoms. See Smith and Wise, *Energy and Empire,* chap. 12.

183. Tolstoy, *James Clerk Maxwell,* 124. Maxwell's comments on his own use of concrete models are instructive: "The former is built up to show that the phenomena are such as can be explained by mechanism. The nature of the explanation is to the true mechanism what an orrery is to the Solar System" (letter to Peter Guthrie Tait, 23 December 1867, Cambridge University Library).

184. The phrase was borrowed from Goethe.

185. Thomson, "Mechanical Representation of Electric, Magnetic, and Galvanic Forces."

186. Maxwell, *Scientific Papers,* 1:156.

187. Ibid., 1:486.

188. See Siegel, *Innovations in Maxwell's Electromagnetic Theory,* 66–70. For a somewhat different, personal, but historically informed view, see Simpson, *Maxwell on the Electromagnetic Field: A Guided Study.*

189. Maxwell, *Scientific Papers,* 1:489.

190. Siegel offers an interesting discussion of this calculation in *Innovation in Maxwell's Electromagnetic Theory,* which sheds light on the seemingly ad hoc nature of Maxwell's choices for the parameters of the medium—the elastic aether—in which he saw light propagating as transverse torsion waves.

191. Maxwell, *Scientific Papers,* 1:50 (italics are his). For details in modern notation, see Hendry, *James Clerk Maxwell,* and Jed Z. Buchwald, *Maxwell to Microphysics* (Chicago: University of Chicago Press, 1985).

192. Maxwell, *Scientific Papers,* 2:306.

193. Siegel, *Innovations in Maxwell's Electromagnetic Theory,* 125–43.

194. See Hendry, *James Clerk Maxwell,* chaps. 5 and 6, and Whittaker, *A History of the Theories of Aether and Electricity,* vol. 1, chap. 8. Riemann had speculated about the possibility of adding to Poisson's equation, $\nabla^2 V + 4\pi\rho = 0$, the term $- 1/c^2 \, \partial^2 V/\partial t^2$, a speculation that was published after his death (*Ann. d. Phys.* 131 [1867]; see Whittaker). See also James Clerk Maxwell, "Note on the Electromagnetic Theory of Light," *Phil. Trans. Roy. Soc.* 158 (1868): 643–57.

195. The displacement current deserves more than a footnote and indeed has been the source of some of the most heated discussions surrounding Maxwell's theory. For details, see Siegel, *Innovations in Maxwell's Electromagnetic Theory.* Briefly, however, this is the situation: Faraday's law in modern notation is written $\nabla \times E + 1/c \, \partial B/\partial t = 0$; the time derivative of the magnetic field gives rise to Faraday's magnetic induction. (One might consult Siegel's footnote 8 to chapter 1 on the delicate issue of translating historical mathematical statements into modern language.) But Ampère's law, $\nabla \times H = 4\pi J/c$, implies $\nabla \cdot J = 0$ because $\nabla \cdot \nabla \times$ is identically 0; this is in conflict with the continuity equation $\nabla \cdot J + \partial r/\partial t = 0$. If a term of the form $\partial E/\partial t$ is added to the righthand side of Ampère's law, then taking the divergence of that law just leads to the continuity equation. A critical part of Siegel's discussion involves how Maxwell (and Faraday before him) viewed what we consider to be

the sources of the electromagnetic field, namely the charges and currents. Siegel argues, as have others, that Maxwell and Faraday thought of these as products of the fields, or, as Siegel says, "epiphenomena" of the fields. This perspective is quite different from the modern one and reminds us that, simply because the equations are the same as those in a modern textbook, the theories do not need to be ontologically the same.

196. See Daniel Siegel, "The Origin of the Displacement Current," *Historical Studies of Physical and Biological Science* 17 (1986): 99–146.
197. Maxwell, *Scientific Papers,* 1:535.
198. Ibid., 1:580.
199. As Goldman says in *The Demon in the Aether,* "Cambridge was luckier than it deserved. It got the greatest of the three very great scientists" (177).
200. William Cavendish, like Maxwell, had been second wrangler and the Smith's Prize winner.
201. James Clerk Maxwell, *Electrical Researches of the Honorable Henry Cavendish* (Cambridge, 1879).
202. See Hendry, *James Clerk Maxwell,* chap. 6, especially 237–38.
203. In his last letter to Tait, written two months before his death, Maxwell wrote: "I have been so seedy that I could not read anything however profound without going to sleep over it" (letter in Maxwell Archives, Cambridge University Library).
204. See C. F. Gauss, *Werke,* 5:199–200.
205. See *Crelle's Journal* 39 (1850): 73–89; 44 (1852): 356–74; 47 (1854): 161–221.
206. See Jungnickel and McCormmach, *Intellectual Mastery of Nature;* Whittaker, *A History of the Theories of Aether and Electricity;* and Siegel, *Innovations in Maxwell's Electromagnetic Theory.*
207. George Simon Ohm, *Die Galvanische Kette, Mathematische Bearbeitet* (Berlin, 1827). See Jungnickel and McCormmach, *Intellectual Mastery of Nature,* 1:55.
208. Jungnickel and McCormmach, *Intellectual Mastery of Nature,* 1:76.
209. Ibid., 1:149.
210. In 1848 Weber moved back to Göttingen.
211. Poisson's equation, in electrostatics or gravitational theory, was $\nabla^2 V = -4\pi\rho$, where ρ is the charge or mass density (see note 194). See Whittaker, *A History of the Theories of Aether and Electricity,* 1:206.
212. See Leo Königsberger, *Hermann von Helmholtz* (New York: Dover, 1965), 24. This is an abridged version of the original three-volume work.
213. Jungnickel and McCormmach, *Intellectual Mastery of Nature,* 1:157; Hermann von Helmholtz, "Über den Stoffverbrauch bei der Muskelaktion," *Archiv für Anatomie, Physiologie, und wissenschaftliche Medcin* (1845): 72–83.
214. Hermann von Helmholtz, "On the Integrals of the Hydrodynamic Equations Which Express Vortex-Motion," *Journal für Reine und Angewandte Mathematik* (1857); see Königsberger, *Hermann von Helmholtz,* 167
215. Königsberger, *Hermann von Helmholtz,* 223–24.
216. When Helmholtz turned entirely to physics in 1871, he cited as his reason the fact that his physiology students were not well versed in mathematics and physics.
217. See Steven Turner's article on Helmholtz in *Dictionary of Scientific Biography.*
218. See, for example, O. Darrigol, "The Electrodynamic Revolution in Germany As Documented by Early German Expositions of 'Maxwell's Theory,'" *Arch. Hist. Exact Sci.* 45 (1993): 189–280.

219. I. Yavetz, "Oliver Heaviside and the Significance of the British Electrical Debate," *Ann. Sci.* 50 (1993): 135–73.
220. H. Hertz, *Electric Waves, Being Research on the Propagation of Electric Action with Finite Velocity Through Space,* trans. D. E. Jones (London: Macmillan, 1893), 21.
221. H. Hertz, "Die ausbreitung der elektrischen Kraft." See his *Electric Waves,* trans. Jones. See also Bruce J. Hunt, *The Maxwellians* (Ithaca, N.Y.: Cornell University Press, 1991), an excellent study of those successors of Maxwell who were largely responsible for the acceptance of his theory.
222. The definitive work in English is Jed Z. Buchwald, *The Creation of Scientific Effects: Heinrich Hertz and Electric Waves* (Chicago: University of Chicago Press, 1994).
223. For details, see Max Born and Emil Wolf, *Principles of Optics* (Oxford: Pergamon, 1980), which, in addition to describing the Fresnel-Kirchhoff diffraction theory, discusses the historical context.
224. Maxwell may not have agreed with this remark.
225. *James Clerk Maxwell, A Commemoration Volume, 1831–1931,* p. 71.

CHAPTER 4   *Heat and Thermodynamics*

1. The Celsius, or Centigrade, scale of Anders dates from about 1742.
2. Pierre Duhem, *Evolution of Mechanics,* trans. Michael Cole (Alphen aan den Rijn: Sijthoff and Noordhoff, 1980), 35; Donald Cardwell, *From Watt to Clausius* (Ithaca, N.Y.: Cornell University Press, 1971). See Galileo's *Il Saggiatore* [The Assayer], in Drake Stillman, trans., *The Discoveries and Opinions of Galileo* (Garden City, N.Y.: Doubleday Anchor, 1957), 217–81, and *Dialogues Concerning Two New Sciences,* trans. Henry Crew and Alfonso de Salvio (New York: Dover, 1954). When Descartes proposed that cold is only the absence of heat, his view was consistent with either the idea of a subtle fluid or the motion of particles.
3. He mentions this in several places—for example, query 8: "Do not all fix'd bodies, when heated beyond a certain degree, emit Light and shine, and is not this emission perform'd by the vibrating motions of their parts?" Query 31 is another example. Isaac Newton, *Opticks,* 4th ed. (1730; reprint, New York: Dover, 1952).
4. Quoted by C. C. Gillispie, *The Edge of Objectivity* (Princeton, N.J.: Princeton University Press, 1960), 372.
5. R. Boyle, *New Experiments Physico-Mechanical, Touching the Spring of the Air, and Its Effects; Made, for the Most Part in a New Pneumatic Engine* (Oxford, 1660), and Boyle, *Works* (London, 1772). In *Kinetic Theory,* vol. 1 (Oxford: Pergamon, 1965), Stephen Brush reprints part of Boyle's *New Experiments.* See also Cardwell, *From Watt to Clausius,* 4; Louis Trenchard More, *The Life and Works of the Honorable Robert Boyle* (Oxford: Oxford University Press, 1944); and Marie Boas, *Robert Boyle and Seventeenth Century Chemistry* (Cambridge: Cambridge University Press, 1958).
6. I call this, variously, the dynamical, motional, mechanical, or atomic theory of heat, in each case meaning that heat is a manifestation of some kind of mechanical motion internal to the substance.
7. While Black's (1728–99) calorimetric measurements in 1757 represent the discovery of latent heat, his disciples William Irvine (1743–87) and Adair Crawford (1748–

95) carried the doctrine of latent heat and heat capacity forward. Crawford, especially, who did not share the reluctance of Black and Irvine to publish, played an important role in the establishment of the material theory of heat. See Cardwell, *From Watt to Clausius,* chap. 2, and Henry Guerlac's article on Black in the *Dictionary of Scientific Biography.*

The latent heat of fusion is the amount of heat required to convert a certain quantity of a solid substance at the melting point (conventionally 1 gram) into a liquid at the same temperature. The latent heat of fusion for ice, for example, is about 80 calories per gram; the latent heat of vaporization is about 540 calories per gram.

8. The quotation is from Antoine Lavoisier and P. S. Laplace, "Mémoire sur la chaleur," delivered to the Academy of Sciences, 18 June 1783; see Duhem, *Evolution of Mechanics,* 55. As Laplace once wrote, "perhaps they both obtain at the same time." On Laplace, see I. Grattan-Guinness's article in the *Dictionary of Scientific Biography* and the accompanying bibliography. See R. Fox, *The Caloric Theory of Gases from Lavoisier to Regnault* (Oxford: Clarendon, 1971). On the appearance of the term *caloric,* see Fox, 6n, and Henry Guerlac, "Chemistry As a Branch of Physics: Laplace's Collaboration with Lavoisier," *Hist. Stud. Phys. Sci.* 7 (1976): 193–276.

9. See, for example, Peter Landsberg, *Thermodynamics* (New York: Interscience, 1961).

10. James Clerk Maxwell, "Tait's 'Thermodynamics,'" *Nature* 17 (1878): 257–59, 278–80.

11. In particular, see Valeriano Magni, *Demonstratio ocularis* (1647).

12. See Cardwell, *From Watt to Clausius,* 19. In modern terms, it takes 540 calories per gram of water to convert it to steam at 100°C. This energy is called the latent heat of vaporization.

13. On this point, see C. Webster, "The Discovery of Boyle's Law, and the Concept of Elasticity of Air in the Seventeenth Century," *Arch. Hist. Exact Sci.* 2 (1965): 441–502; C. Webster, "Richard Towneley and Boyle's Law," *Nature* 197 (1963): 226–28; and I. B. Cohen, "Newton, Hooke, and 'Boyle's Law,'" *Nature* 204 (1964): 618–21.

14. Mariotte did acknowledge Pascal; see, for example, M. S. Mahoney's article on Mariotte in the *Dictionary of Scientific Biography.*

15. This is the discovery that pressure of a gas is directly proportional to its temperature if the volume is held constant.

16. According to Cardwell, Peter Guthrie Tait was responsible for Charles's receiving credit for first conceiving this law, something that Cardwell rejects as unjustified. See *From Watt to Clausius,* 130. Charles's contributions were mainly to ballooning; for example, he introduced the use of hydrogen.

17. Gay-Lussac established that the ratio of densities of a gas at two temperatures (θ, θ') was equal to the inverse ratio of some universal function of the temperature for the two gases: $\rho'/\rho = f(\theta)/f(\theta')$. He cited and criticized Charles's work in his "Sur la dilatation des gaz et des vapeurs" (*Ann. Chim.* 48 [1802]: 137–75). Gay-Lussac also carried out some of the most important early experiments on the specific heats of gases, including an experiment with connecting flasks in which a gas contained in one was allowed to expand into the other, which had been evacuated. In this experiment, which foreshadowed Joule's experiments of the 1840s, he showed that the change in temperature of the gas was directly proportional to the change in pressure and that the heat absorbed and evolved in the two flasks was the same. Gay-

Lussac also established the law of combining volumes—that "gases combine in very simple proportions"—which was the foundation for the chemical theory of the atom. See Maurice Crosland's article on Gay-Lussac in the *Dictionary of Scientific Biography.*

18. This result is only approximately true, near 0°C.

19. Jacob Hermann, *Phoronomia* (1716); Daniel Bernoulli, *Hydrodynamica* (1738). See, for example, Cardwell, *From Watt to Clausius,* 24–25.

20. In *Kinetic Theory,* Brush notes that early eighteenth-century figures such as Lomonosov, De Luc, and Le Sage helped keep Bernoulli's kinetic theory alive.

21. Edmund Halley may have first remarked on this phenomenon; see Halley, "An Account of Several Experiments Made to Examine the Nature of the Expansion and Contraction of Fluids by Heat and Cold, in Order to Ascertain the Divisions of the Thermometer, and to Make That Instrument, in All Places, without Adjusting by a Standard," *Phil. Trans. Roy. Soc.* 17 (1692): 650.

22. See Cardwell, *From Watt to Clausius,* 37–38.

23. Black made most of his discoveries by 1761. See Guerlac's article on Black in the *Dictionary of Scientific Biography* and Cardwell, *From Watt to Clausius,* 34–42.

24. See J. L. Heilbron, *Electricity in the 17th and 18th Centuries* (Berkeley: University of California Press, 1979).

25. Laplace and Lavoisier, "Mémoire sur la chaleur."

26. Published as Antoine Lavoisier, *Traité élémentaire de chimie* (Paris, 1789).

27. Fox, *The Caloric Theory of Gases.*

28. See, for example, the discussion by Stephen G. Brush, *Statistical Physics and the Atomic Theory of Matter from Boyle and Newton to Landau and Onsager* (Princeton, N.J.: Princeton University Press, 1983), 42–45.

29. Carl Wilhelm Scheele, "On Air and Fire," in *Collected Papers of Carl Wilhelm Scheele,* trans. L. Dobbin (London: Bell, 1931); H. B. de Saussure, Marc-Auguste Pictet. In his "Essai sur le feu," Pictet accepts the materiality of heat.

30. William was the father of John Herschel. He discovered Uranus in 1781.

31. R. Clausius, "On the Mean Lengths of the Paths Described by the Separate Molecules of Gaseous Bodies," originally published as "Über die mittlere Länge der Wege, welche bei der Molecularbewegung gasförmigen Körper von den einzelnen Molecülen zurüchgelegt werden," in *Ann. d. Phys.* 105 (1858): 239–58 and translated in *Phil. Mag.,* series 4, 17 (1859): 81–91; James Clerk Maxwell, "Illustrations of the Dynamical Theory of Gases," *Phil. Mag.* 19 (1860): 19–32; 20 (1860): 21–37.

32. We refer here to the patents of Watt and Boulton; both were members of the Lunar Society of Birmingham (see chapter 2).

33. The p-V (indicator) diagram, in which pressure is plotted versus volume, gives a visual and quantitative view of a thermodynamic process or cycle. The area under a p-V curve, specifically $\int p dV$, represents work done. See figure 4–1.

34. The engine was used for Professor Anderson's course in natural philosophy at the University of Glasgow.

35. Black left Glasgow for Edinburgh in 1764.

36. The indicator diagram was kept such a secret that, as late as 1826, experienced engineer John Farey only learned of it on a trip to Russia; the information evidently came from Boulton and Watt engineers working there. See Cardwell, *From Watt to Clausius,* 220.

37. Daniel Bernoulli had also emphasized it (see A. J. Pacey and S. J. Fisher, "Daniel Bernoulli and the Vis Viva of Compressed Air," *Brit. J. Hist. Sci.* 3 [1967]: 388–92) as did William Gilbert.

38. Cardwell, *From Watt to Clausius,* 150.

39. L. J. Henderson, recounted by Gillispie, *The Edge of Objectivity,* 357; Robert Schofield, *The Lunar Society of Birmingham* (Oxford: Oxford University Press, 1963).

40. In 1753, Euler conceived the idea of an equation of state relating the pressure, volume, and temperature of a fluid body; see Clifford Truesdell, *Rational Thermodynamics,* 2d ed. (New York: Springer-Verlag, 1984), 3.

41. Davy was from Penzance, Cornwall, which had heavy mining activity, especially tin.

42. Fox, *The Caloric Theory of Gases.* A crucial factor in the decline of the caloric theory seems to have been the electrochemical theory of Berzelius.

43. Herman Boerhaave, *Elementa chemiae* (Leiden, 1732); published in English as *Elements of Chemistry,* trans. Timothy Dallowe, 2 vols. (London, 1735), and *A New Method of Chemistry,* trans. Peter Shaw (London, 1741). Boerhaave asked whether fire "be originally such, formed thus by the Creator himself at the beginning of things; or whether it be mechanically producible from other bodies, by inducing some alteration in the particles thereof" (from the 3d ed., trans. Shaw [London, 1753], 206). He also noted: "When a smithy briskly hammers a piece of iron, the metal thereby becomes hot; yet there is nothing to make it so, except the forcible motion of the hammer impressing a vehement, vigorously determined agitation, on the small part of iron" (ibid.). Boerhaave's view partakes of both the imponderable fluid and vibratory theories. The particle of fire vibrated, as did the particles of ponderable matter. See Henry Steffens's discussion of Thompson and Boerhaave in *James Prescott Joule and the Concept of Energy* (New York: Science History Publications, 1979), 55. James Watt was also apparently influenced by Boerhaave.

44. See, for example, Sanford Brown, *Benjamin Thompson, Count Rumford* (Cambridge: MIT Press, 1979).

45. Thompson's paper, "An Inquiry Concerning the Source of Heat Which Is Excited by Friction," appears in *Phil. Trans. Roy. Soc.* 80 (1798) and *The Collected Works of Count Rumford,* ed. Sanborn C. Brown (Cambridge: Harvard University Press, 1970).

46. Cardwell, *From Watt to Clausius,* 95–107. Cardwell gives an excellent summary of Thompson's contributions and the skeptical reaction he received. Although he is rather too sympathetic to Thompson's critics, he does correctly point out that the scientist's experiments *proved* very little.

47. In *The Mercurial Chemist* (London: Methuen, 1963), Anne Treneer refers to Davy's experiments on heat, performed when he was about nineteen years old and before Rumford's results were published (*The Mercurial Chemist,* 36). See also Treneer's discussion on 37; E. N. da C. Andrade, "Two Historical Notes: Humphrey Davy's Experiments on the Frictional Development of Heat," *Nature* 135 (1935): 359–60; and McKie, "Davy's Experiments on the Frictional Development of Heat," *Nature* 135 (1935): 878. For the rest of his life Davy regretted having published these early essays.

48. On Davy's experiment, see Andrade, "Two Historical Notes." Thomas Beddoes (1760–1808) was a member of a group of dissenters (several of whom were doctors) that included Joseph Priestly. Beddoes was a member of the Lunar Society of

Birmingham, founded the Pneumatic Institution, and installed nineteen-year-old Humphrey Davy as its superintendent. See J. E. Stock, *Memoirs of the Life of Thomas Beddoes, M.D.* (London, 1811).

49. On the other hand, it seems evident that Davy's commitment to the motional theory of heat was largely philosophical and a priori and that the material theory of heat could hardly be ruled out by his experiments. By the 1820s, Davy himself did not seem to consider the matter settled. It is somewhat beside the point to note that Davy's experiment actually proves little (see Andrade, "Two Historical Notes"). In *James Prescott Joule and the Concept of Energy,* Steffens notes that Joule was less interested in the details of the experiments of Thompson and Davy than in their implications for the material theory of heat. As we shall see, the evolution of Carnot's ideas between 1824 and his death in 1832 is especially interesting.

50. In notes published in 1878, long after his death in 1832, Sadi Carnot wrote that "heat is the result of a motion." See my discussion later in this chapter.

51. See J. Herapath, *Mathematical Physics* (London: Whittaker, 1847); reprinted in Stephen G. Brush, ed., *Selected Papers by John Herapath* (New York: Johnson Reprint Corporation, 1972).

52. Or $pV = c(F + 448)$; this is a form of the equation of state for an ideal gas. Attempts to locate absolute zero had been made by Irvine, Crawford, and Dalton, among others. See, for example, Robert Fox, "Dalton's Caloric Theory," in D.S.L. Cardwell, ed., *John Dalton and the Progress of Science* (Manchester: Manchester University Press, 1968).

53. J. Herapath, "On Mr. Tredgold's 'Refutation of Mr. Herapath's Theory,'" *Ann. Phil.* 2 (1812): 303. See Cardwell, *From Watt to Clausius,* 148.

54. The work was based on earlier papers by Fourier. One might say that the very different work of Carnot and Fourier, both done during the 1820s, began the process of giving the science of heat a theoretical foundation. On Fourier, see I. Grattan-Guinness with J. R. Ravetz, *Joseph Fourier, 1768–1810* (Cambridge: MIT Press, 1972).

55. It is true, nonetheless, that Fourier believed in the caloric theory, even though his theory was independent of it. See Brush, *Statistical Physics and the Atomic Theory of Matter,* 42.

56. Norton Wise, "The Maxwell Literature and British Dynamical Theory," *Hist. Stud. Phys. Sci.* 13 (1982): 175–205.

57. See Jerome Ravetz and I. Grattan-Guinness's article on Fourier in the *Dictionary of Scientific Biography.*

58. He was reproached for "lack of rigor and generality."

59. See any book on mathematical physics for details: for example, George Arfken, *Mathematical Methods for Physicists* (Orlando, Fla.: Academic Press, 1985). Fourier had introduced what is now called the Fourier integral in the previous year's competition. He presented the Fourier series in a paper on the problem of heat diffusion, delivered to the Academy of Sciences in 1807.

60. Specific heats were first being measured in this period. François Delaroche and Jacques Etienne Bérard were awarded the prize of the French Institute in 1811 for their measurements of the specific heats of several gases, including air, hydrogen, and oxygen.

61. In *A New System of Chemical Philosophy* (London, 1808), John Dalton wrote that "the specific heats of equal *bulks* of elastic fluids, are directly as their specific gravi-

ties, and inversely as the weights of their atoms." See L. Pearce Williams and Henry John Steffens, *The History of Science in Western Civilization,* vol. 3, *Modern Science, 1700–1900* (Washington: University Press of American), 164–72, quotation on 72. The specific heats at constant volume (c_v) and constant pressure (c_p) represent the amount of heat required to raise the temperature of 1 gram of a substance 1 degree Celsius (with either the volume or pressure held constant). For a monatomic gas, the specific heat dQ/dt has the value (at constant volume) of 3/2R, where R is the universal gas constant, per mole. This comes from the fact that the internal energy of the gas is 3/2kT per atom and that dQ/dt = $\partial U/\partial$ t at constant volume. For diatomic or more complex molecules, which can vibrate and rotate, the specific heat proves to depend on temperature, an issue we address later in this chapter and in chapter 7.

62. Actually, although Laplace did present the correct formula in 1816, taking into account the heat gained or lost in adiabatic compression or expansion, he provided a derivation only in volume 5 of the *Traité de mécanique céleste.* Adiabatic processes are those in which no heat (caloric) is transferred. An example is the compression of a gas in an insulated cylinder. The fact that the temperature of a gas can be raised by doing work on it rather than by transferring heat to it was conceptually important in clarifying the difference between heat and temperature.

63. Cardwell, *From Watt to Clausius,* 139–43.

64. Brush, *Kinetic Theory,* 13.

65. Fox, *The Caloric Theory of Gases,* and Cardwell, *From Watt to Clausius,* 58–59. Erasmus Darwin explained it in "Frigorific Experiments on the Mechanical Expansion of Air," *Phil. Trans. Roy. Soc.* 16 (1788): 43.

66. Sadi Carnot, "Reflexions sur la puissance motrice du feu et sur les machines propres à développer cette puissance" (Paris, 1824). A good modern translation is Sadi Carnot, *Reflexions on the Motive Power of Fire,* ed. Robert Fox (Manchester: Manchester University Press, 1986). See also E. Mendoza, ed., *Reflections on the Motive Power of Heat by Sadi Carnot, and Other Papers on the Second Law of Thermodynamics by E. Clapeyron and R. Clausius* (New York: Dover, 1960), and W. F. Magie, ed., *The Second Law of Thermodynamics, Memoirs by Carnot, Clausius and Thomson* (New York: Harper Brothers, 1899).

   Joseph Larmor called Carnot's argument leading to the idea that all reversible cyclic thermal operations, acting between the same two temperatures, have the same efficiency "perhaps the most original in physical science"; Joseph Larmor, "On the Nature of Heat," *Proc. Roy. Soc.* 94 (1918): 326.

67. In modern parlance, the first law of thermodynamics is the law of conservation of energy specifically applied to thermodynamic systems.

68. He was speaking in the context of a discussion of perpetual motion. See Mendoza, *Reflections,* 12.

69. Referring to the conservation of heat. See Mendoza, *Reflections,* 19, and Magie, *The Second Law of Thermodynamics,* 20.

70. Gillispie considers the interesting question of whether Carnot could have arrived at his important results without a belief in the caloric. See Gillispie, *The Edge of Objectivity,* 369.

71. Mendoza, *Reflections,* 63, 67.

72. See the *Dictionary of Scientific Biography* articles on the Carnots: C. C. Gillispie's on Lazare, and J. F. Challey's on Sadi.

73. Apparently he died of cholera after suffering from scarlet fever.
74. "Whenever there is a difference of temperature the production of motive power is possible" (Carnot, "Reflections," in Magie, *The Second Law of Thermodynamics,* 10).
75. "The motive power of heat is independent of the agents employed to develop it; its quantity is determined solely by the temperatures of the bodies between which, in the final result, the transfer of caloric occurs" (ibid., 20).
76. We speak here in the sense of thermodynamics as opposed to simply the theory of heat. As I have already mentioned, the work of Fourier and others, including Poisson, dealt with mathematical problems of heat conduction but not the conversion of heat into mechanical work, which is Carnot's great contribution.
77. E. Clapeyron, "Mémoire sur la puissance motrice de la chaleur," *Journal de l'Ecole Polytechnique* 14 (1834): 153–90; translated as "On the Motive Power of Heat" in *Taylor's Scientific Memoirs* 1 (1837): 347–76. See Mendoza, *Reflections,* 73–105.
78. Both James and William Thomson became interested in Clapeyron and Carnot in 1844, evidently from the translation of Clapeyron's paper in *Taylor's Scientific Memoirs.*
79. The mechanical equivalent of heat was first determined by Mayer in about 1842. Joule's various experiments yielded values of approximately 740 to 860 foot-pounds of work per British thermal unit (Btu). As Joule expressed it, "the quantity of heat capable of increasing the temperature of a pound of water . . . by $1°$ Fahr. requires for its evolution the expenditure of a mechanical force represented by the fall of 772 lb. through the space of one foot" (J. P. Joule, "On the Mechanical Equivalent of Heat," *Trans. Roy. Soc. London* 140 [1850]: 61). Several excerpts from Joule's reports on his experiments appear in W. F. Magie, *A Source Book in Physics* (Cambridge: Harvard University Press, 1935).

    The Btu is the amount of heat required to raise the temperature of 1 pound of water $1°F$. It is related to the metric system unit called the calorie (the amount of heat required to raise 1 gram of water $1°C$): 1 Btu = 251.997 cal (or 1 Btu = 1055.06 Joules; 1 cal = 4.1868 J). In Joule's units, the modern value for the mechanical equivalent of heat, which is seen only as the conversion factor from energy units typically used in mechanics to those commonly used in thermodynamics, is 778.6 ft-lbs/Btu.
80. For example, see J. P. Joule, "On the Heat Evolved by Metallic Conductors of Electricity," *Phil. Mag.,* series 3, 19 (1841): 260; and Joule, "On the Caloric Effects of Magneto-Electricity, and on the Mechanical Effects of Heat," read on 21 August 1843 at the BAAS meeting in Cork, Ireland, and published in *Phil. Mag.,* series 3, 23 (1843): 263–76, 347–55, 435–43. See my discussion later in this chapter.
81. Joule published his first papers in Sturgeon's journal *Annals of Electricity.*
82. See Leon Rosenfeld's article on Joule in the *Dictionary of Scientific Biography.* Rosenfeld says that Joule never imagined anything more complicated than a linear or quadratic relationship between physical quantities.
83. Joule, "On the Heat Evolved by Metallic Conductors of Electricity," 260.
84. Joule, "On the Caloric Effects of Magneto-Electricity, and on the Mechanical Value of Heat," 263, 347, 435. In the paper, he gave the mechanical equivalent of heat: the amount of work that must be expended to raise 1 pound of water by 1 degree Fahrenheit at 838 foot-pounds. The paper attracted little attention. See Cardwell, *Watt to Clausius,* 232–33.

85. He presented his results at the 1845 BAAS meeting in Cambridge. See J. P. Joule, "On the Changes of Temperature Produced by the Rarefaction and Condensation of Air," *Phil. Mag.* 26 (1845): 369–83.

86. The apparatus consisted of two vessels connected by a stopcock and submerged in water. One vessel was filled with air, and the other was evacuated. According to Joule, when the stopcock was opened, there was no mechanical effect outside the vessels, and the temperature of the water remained unchanged (except for a slight effect), thus showing him that no heat was transferred. Thomson, however, focused on the slight change in temperature. See Crosbie Smith and Norton Wise, *Energy and Empire* (Cambridge: Cambridge University Press, 1989).
    Joule believed in conservation of vis viva, even to the point of advancing a theological argument.

87. This took place before he met Thomson. See D.S.L. Cardwell, *James Joule: A Biography* (Manchester: Manchester University Press, 1989), 297.

88. J.N.P. Hachette, *An Elementary Treatise on Mechanics,* 2d ed. (London, 1824); originally published in French as *Traité élémentaire des machines* (Paris, 1811). See Cardwell, *From Watt to Clausius,* 164, 237.

89. J. P. Joule, "Some Remarks on Heat and the Constitution of Elastic Fluids," *Phil. Mag.,* series 4, 14 (1848): 211. Joule apparently learned about the kinetic theory from Herapath's *Mathematical Physics,* published one year earlier, in 1847.

90. The St. Ann's Church lecture was published in the *Manchester Courier,* 12 May 1847.

91. Thomson also became godfather to Joule's daughter.

92. William Thomson, "On the Dynamical Theory of Heat," in *Mathematical and Physical Papers,* 1:181; presented to the Royal Society of Edinburgh in Manchester in March 1851 and originally published in eight parts in *Philosophical Magazine* (1852–56).

93. William Thomson, "On an Absolute Thermometric Scale, Founded on Carnot's Theory of the Motive Power of Heat, and Calculated from the Results of Regnault's Experiments on the Pressure and Latent Heat of Steam," *Phil. Mag.,* series 3, 33 (1848): 313–17; Thomson, *Mathematical and Physical Papers,* 1:100–106.

94. Letters from J. P. Joule to William Thomson, 6 and 27 October 1848, Cambridge University Library. The Joule-Thomson correspondence in the Cambridge University Library includes more than thirty letters, all essentially scientific in character.

95. William Thomson, "On the Dynamical Theory of Heat."

96. T. S. Kuhn, "The Conservation of Energy As an Example of Simultaneous Scientific Discovery," in *Critical Problems in the History of Science,* ed. M. Clagett (Madison: University of Wisconsin Press, 1959).

97. Published in German as R. Clausius, "Über die bewegende Kraft der Wärme, und die Gesetze, welche sich daraus für die Wärmelehre selbst ableiten lassen," *Ann. d. Phys.* 79 (1850): 368–97, 500–524; published in English as "On the Moving Force of Heat and the Laws Regarding the Nature of Heat Itself Which Are Deducible Therefrom" in *Phil. Mag.,* series 4, 2 (1851): 1–21, 102–19. Note the different versions of the English title, including "On the Motive Power of Heat, and on the Laws Which Can Be Deduced from It for the Theory of Heat." See the translation in Magie, *The Second Law of Thermodynamics,* 65–107.

98. See, for example, E. E. Daub's article on Clausius in the *Dictionary of Scientific Biography.*

99. Our knowledge is based on a letter to Rankine in that month. See Smith and Wise, *Energy and Empire,* 320. Much of Thomson's paper is reprinted in Magie, *The Second Law of Thermodynamics.*

100. I quote (without comment) Gillispie's evaluation: "Clausius' was a subtler mind, less concrete than Joule's or Thomson's and less comprehensive, perhaps, than Helmholtz's, but the supplest of their generation in the abstract reaches of physics" (*The Edge of Objectivity,* 396). See Mendoza, *Reflections.*

101. Carnot's "Reflections" sold few copies and vanished almost without a trace.

102. Clausius, "On the Motive Power of Heat," in Magie, *The Second Law of Thermodynamics,* 65. For details of his mathematical reasoning, see E. Yagi, "Clausius' Mathematical Method and the Mechanical Theory of Heat," *Hist. Stud. Phys. Sci.* 15 (1984): 177–95.

103. Thomson, "An Account of Carnot's Theory of the Motive Power of Heat," *Mathematical and Physical Papers,* 1:119.

104. Clausius, "On the Motive Power of Heat," in Magie, *The Second Law of Thermodynamics,* 68.

105. Q is the heat; hence, dQ is the differential (or incremental) heat added, U the internal energy, and p and V the pressure and volume.

106. See Magie, *The Second Law of Thermodynamics,* 102–4. Here p, V, and t stand for pressure, volume, and temperature, respectively.

107. Clausius, "On the Motive Power of Heat," in ibid., 90.

108. Thomson, "On the Dynamical Theory of Heat," *Mathematical and Physical Papers,* 1:181. We give Thomson's version of the Clausius formulation of the second law because Clausius does not give a succinct statement of it in his original paper.

109. Legions of university students have been and are asked to establish this equivalence in their first physics course.

110. R. Clausius, "Ueber eine veranderte Form des zweiten Hauptsatzes der mechanischen Wärmetheorie," *Ann. d. Phys.* 93 (1854): 481–506. Clausius introduced the term *entropy* in "Über verschiedene für die Anwendung bequeme Formen der Hauptgleichungen der mechanischen Wärmetheorie," *Ann. d. Phys.* 125 (1865): 353–400; translated as "On a Modified Form of the Second Fundamental Law in the Mechanical Theory of Heat," in *Phil. Mag.,* series 4, 12 (1856): 81–98.

111. Thomson, "On the Dynamical Theory of Heat," in *Mathematical and Physical Papers,* 1:232–91. See Cardwell, *From Watt to Clausius,* 265–67.

112. Irreversible processes are those in which there is dissipation. That is, if the system is reversed so that it retraces its path in a p-V diagram, it will not return to its original state. In general, macroscopic processes are irreversible. At the microscopic level, where (almost) all processes satisfy time-reversal invariance, irreversibility is not an issue. Dissipation was an important preoccupation of Thomson.

113. Cardwell, *From Watt to Clausius,* 263–76

114. See, for example, Smith and Wise, *Energy and Empire,* chap. 9, especially 283, 305–8. The chapter discusses Joule's commitment to heat as motion (that is, vis viva) and the conclusion that there can be no work without the loss of heat, which contrasts with Thomson's lack of commitment to any theory of heat but instead to heat as a state of a body and the idea that work is accomplished by a fall in intensity. In "On the Theory of Electromagnetic Induction" (*BAAS Report* 18 [1848]: 9–10), Thomson commented on Joule's work: "Some known experiments with magneto-electric machines seem to indicate an actual conversion of mechanical effect

into caloric. No experiment however is adduced in which the converse is exhibited; but it must be confessed that as yet much is involved in mystery with reference to these fundamental questions of Natural Philosophy" (see Smith and Wise, *Energy and Empire,* 310).

115. Joule, "On the Calorific Effects of Magneto-Electricity, and on the Mechanical Equivalent of Heat," and "On the Changes of Temperature Produced by the Rarefaction and Condensation of Air."

116. See the letter from Thomson to his father, dated 1 July 1847, in the Cambridge University Library. Quoted in Smith and Wise, *Energy and Empire,* 302.

117. Smith and Wise, *Energy and Empire,* 296–99.

118. In "Account of Carnot's Theory of the Motive Power of Heat," Thomson quoted Carnot: "In our demonstrations we tacitly assume that after a body has experienced a certain number of transformations, if it be brought identically to its primitive physical state as to density, temperature, and molecular constitution, it must contain the same quantity of heat which it initially possessed. . . . This fact has never been doubted. . . . To deny it would be to overturn the whole theory of heat, in which it is a fundamental principle" (*Mathematical and Physical Papers,* 1:115).

119. Thomson, "On an Absolute Thermometric Scale." In a footnote he remarks that he had not then seen Carnot's work but learned of it from Clapeyron's paper, which had been translated in *Taylor's Scientific Memoirs.* When "On an Absolute Thermometric Scale" was reprinted in *Mathematical and Physical Papers,* Thomson said that he had since obtained a copy of Carnot's work from Professor Lewis Gordon. In another footnote, to his own comment that "the conversion of heat (or *caloric*) into mechanical effect is probably impossible," he said: "This opinion seems to be nearly universally held by those who have written on the subject. A contrary opinion however has been advocated by Mr. Joule of Manchester; some very remarkable discoveries which he has made with reference to the *generation* of heat by the friction of fluids in motion, and some known experiments with magneto-electric machines, seeming to indicate an actual conversion of mechanical effect into caloric. No experiment however is adduced in which the converse operation is exhibited; but it must be confessed that as yet much is involved in mystery with reference to these fundamental questions of natural philosophy" (*Mathematical and Physical Papers,* 1:102n).

On 20 March 1852, in the *Philosophical Magazine* reprint of "On the Dynamical Theory of Heat," Thomson commented that he had just learned of Helmholtz's "On the Conservation of Force" (1847), "having seen it for the first time on the 20th of January of this year." This may have been because *Annalen der Physik* had rejected the paper.

120. See Crosbie Smith, "Natural Philosophy and Thermodynamics: William Thomson and 'The dynamical theory of heat,'" *Brit. J. Hist. Sci.* 9 (1976): 293–319.

121. Smith and Wise shed light on the evolutions of Thomson's ideas, quoting from William Smith's notes of Thomson's lectures in natural philosophy for the 1849–50 academic session (*Energy and Empire,* 324–25). See the letter from Joule to Thomson, dated 26 March 1850, in the Cambridge University Library.

122. W.J.M. Rankine, "On the Mechanical Action of Heat, Especially in Gases and Vapors," *Trans. Roy. Soc. Edin.* 20 (1850): 147–90. See also Keith Hutchison, "W.J.M. Rankine and the Rise of Thermodynamics," *Brit. J. Hist. Sci.* 14 (1981): 1–26.

123. William Thomson, "On a Remarkable Property of Steam Connected with the Theory of the Steam Engine," *Phil. Mag.* (November 1850); *Pogg. Ann.* 81 (1850). In his March 1851 paper, "On the Dynamical Theory of Heat," Thomson claimed to have arrived at the conclusions "about a year ago" (*Mathematical and Physical Papers,* 1:181).

124. The nonscalding property (that is, the fact that the steam was not "wet") meant that condensation did not in fact occur.

125. Here, the heat consisted of the vis viva of the vortices.

126. Letter from Thomson to Joule, 15 October 1850.

127. See, for example, Crosbie Smith, "William Thomson and the Creation of Thermodynamics," *Arch. Hist. Exact Sci.* 16 (1976): 280–88. Thomson also commented: "This very remarkable conclusion was first announced by Mr. Rankine, in his paper communicated to this Society on the 4th of February last year. It was discovered independently by Clausius, and published in his paper in *Annalen der Physik* in the months of April and May of the same year" (*Mathematical and Physical Papers,* 1:209). Had Carnot's 1832 paper been published, the pace of Thomson's work on the problem might have been very different. Cardwell says that it is surprising that Thomson was so long in coming around to the views of Joule, for whom he had great personal affection, and wonders if this delay was due to Thomson's admiration for French writers.

128. Letter from Joule to Thomson, dated 6 February 1851, in the Cambridge University Library.

129. Thomson, "On the Dynamical Theory of Heat," in *Mathematical and Physical Papers,* 1:181.

130. Ibid.

131. Ibid.

132. Letter from Joule to Thomson, dated 8 March 1851, in the Cambridge University Library.

133. Thomson, "On the Dynamical Theory of Heat." In other words, Thomson refrains from any generalizing or philosophical comments. But Thomson usually kept philosophy out of his papers, which are very businesslike—full of tables and drawings of experimental apparatus.

134. Smith and Wise, *Energy and Empire,* 330. See also E. Daub, "Entropy and Dissipation," *Hist. Stud. Phys. Sci.* 2 (1970): 321–54, and William Thomson, "On a Universal Tendency in Nature for the Dissipation of Mechanical Energy," *Phil. Mag.,* series 4, 4 (1852): 304.

135. William Thomson, "On the Age of the Sun's Heat," *Macmillan's Magazine* 5 (1862): 288–93. See also Smith and Wise, *Energy and Empire,* 500–501.

136. Part 1 of the *Treatise* was published in 1867, part 2 not until 1883, with the help of George Darwin. The smaller *Elements of Natural Philosophy* was published in 1872 and translated into German by Helmholtz and Wertheim.

137. In *Energy and Empire,* Smith and Wise devote chapters 15 through 17 to Thomson's role in these fascinating issues.

138. M. J. Klein, "Gibbs on Clausius," *Hist. Stud. Phys. Sci.* 1 (1969): 127–49; J. W. Gibbs, "Rudolph Julius Emanuel Clausius," *Proc. Am. Acad. Arts. Sci.* 16 (1889): 458–65.

139. James Clerk Maxwell, *Theory of Heat* (London: Longman's, 1871). See Klein, "Gibbs on Clausius"; R. Clausius, "A Contribution to the History of the Mechani-

cal Theory of Heat," *Phil. Mag.,* series 4, 43 (1872): 106; Clausius, "A Necessary Correction of One of Mr. Tait's Remarks," *Phil. Mag.,* series 4, 44 (1872): 117; and P. G. Tait, *Sketch of Thermodynamics* (Edinburgh: Edmunston and Douglas, 1868; 2d ed., 1877).

140. Letter from Maxwell to Tait, dated 1 December 1873. The letters between the two men are wonderful documents in both the history of nineteenth-century science and of nineteenth-century letters in general. Maxwell signed himself "dp/dt" and Tait "T′" (to distinguish between himself and Thomson). In a letter dated 10 March 1873, Maxwell advised Tait that "to speak familiarly of a 2nd Law of thermodynamics as of a thing known for some years, to men of culture who have never even heard of a 1st Law, may arouse sentiments unfavorable to perfect attention." Both letters are in the collection of the Cambridge University Library.

CHAPTER 5   *Energy and the Energy Principle*

1. A reader might well ask, "What are some other global principles of physics?" The answer includes the following conservation laws or invariance principles, some of which are now known to have exceptions: conservation of charge, conservation of angular momentum, parity conservation, invariance under time reversal, and CPT invariance. In 1957 the theoretical work of Lee and Yang and the experiments of Wu showed parity conservation to be violated in weak interactions, and a team led by Val Fitch and James Cronin in 1963–64 showed time-reversal invariance to be violated in the two-pi decay of the kaon.

2. The roots of the idea of conservation of energy lie in classical antiquity. Nevertheless, while any purely mechanical theory of the world (such as that of the Greek atomists) implies conservation of motion, the idea did not play a significant role in physics until the seventeenth century.

   Conservation of motion in the seventeenth century was equivalent to our conservation of momentum or the law of inertia. From its generalization, and in the absence of net external forces, Newton's second law (the conservation of vis viva) can be derived. In the presence of forces, the conservation of mechanical energy can be obtained if one introduces a properly understood concept of external work. By introducing the idea of a potential function, one can obtain the result that any change in the kinetic plus the potential energy must equal external work done on the system. For examples, see any introductory physics text.

   As we have seen, Green introduced the term *potential*. Rankine, however, coined *potential energy* in 1853 as an alternative to Thomson's *actual energy*. Thomson introduced the term *kinetic energy* (for vis viva) in the 1870s.

3. Readers familiar with modern physics will know that the idea of conservation of energy was generalized by Einstein in 1905 to conservation of mass-energy.

4. On occasion, a crisis in science has motivated one brave soul or another to propose the violation of energy conservation—for example, Bohr, Kramers, and Slater in 1924 and Bondi, Gold, and Hoyle in their steady-state theory of cosmology after World War II. Modern physics does not necessarily forbid violations of energy conservation on a time scale that makes such violation fundamentally undetectable. So when we talk about the establishment of conservation by Joule and Helmholtz (and others), we should understand the statement in this context.

5. My discussion in this chapter owes much to Y. Elkana, *The Discovery of the Con-*

servation of Energy (Cambridge: Harvard University Press, 1974); Wilson L. Scott, *The Conflict between Atomism and Conservation Theory, 1644–1860* (London: Macdonald, 1970); and T. S. Kuhn, "Energy Conservation As an Example of Scientific Discovery," in *Critical Problems in the History of Science,* ed. M. Clagett (Madison: University of Wisconsin Press, 1969), 321–56.

6. Elkana, *Discovery,* 25. On issues in seventeenth- and eighteenth-century mechanics, see René Dugas, *A History of Mechanics* (New York: Dover, 1988); Clifford Truesdell, *Essays in the History of Mechanics* (New York: Springer-Verlag, 1968); Stillman Drake and I. E. Drabkin, *Mechanics in Sixteenth-Century Italy* (Madison: University of Wisconsin Press, 1969); and Richard Westfall, *Force in Newton's Physics* (London: Macdonald, 1971).

7. These studies were conducted by Lazare Carnot among others.

8. According to Elkana, the factor 1/2 was introduced by Coriolis in "Calcul de l'effet de machines" in 1829 (*Discovery,* 41). *Vis,* in Latin, means "force." On the concept of force, see Max Jammer, *Concepts of Force: A Study in the Foundations of Dynamics* (Cambridge: Harvard University Press, 1957); Westfall, *Force in Newton's Physics;* and I. B. Cohen, "Newton's Second Law and the Concept of Force in the *Principia,*" in *The Annus Mirabilis of Sir Isaac Newton, 1666–1966,* ed. R. Palter (Cambridge: MIT Press, 1970), 244–57.

9. J. Bernoulli, *Opera Johannis Bernoulli* (Geneva, 1742), 3:239, 243.

10. Dugas, *A History of Mechanics.* The theorem of vis viva conservation emerged from Huygens's studies of collision theory and the compound pendulum in the 1650s and 1660s, using the principle that a falling body cannot rise higher than its initial position, and (according to Gabbey) Leibniz's subsequent criticisms of Descartes's views on motion (1686). See Alan Gabbey, "Force and Inertia in Seventeenth-Century Dynamics," *Stud. Hist. Phil. Sci.* 2 (1971): 1–67.

11. For a discussion of perpetual motion, see Elkana, *Discovery,* 28ff.

12. Elkana emphasizes, however, the gulf between the results of analytical mechanics and experimental practice (ibid., 59). On Lagrangian mechanics, see H. Goldstein, *Classical Mechanics,* 2d ed. (Reading, Mass.: Addison-Wesley, 1980).

13. For more on this history, see Westfall, *Force in Newton's Physics;* Elkana, *Discovery;* Dugas, *A History of Mechanics;* Jammer, *Concepts of Force;* and Cohen, "Newton's Second Law."

14. Momentum, or quantity or force of motion, was thought to be conserved in collisions; Huygens understood it as having the properties of a vector. In Euler's *Mechanica, vis inertiae* was clear and unambiguous—and Newtonian.

15. Michael Faraday, "On the Conservation of Force," *Phil. Mag.* 13 (1857): 225–39; see also *Proceedings of the Royal Institute,* 27 February 1857.

16. Hermann von Helmholtz's landmark paper was titled "Die Erhaltung der Kraft," translated as "On the Conservation of Force." See John Tyndall, ed., *Scientific Memoirs, Natural Philosophy* (London, 1853), 114. By using the term *force* rather than *energy,* which Young had introduced into physics forty years earlier, he passed up an opportunity to end the confusion. Elkana's discussion of this issue is especially interesting (see *Discovery*). This is not to say that Helmholtz himself was unclear about the distinction between the physical ideas; *Kraft* was the term he used for "energy." To some extent, then, it was a linguistic or etymological problem rather than a conceptual one. Others, however, were confused.

17. Poncelet introduced the term *travail* in 1826, but he was anticipated by Jean Ber-

noulli. The German is *arbeit.* See Elkana's discussion in *Discovery* and Kuhn, "Energy Conservation," note 48, p. 349. Common usage of *work* can be traced to Clausius, in about 1850; in 1852 Thomson (following Young) popularized the use of *energy.*

18. But it is interesting to note, as Kuhn does ("Energy Conservation," 353, note 76), that none of the pioneers of energy conservation mention the eighteenth-century traditions.

19. T. Young, *A Course of Lectures on Natural Philosophy and the Mechanical Arts,* 2 vols. (London, 1807), 1:59.

20. Of course, living force (kinetic energy) is not by itself conserved. But as it became clear that energy was converted from one form to another rather than being lost, one could think of vis viva as being indestructible.

21. Kuhn names twelve scientists with some claim to being among the first to state the principle of energy conservation ("Energy Conservation"). Others also could be named.

22. Benjamin Thompson, *The Complete Works of Count Rumford,* ed. G. E. Ellis, 4 vols. (Boston: American Academy of Arts and Sciences, 1870), 3:172.

23. See especially Scott, *Conflict.*

24. See Kuhn, "Energy Conservation," 332

25. See Elkana, *Discovery,* 70.

26. Sadi Carnot, "Reflections on the Motive Power of Heat," in *The Second Law of Thermodynamics, Memoirs by Carnot, Clausius and Thomson,* ed. W. F. Magie (New York: Harper Brothers, 1899), 12–13. See chapter 3, note 66, for English translations of Carnot's work.

27. Sadi Carnot, "Posthumous Remarks," in *Reflections on the Motive Power of Heat by Sadi Carnot, and Other Papers on the Second Law of Thermodynamics,* ed. E. Mendoza (Gloucester, Mass.: Smith, 1977), 67.

28. Although I say "unfortunately," I know that what is important is how the theory of heat evolved, not how it *might* have.

29. Marc Seguin was born in 1786, Sadi Carnot ten years later, five years after Faraday. Carl Friedrich Mohr was born in 1804, Karl Holtzmann and William Grove in 1811, Ludwig Colding and Julius Robert Mayer in 1814–15. Only Joule (1818) and Helmholtz (1821) were born after 1815.

30. Some may argue that I am merely stating the obvious—namely, that these figures came after those who preceded them. Let me clarify the point. When these early hints of energy conservation were surfacing in the early 1840s, Maxwell was a teenager, Thomson was about to go up to Cambridge, and Helmholtz was barely in his twenties.

31. See Kuhn, "Energy Conservation," 337.

32. Studies of animal heat were important and suggestive, especially to von Liebig and Mayer. It is not entirely an accident that it fell to Helmholtz, who was a physiologist as well as a physicist, to frame energy conservation in mathematical terms.

33. It was also published in Mohr and von Liebig's *Annalen der Pharmacie.*

34. Friedrich Mohr, "Ueber die Natur der Wärme," *Zeitschrift für Physik* 5 (1837): 442.

35. Johann Gottlieb Fichte was a German idealist and Kantian philosopher.

36. Kuhn, "Energy Conservation," 336.

37. Mendoza, ed., *Reflections on the Motive Power of Heat by Sadi Carnot,* xxi.

38. Elkana, *Discovery,* 144. See J. R. Mayer, "Bermerkungen über die Kräfte der

unbelebten Natur," *Liebig's Annalen der Chemie und Pharmacie* 42 (1842): 233–40; translated as "Remarks on the Forces of Inorganic Nature" in *Phil. Mag.* 24 (1862): 371–77. For a biography of Mayer, see Kenneth L. Caneva, *Robert Mayer and the Conservation of Energy* (Princeton: Princeton University Press, 1993).

39. Mayer, "Bemerkungen über die Kräfte der unbelebten Natur." As many authors have pointed out since Thomson in 1851, Mayer's result was based on the assumption that the work expended in compressing a gas is precisely the mechanical equivalent of the heat evolved, a result that is only approximately true.

40. In "The Conservation of Force," Helmholtz presented values obtained by Holtzmann, Joule, and Mayer and expressed skepticism at some of Joule's numbers. In an appendix added in 1881, he offered a spirited defense of Mayer. See Russell Kahl, ed., *Selected Writings of Hermann von Helmholtz* (Middletown, Conn.: Wesleyan University Press, 1971). The mechanical equivalent of heat was expressed in two ways: (1) as the weight in pounds that could be raised 1 foot if 1 Btu of thermal energy (that required to raise 1 pound of water 1°F) were converted into mechanical work, and (2) the height (in meters) to which 1 gram of matter could be raised if 1 calorie of thermal energy (that required to raise 1 gram of water 1°C) were converted into mechanical work. Mayer's 1842 value of 365 meters (or 365 kg-m/kcal) corresponds to 665 pounds (665 ft-lbs/Btu).

41. Ludwig Colding, "On the History of the Principle of the Conservation of Energy," *L. Phil. J.* (1864). See Per F. Dahl, trans., *Ludwig Colding and the Conservation of Energy* (New York: Johnson Reprint Corporation, 1972).

42. Marc Seguin, "Observations on the Effects of Heat and of Motion," letter to John Herschel, 12 September 1822, published in *Edin. Phil. J.* 10 (1824): 280–83. See D.S.L. Cardwell, *James Joule: A Biography* (Manchester: University of Manchester Press, 1989), 61.

43. Marc Seguin, *De l'influence des chemins de fer* (Paris, 1839). See Cardwell, *James Joule,* 61, 67, 98.

44. Joule first published these descriptions in William Sturgeon's *Annals of Electricity* (1842), then in the *Philosophical Magazine* (1841–45). See James Joule, *The Joint Scientific Papers of J. P. Joule,* 2 vols. (London, 1884–87). As I have already noted, Joule's early value of 770 pounds for the mechanical equivalent of heat was very close to the correct value of 777.

45. Joule, *Scientific Papers,* 2:215. On Joule, see Henry Steffens, *James Prescott Joule and the Concept of Energy* (Folkestone, England: Dawson, 1979), and Cardwell, *James Joule.*

46. Quoted in Elkana, *Discovery,* 116. They next met by accident near Chamonix in the French Alps, where Joule, with thermometer in hand, was trying to measure the elevated temperature in waterfalls (William Thomson, *Mathematical and Physical Papers* [London, 1872], 2:215).

47. The lecture was reported in the *Manchester Courier,* 12 May 1847.

48. Kahl, *Selected Writings of Hermann von Helmholtz,* 3–55.

49. Elkana, *Discovery,* 124. By the sum of tensions Helmholtz meant the integral $\int \phi dr$, where ϕ was the intensity of the force $F = -\nabla\phi$, where for an inverse-square force $F_i = -(r_i/r)\phi$.

50. See Merz, *European Thought in the Nineteenth Century* (1896–1914), 2:113–14. Mayer's claims, versus those of Joule, were also disputed by Tait (especially) and Thomson. In 1862 Tyndall credited Mayer with having arrived at the mechanical

equivalent of heat in 1842, a year before Joule, and met an angry response from both Thomson and Tait.

51. Tait's book is one of the documents stemming from the controversies over priority in both this issue and the matter of the second law of thermodynamics that raged in the 1860s and 1870s. Tait, who was something of a bulldog, was directly involved in both. See Peter Guthrie Tait, *Recent Advances in Physical Science* (1876) and Merz, *European Thought.*

52. See Kuhn, "Energy Conservation."

53. Concerning their book, Tait wrote to Thomson on 12 December 1861: "The next generation will thank us." He also predicted, thirteen days later, that "we are certain to be pirated in America" (letters in the Cambridge University Library).

CHAPTER 6   *Atomism*

1. Sources for the history of atomism include L. L. Whyte, *Essay on Atomism: From Democritus to 1960* (Middletown, Conn.: Wesleyan University Press, 1961); A. G. van Melsen, *From Atomos to Atom: The History of the Concept Atom,* trans. H. J. Koren (Pittsburgh: Duquesne University Press, 1952); David M. Knight, *Atoms and Elements* (London: Hutchinson, 1967); E. MacKinnon, *Scientific Explanation and Atomic Physics* (Chicago: University of Chicago Press, 1992); Arnold Thackray, *Atoms and Powers: An Essay on Newtonian Matter Theory and the Development of Chemistry* (Cambridge: Harvard University Press, 1970); and B.F.J. Schonland, *The Atomists (1805–1933)* (Oxford: Clarendon, 1968).

2. MacKinnon, *Scientific Explanation,* 14.

3. A. Koyré, *From the Closed World to the Infinite Universe* (Baltimore: Johns Hopkins University Press, 1957), 278.

4. William Whewell, *The Philosophy of the Inductive Sciences, Founded upon Their History,* 2 vols., 2d ed. (London, 1847), 1:427.

5. Whyte, *Essay on Atomism,* 48–49. On atomism in the early period in England, see R. H. Kargon, *Atomism in England from Hariot to Newton* (Oxford: Clarendon, 1966).

6. See Isaac Newton, query 31, in *Opticks:*

   All these things being consider'd, it seems probable to me, that God in the Beginning form'd Matter in solid, massy, hard, impenetrable, moveable Particles, of such Sizes and Figures, and with such other Properties, and in such Proportion to Space, as most conduced to the End for which he form'd them and that these primitive Particles being Solids, are incomparably harder than any porous bodies compounded of them; even so very hard, as never to wear or break in pieces. . . . And therefore, that Nature may be lasting, the Changes of corporeal Things are to be placed only in the various Separations and new Associations and Motions of these permanent Particles.

   From *Opticks,* 4th ed. (1931; reprint, New York: Dover, 1952), book 3, part 1, 389.

7. L. Euler, "General Principles of the State of Equilibrium of Fluids" (1755); published in his collected works, *Opera omnia,* series 2, vol. 12. See MacKinnon, *Scientific Explanation,* 39.

8. Immanuel Kant, *Metaphysical Foundations of Natural Science,* quoted in John Hendry, *James Clerk Maxwell and the Theory of the Electromagnetic Field* (Bristol, England: Hilger, 1986), 15.

9. He was also known as Rogerio Josepho Boscovich and Rudjer Josip Boskovic.

Boscovich was born at Ragusa in Dalmatia. See "Life of Roger Joseph Boscovich," by Branislav Petroniević, appended to R. J. Boscovich, *A Theory of Natural Philosophy* [originally titled *Philosophiae naturalis theoria*], trans. J. M. Child (London: Open Court, 1922), based on the edition published in Venice in 1763. The book was first published in Vienna in 1758.

10. Boscovich, *A Theory of Natural Philosophy,* 10.
11. Ibid., 37. Note the similarity to the point centers of G. B. Vico (1668–1744), which he worked on around 1710. Also compare the work of E. Swedenborg (1688–1772), a Swedish physician. See Whyte, *Essay on Atomism,* 53
12. This separation is evident to anyone who has read documents from the nineteenth century. For elaboration, see Wilson L. Scott, *The Conflict between Atomism and Conservation Theory, 1644–1860* (London: Macdonald, 1970), chapter 9; and Alan Rocke, "Atoms and Equivalents: The Early Development of the Chemical Atomic Theory," *Hist. Stud. Phys. Sci.* 9 (1978): 225–63.
13. See the discussion in Scott, *Conflict,* chap. 9.
14. We discuss Davy's complex and shifting views later in the chapter. On Wollaston, see D. C. Goodman, "Wollaston and the Atomic Theory of Dalton," *Hist. Stud. Phys. Sci.* 1 (1969): 37–59.
15. This law was variously known as the law of definite proportions, the law of constant proportions, the law of fixed proportions, and the law of constant composition. The law of combining proportions is an expression of the fact that elements combine or recombine to form chemical compounds in definite ratios by weight. We know now that this results from the atomic weights of the atoms and their electronic properties. Richter was a student of Kant.
16. Ernst von Meyer, *A History of Chemistry,* trans. G. McGowan (London, 1891). Fischer published this table in his translation of Berthollet's *Researches on the Laws of Affinity;* Berthollet later adopted Richter's numbers. For a good discussion of what various chemists of the early nineteenth century meant by proportions or equivalents (or atoms), see Rocke, "Atoms and Equivalents."
17. Thackray, *Atoms and Powers,* 214–15.
18. Arianism was a heretical view that denied the divinity of Christ. See R. Schofield, *A Scientific Autobiography of Joseph Priestley, 1733–1804: Selected Scientific Correspondence* (Cambridge: MIT Press, 1966).
19. Like Thompson, Priestley had encountered Boerhaave's *Elementa chemiae* and s'Gravesande's *Elements of Natural Philosophy.*
20. See Thackray, *Atoms and Powers,* chap. 6. Priestley's support for this doctrine is clear in *History and Present State of Discoveries Relating to Vision, Light, and Colours* (London, 1772) and *Disquisitions Relating to Matter and Spirit* (London, 1777).
21. Thackray, *Atoms and Powers,* 189–90. See Priestley, *Disquisitions Relating to Matter and Spirit.*
22. Thackray, *Atoms and Powers,* 249–50, and references therein.
23. R. Schofield, *The Lunar Society of Birmingham: A Social History of Provincial Science and Industry in Nineteenth-Century England* (Oxford: Clarendon, 1963).
24. Nevertheless, the Lunar Society still exists today.
25. Arthur Donovan, *Antoine Lavoisier: Science, Administration, and Revolution* (Oxford: Blackwell, 1993). See also Henry Guerlac's article on Lavoisier by in the *Dictionary of Scientific Biography.*

26. For details, see von Meyer, *A History of Chemistry.*

27. Lavoisier tried to bring order to the chaos that resulted from different names' being used in different countries for the same substances by writing, with de Moreau, Berthollet, and de Fourcroy, the *Method of Chemical Nomenclature* (Paris, 1787; English translation, London: James St. John, 1788).

28. See Whyte, *Essay on Atomism,* 59.

29. His father was a small landowner and a weaver. These biographical details are largely from Arnold Thackray's article on Dalton in the *Dictionary of Scientific Biography;* Arnold Thackray, *John Dalton: Critical Assessments of His Life and Science* (Cambridge: Harvard University Press, 1972; and Thackray, *Atoms and Powers.* See also Arnold Thackray, "The Emergence of Dalton's Chemical Atomic Theory, 1801–08," *Br. J. Hist. Sci.* 3 (1966): 1–23; Thackray, *John Dalton: Critical Assessments of His Life and Science* (Cambridge: Harvard University Press, 1972); and Donald Cardwell, ed., *John Dalton and the Progress of Science* (Manchester: Manchester University Press, 1968).

30. Dalton kept daily meteorological records from 1787 until the day before his death in 1844. Some of his meteorological studies involved taking an altimeter to high altitudes in the Lake District and bringing back air samples.

31. John Dalton, *A New System of Chemical Philosophy* (London, 1808–10), vol. 1, part 1, 141–43. The first and second parts of volume 1 were published in 1808 and 1810, respectively. Volume 2, part 1, was published in 1827. Chapter 1, which is devoted to heat, begins: "The most probable opinion concerning the nature of caloric, is, that of its being an elastic fluid of great subtilty, the particles of which repel one another, but are attracted by other bodies."

32. John Dalton, "On the Absorption of Gases by Water and Other Liquids," *Mem. Proc. Manchester Lit. Phil. Soc.,* series 2, 1 (1805): 244–58.

33. The modern value would be about 16.

34. This was Boerhaave's theory of fire as an imponderable but hard, granular substance distinct from the elements. See Scott, *Conflict,* 188.

35. Dalton taught both of the Joule brothers.

36. Quoted in Scott, *Conflict,* 103, from L. K. Nash, "The Origins of Dalton's Chemical Atomic Theory," *Isis* 47 (1956): 101–16.

37. Scott, *Conflict,* 200

38. The terms were introduced, respectively, by Wollaston and Davy. See William Wollaston, "A Synoptic Scale of Chemical Elements," *Phil. Trans. Roy. Soc.* 104 (1814): 1–22. Rocke ("Atoms and Equivalents") says that *chemical equivalents* had earlier been used by Henry Cavendish.

39. Scott, *Conflict,* 203; W. V. Farrar, "Dalton and Structural Chemistry," in D. S. Cardwell, ed., *John Dalton and the Progress of Science* (Manchester: Manchester University Press, 1968), 290.

40. Avogadro (1776–1856), an Italian chemist, modified the view that prevailed following Dalton (that equal numbers of atoms of a gas occupied equal volumes) to mean that equal numbers of *molecules* occupied equal volumes.

41. Wollaston wrote in January 1808: "The inquiry which I had designed appears to be superfluous, as all the facts that I have observed are but particular instances of the more general observation of Mr. Dalton" (see Thackray, *Atoms and Powers,* 276). Wollaston converted Gilbert, who in turn convinced Davy, according to Thomas

Thomson in *History of Chemistry* (London, 1831), but to the law of proportions, or equivalents, not to atoms.

42. John Herschel, *Preliminary Discourse on the Study of Natural Philosophy* (London, 1830). See Knight, *Atoms and Elements,* 36.

43. Knight borrowed this phrase from Davy. When presenting Dalton with the medal of the Royal Society, Davy referred to his establishment of a simple, universal principle that laid the foundation for future work, saying, "His merits in this respect resemble those of Kepler in astronomy" (see Thackray, *Atoms and Powers,* 277).

44. See Knight's discussion in *Atoms and Elements,* chap. 2.

45. Humphrey Davy, Dialogue 5, "The Chemical Philosopher," in *Consolation in Travel: The Last Days of a Philosopher* (London, 1830), 250.

46. L. Pearce Williams, *Michael Faraday: A Biography* (London: Chapman and Hall, 1965), 79–80.

47. From a course of lectures delivered at the Royal Institution in 1802, in Humphrey Davy, *Collected Works* (London, 1839), 2:333.

48. Letter from Davy to Berzelius, 24 March 1811, in H. G. Soderbaum, ed., *Jac. Berzelius Bref* (Uppsala, 1912–32), vol. 1, part 2, p. 23; quoted in Rocke, "Atoms and Equivalents," 236.

49. Davy, *Consolations in Travel,* 247.

50. Davy, *Collected Works,* 4:83.

51. In "Atoms and Equivalents," Rocke recounts an interesting story involving Wollaston, Davy, Thomas Thomson, and Davies Gilbert at the Crown and Anchor Inn in London, in which Wollaston supposedly convinced Gilbert of the merits of atomism, who then persuaded Davy (235).

52. Humphrey Davy, *Elements of Chemical Philosophy* (London, 1812), vol. 1, part 1, 112–13.

53. Rocke, "Atoms and Equivalents," 259.

54. Williams, *Michael Faraday,* 77–78

55. Ibid., 87. See also Thomson's *History of Chemistry.* Thomson was editor of *Annals of Philosophy.*

56. See Williams, *Michael Faraday,* chaps. 2–4.

57. Michael Faraday, "A Speculation Touching Electrical Conduction and the Nature of Matter," in *Experimental Researches in Electricity,* 3 vols. (London, 1839–55), 2:284–93.

58. Ibid., 2:289–91. In 1836 Faraday embraced a new theory of point atoms with attractive and repulsive powers, proposed by O. F. Mossotti (*Experimental Researches,* 2:293).

59. Williams, *Michael Faraday,* 256.

60. Faraday, *Experimental Researches,* 1:256.

61. Ibid., 2:236.

62. See ibid., 2:289–93. See also Williams, *Michael Faraday,* 379.

63. Faraday, *Experimental Researches,* 3:414.

64. Claude Louis Berthollet (1748–1822), one of the great scientific figures of France at the end of the century, fails to get his due in this narrative. He contributed importantly to French chemistry from about 1780 on, especially in his attempts to understand chemical affinity. A teacher and a civil servant, he was a central member of the Society of Arcueil (see chapter 3), which met at his home in Arcueil near Paris. See Thackray, *Atoms and Powers,* chap. 7.

65. Dulong was a chemist, Petit a physicist. For more about them, see von Meyer, *A History of Chemistry,* and the articles on Dulong and Petit in the *Dictionary of Scientific Biography.* Their results concerning specific heat and atomic weight appear in Pierre Louis Dulong and Alexis Thérèse Petit, "Recherches sur quelques points importants de la théorie de la chaleur," *Ann. Chim. Phys.* 10 (1819): 395–413.

66. Note that French luminaries such as Buffon and Macquer saw in gravitation an explanation of chemical affinity.

67. Berzelius encountered Richter's *Über die neuern Gegenstände der Chymie* (Breslau, 1791–1802) in 1807, and it "became the basis for the direction of my scientific endeavors during the greater part of my most active years of life, i.e. for my work on chemical proportions" (from Jacob Berzelius, *Autobiographical Notes,* trans. O. Larsell, trans. (Baltimore: Williams and Wilkins, 1934; quoted in Rocke, "Atoms and Equivalents," 247). Berzelius first met Davy in 1812, and the two corresponded.

68. According to von Meyer, the distinction between organic and inorganic compounds was first made by Bergman, about 1780 (von Meyer, *A History of Chemistry,* 233).

69. Rocke, "Atoms and Equivalents," 250.

70. See van Melsen, *From Atomos to Atom.*

71. Whyte, *Essay on Atomism.*

72. William Whewell, *History of the Inductive Sciences from the Earliest to the Present Time* (London: Porter, 1847), 1:422–23. On Whewell, see Menachem Fisch and Simon Schaeffer, eds., *William Whewell: A Composite Portrait* (Oxford: Clarendon, 1991), and Menachem Fisch, *William Whewell, Philosopher of Science* (Oxford: Clarendon, 1991).

73. Whewell, *Philosophy of the Inductive Sciences,* 1:425.

74. Ibid., 1:434–35.

75. J. B. Dumas, *Leçons sur la philosophie chimique* (Paris, 1837). See Rocke, "Atoms and Equivalents," 261.

76. Alexander Williamson, "On the Atomic Theory," *J. Chem. Soc.* 22 (1869): 331. See Rocke, "Atoms and Equivalents," 225.

77. Whewell, *Philosophy of the Inductive Sciences,* 1:425. This paragraph leans heavily on Schonland, *The Atomists.*

78. The van der Waals equation of state ($P = a/V^2$)($V - b$) = nRT for a real gas is a modification of the equation of state pV = nRT for an ideal gas. The parameters a and b are related to the intermolecular force law and molecular sizes, respectively. Van der Waals derived this famous equation in his dissertation in 1873 in Leiden. Berthelot proposed a somewhat similar equation of state, ($P + a/V^2T$)($V - b$) = RT, and there are dozens of others.

79. The estimate was based on the transport properties (for example, viscosity); see R. D. Present, *Kinetic Theory* (New York: McGraw-Hill, 1958), 43–45.

80. Crosbie Smith and Norton Wise, *Energy and Empire* (Cambridge: Cambridge University Press, 1989), 164.

81. Ibid., 229–30.

82. This specifically refers to the vortex atom hypothesis of W.J.R. Rankine.

83. William Thomson, "On Vortex Atoms," in *Mathematical and Physical Papers,* 4:1; James Clerk Maxwell, "Atom," in *Encyclopaedia Britannica,* 9th ed. (1875), 3:36–49.

84. Pierre Duhem, *L'Evolution de la mécanique* (Paris, 1903); published in English as *The Evolution of Mechanics,* trans. Michael Cole (Alphen aan den Rijn: Sijthoff and Noordhoff, 1980).

85. William Thomson, "The Size of Atoms," *Nature* 1 (1870): 551–53.

86. From Helmholtz's *Wirbelbewegungen* (vortex motion). For details, see Smith and Wise, *Energy and Empire,* especially 417–25.

87. A longer version of the paper was presented to the Manchester Literary and Philosophical Society in October 1847.

88. See D.S.L. Cardwell, *James Joule: A Biography* (Manchester: University of Manchester Press, 1989), chap. 5, for details. Joule had a copy of Herapath's work and, according to Cardwell, took note of it in a letter to John Playfair in 1846.

89. As Stephen Brush says, "quite apart from its success in explaining the properties of gases, we may regard the kinetic theory's most significant achievements as being the 'discovery' of the physical atom and the elucidation of its properties" (Stephen G. Brush, *Kinetic Theory* [Oxford: Pergamon, 1965], 1:43).

90. Maxwell, "Atom," in *Encyclopaedia Britannica;* reprinted in W. D. Niven, ed., *The Scientific Papers of James Clerk Maxwell* (Cambridge, 1890), 2:445–84.

91. James Clerk Maxwell, "Molecules," *Nature* 8 (1873): 437–41. See also Maxwell, "On the Dynamical Evidence of the Molecular Constitution of Bodies," *J. Chem. Soc.* 13 (1875): 493–508, reprinted in Niven, ed., *Scientific Papers,* 2:418–30.

92. Hermann von Helmholtz, "Über die erhaltung der kraft" (Berlin, 1847); published in English as "The Conservation of Force," trans. John Tyndall, in *Taylor's Scientific Memoirs* (1853): 114. See also Russell Kahl, ed., *Selected Writings of Hermann von Helmholtz* (Middletown, Conn.: Wesleyan University Press, 1971), 3–55.

93. Ernst Mach, *The Science of Mechanics,* trans. Thomas J. McCormack (LaSalle, Ill.: Open Court, 1942), 492.

94. Specifically, there was evidence from Einstein's explanation of Brownian motion and the interpretation of Perrin's experiments in terms of that model. See Albert Einstein, "On the Motion Required by the Molecular Kinetic Theory of Heat of Particles Suspended in Fluids at Rest," *Ann. d. Phys.* 17 (1905): 549; M. von Smoluchowski, "Zur Kinetischen Theorie der Brownischan Molekularbewegung und der Suspensionen," *Ann. d. Phys.* 21 (1906): 756; and J. Perrin, "Mouvement Brownien et réalité moléculaire," *Ann. Chim. Phys.* 18 (1909): 1. See the article on Ostwald by E. N. Hiebert and H. G. Körber in the *Dictionary of Scientific Biography.*

95. See van Melsen, *From Atomos to Atom,* 156.

96. On H. A. Lorentz and electron theories, see R. McCormmach, "Einstein, Lorentz, and the Electron Theory," *Hist. Stud. Phys. Sci.* 2 (1970): 41–87, and McCormmach, "H. A. Lorentz and the Electromagnetic View of Nature," *Isis* 61 (1970): 459–97. Others, including Wiechert and Larmor, also developed electron theories based on a discrete or molecular view of electricity.

97. Thomson was not the first to claim that cathode rays were not electromagnetic in nature, but he was able to show not only that they were charged, which was relatively easy, but that they represented the quantum of electricity.

98. This was one of the last of Lenard's final important contributions to physics; for he became increasingly embittered, perhaps as a result of being beaten to a number of fundamental discoveries by Röntgen, Thomson, and others. He railed against Thomson and Rutherford for "falsifying scientific history in favor of England" and later attacked Einstein for his "non-Aryan" physics.

99. See note 94.

100. The role of those doubts in his suicide is still controversial.

CHAPTER 7 *The Kinetic Theory of Gases and Statistical Mechanics*

1. Recall, however, that in chemistry the atomic theory often manifested itself in the form of combining proportions or equivalent weights rather than atoms per se.
2. Quoted in Crosbie Smith and Norton Wise, *Energy and Empire* (Cambridge: Cambridge University Press, 1989), 380.
3. This influence was exerted largely through Newton's *Opticks,* especially the famous query 31.
4. Such models were employed well into the nineteenth century. For example, consider Ampère's attempts to derive the general relation between pressure, volume, and temperature of a gas using a model in which "more or less static atoms [were] held apart by the repulsive effects of the caloric" (Donald Cardwell, *From Watt to Clausius* [Ithaca, N.Y.: Cornell University Press, 1971], 136).
5. Of course, it is true that, in the end, all properties of gases depend on their microscopic structure. But thermodynamics, starting from the empirical ideas of energy conservation and irreversibility, along with the assumption of certain mathematical properties, and with the addition of certain other empirical results such as those pertaining to specific heats, is indeed independent of assumptions about the underlying structure of a gas.
6. See R. Fox, *The Caloric Theory of Gases from Lavoisier to Regnault* (Oxford: Clarendon, 1971), 32–39, and Arthur Donovan's article on Crawford in the *Dictionary of Scientific Biography.* Crawford's first experiments, performed in 1777, were described in "Experiments and Observations on Animal Heat, and the Inflammation of Combustible Bodies; Being an Attempt to Resolve These Phaenomena into a General Law of Nature" (London, 1779).
7. Specific heat, or specific heat capacity, measured in units of energy per degree Celsius (typically calories per degree Celsius) or per degree per gram, measure the temperature change of a body when heat is added or taken away from it. Specific heats at constant volume or pressure represent what is obtained when either the volume or the pressure is held constant during the temperature change.
8. William Irvine (1743–87) was a student of Black whose work was published posthumously in 1805.
9. Sadi Carnot, "Reflections on the Motive Power of Heat," in W. F. Magie, ed., *The Second Law of Thermodynamics, Memoirs by Carnot, Clausius and Thomson* (New York: Harper Brothers, 1899), 26.
10. For the work of Pierre Louis Dulong (1785–1838) and Alexis Thérèse Petit (1791–1820), see their "Rercherches sur quelques points importants de la théorie de la chaleur," *Ann. Chim. Phys.* 10 (1819): 395–413. Here the "relative weights" of the atoms of a number of solids were related to oxygen, and the resulting products came out to about 0.38. Multiplying by 16 gives a value near the accepted figure of 3R, where R is the universal gas constant. Dulong's paper on the specific heat of gases was "Recherches sur la chaleur spécifique des fluides élastiques," *Ann. Chim. Phys.* 41 (1829): 113–58.
11. This is only approximately true, an issue that proved crucial toward the end of the century. See the discussion later in this chapter and in chapter 8.
12. Daniel Bernoulli, *Hydrodynamica* (Strasbourg, 1738); John Herapath, "A Mathematical Inquiry into the Causes, Laws and Phenomenae of Heat, Gases, Gravitation, Etc." *Ann. Phil.,* series 2, 1 (1821): 273–93, 340–51, 401–6. See also John

Herapath, *Mathematical Physics* (London: Whittaker, 1847); reprinted in Stephen G. Brush, ed., *Selected Papers by John Herapath* (New York: Johnson Reprint Corporation, 1972). Clifford Truesdell has suggested that Euler should deserve credit, as a founder of the kinetic theory, for his conclusion in 1727 that pV is proportional to the vis viva of a gas (*Essays in the History of Mechanics* [New York: Springer-Verlag, 1968], 274–75). But as Brush observes, the motion that Euler imagined was rotational rather than translational (see Stephen G. Brush, *Statistical Physics and the Atomic Theory of Matter from Boyle and Newton to Landau and Onsager* [Princeton, N.J.: Princeton University Press, 1983], 27). We should also mention John Waterston, whom we discuss later in the chapter. A good discussion of Herapath and Waterston can be found in Clifford Truesdell, "The Establishment Stifles Genius: Herapath and Waterston," in *An Idiot's Fugitive Essays on Science* (New York: Springer-Verlag, 1984), 380–402.

13. Translations from W. F. Magie, *A Source Book in Physics* (Cambridge: Harvard University Press, 1935), 247–51, and Stephen G. Brush, *Kinetic Theory* (Oxford: Pergamon, 1965), 1:57–65. See the discussion of early kinetic theory in Brush, *Statistical Physics,* 27–28. See also Stephen G. Brush, *The Kind of Motion We Call Heat: A History of the Kinetic Theory of Gases in the 19th Century* (Amsterdam: North-Holland, 1976).

14. Brush, *Kinetic Theory,* 1:57–65.

15. Herapath, *Mathematical Physics;* Brush, *Statistical Physics.*

16. See Brush, *Statistical Physics,* 38–39.

17. The paper Herapath submitted to the *Philosophical Transactions* of the Royal Society in 1820 was rejected, largely because of Humphrey Davy, whose Boscovichean atomic ideas were very different from Herapath's. The paper was published the next year in *Annals of Philosophy.*

18. Stephen G. Brush, "Foundations of Statistical Mechanics, 1845–1915," *Arch. Hist. Exact Sci.* 4 (1967): 145–83. See also Brush, *The Kind of Motion We Call Heat,* and Brush's article on Waterston in the *Dictionary of Scientific Biography.*

19. Brush, *The Kind of Motion We Call Heat;* Brush, *Statistical Physics,* 45. Waterston's abstract of the paper "On the Physics of Media That Are Composed of Free and Perfectly Elastic Molecules in a State of Motion" appears in *Proc. Roy. Soc.* 5 (1846): 604. The original paper was published in *Phil. Trans. Roy. Soc.* A183 (1893): 5–79, as a result of Rayleigh's intervention and with an introduction by him. See also J. S. Haldane, ed., *The Collected Papers of John Waterston* (Edinburgh: Oliver and Boyd, 1928).

Although S. Tolver Preston informed Maxwell of Waterston's early work in 1876, Maxwell apparently never pursued the matter. Had he done so, Waterston might have been rescued from obscurity before his death in 1883. See Stephen G. Brush, "James Clerk Maxwell and the Kinetic Theory of Gases: A Review Based on Recent Historical Studies," *Am. J. Phys.* 39 (1971): 631–40.

See E. Daub, "Waterston's Influence on Krönig's Kinetic Theory of Gases," *Isis* 60 (1971): 512–15, and Brush, *Statistical Physics,* 45–46—in particular, his comments on the negative influence of scientific institutions, especially the Royal Society, in the context of Waterston.

20. In modern form this is stated in terms of the number of quadratic terms in x and p in the Hamiltonian.

21. James Clerk Maxwell, "Illustrations of the Dynamical Theory of Gases," *Phil. Mag.*, series 4, 19 (1860): 19–32; 20 (1860): 21–37.

22. Joule presented his results in a paper to the Manchester Literary and Philosophical Society. See Brush, *Kinetic Theory*, 1:23.

23. R. Clausius, "Über die Art der Bewegung, welche wir Wärme nennen," *Ann. d. Phys.* 100 (1857): 353–80; English translation in *Phil. Mag.*, series 4, 14 (1857): 108–27. The title is often rendered as "The Kind of Motion We Call Heat," which is also the title of Brush's book. Also see R. Clausius, "Grundzuge einer theorie der gase," *Ann. d. Phys.*, series 2, 99 (1856): 315–22.

24. James Clerk Maxwell, "On the Dynamical Theory of Gases," *Phil. Mag.*, series 4, 32 (1866): 390–93; 35 (1868): 129–45, 185–217. See also Maxwell, *The Scientific Papers of James Clerk Maxwell*, ed. W. D. Niven (Cambridge: Cambridge University Press, 1890), 2:26–78. This paper has been reprinted in Brush, *Kinetic Theory*, vol. 2, and in Elizabeth Garber, Stephen G. Brush, and C.W.F. Everitt, eds., *Maxwell on Molecules and Gases* (Cambridge: MIT Press, 1986).

25. The German title was "Ueber die mittlere Länge der Wege, welche bei Molecularbewegung gasförmigen Körper von den einzelnen Molecülen zuruchgelegt werden, nebst einigen anderen Bemerkungen über die mechanischen Wärmtheorie," *Ann. d. Phys.* 105 (1858): 239–58. The English translation is from *Phil. Mag.* 17 (1859): 81–91.

26. The paper was read at the August 1859 meeting of the BAAS. See especially Elizabeth W. Garber, "Clausius and Maxwell's Kinetic Theory of Gases," *Hist. Stud. Phys. Sci.* 2 (1970): 299–319.

27. As Clausius put it, "Kronig assumes that the molecules of gas do not oscillate about definite positions of equilibrium, but that they move with constant velocity in straight lines until they strike against other molecules, or against some surface which is to them impermeable. I share this view completely, and I also believe that the expansive force of the gas arises from this motion. On the other hand, I am of opinion that this is not the only motion present" (quoted from Brush, *Kinetic Theory*, 1:113).

28. A mole is one gram molecular weight, or the molecular weight in grams. Technically, we speak here of *heat capacities* c_p and c_v rather than *specific heats*, which are obtained by dividing by the number of moles, v.

29. R. Clausius, "On the Motive Power of Heat," *Phil. Mag.*, series 4, 2 (1851): 1–21, 102–19; and *Pogg. Ann.* 79 (1850): 376–97, 500–24. For a translation, see Magie, ed., *Second Law of Thermodynamics*, 65–107.

30. *Ann. Chim. Phys.* 41 (1829):113–58. See Magie, ed., *Second Law of Thermodynamics*, 86.

31. Translated by Brush, *Kinetic Theory*, 1:111.

32. Clausius, "On the Kind of Motion We Call Heat." See also Brush, *Kinetic Theory*, 1:131.

33. Christoph H. D. Buys-Ballot, "Über die art der bewegung, welche wir Wärme und Electrizitat nennen," *Ann. d. Phys.* 103 (1858): 240.

34. Translated by Brush, *Kinetic Theory*, 1:137.

35. See the translation in ibid., 1:145.

36. Maxwell paralleled the argument used by Herschel in 1850; see Brush, *Statistical Physics*, 62. The later justification was based on the assumption that the velocities of pairs of colliding molecules were statistically independent rather than the ve-

locity components of one molecule. It was this latter assumption that Maxwell believed might "appear precarious." The 1860 paper was actually delivered in 1859, and the 1867 paper was presented in 1866. See Maxwell, "Illustrations of the Dynamical Theory of Gases."

37. Garber, "Clausius and Maxwell's Kinetic Theory of Gases."

38. R. Clausius, "On the Conduction of Heat by Gases," translated in *Phil. Mag.* 23 (1862): 417–35, 512–34. On Clausius's criticism, see Brush, *Kinetic Theory;* Brush, *Statistical Physics;* or Garber, "Clausius and Maxwell's Kinetic Theory of Gases."

39. Maxwell, "On the Dynamical Theory of Gases."

40. Which depend on the transport of momentum and energy through a gas. The paper referred to was read in 1866 but published in *Phil. Trans.* 157 (1867): 49–88. In it Maxwell gave a detailed calculation of the transport properties, including viscosity; but already in 1860, in his early description of viscosity, he had found the surprising result that the coefficient of viscosity for a gas is independent of density or pressure. For details, see R. D. Present, *Kinetic Theory* (New York: McGraw-Hill, 1958), 41, 221–24. Many of the important early experiments on diffusion were those of Thomas Graham. See his *Elements of Chemistry* (London, 1846).

41. It turns out that $a^2 = 2kT/m$. The expression $v^2 \exp(-v^2/a^2)$ is known as the Maxwell-Boltzmann distribution. Alternatively, the probability that a molecule had a velocity in the x-direction between x and some slightly different value x + dx was proportional to $\exp(-v_x^2/a^2)dv_x$.

42. This provoked Clausius's criticism, the validity of which Maxwell acknowledged in "On the Dynamical Theory of Gases."

43. W. D. Niven, ed., *The Scientific Papers of James Clerk Maxwell* (Cambridge, 1890), 409; originally published as "Illustrations of the Dynamical Theory of Gases" in *Phil. Mag.* 19 and 20.

44. James Clerk Maxwell, "On the Dynamical Evidence of the Molecular Constitution of Bodies," *Nature* 11 (1875): 357–59, 374–77; reprinted in Niven, *Scientific Papers,* 2:418–38. In that interesting paper, Maxwell commented on the structure of the aether: "Among the properties of a gas, it will have that established by Dulong and Petit, so that the capacity for heat of unit of volume of the aether must be equal to that of unit of volume of any ordinary gas at the same pressure. Its presence, therefore, could not fail to be detected by our experiments on specific heats, and we may therefore assert that the constitution of the aether is not molecular" (*Scientific Papers,* 2:433).

It was by this time supposed that molecules (or atoms) had some internal degrees of freedom; and information had been accumulating independent of that from the specific heats, specifically from increasingly detailed studies of atomic spectral lines (see chapter 8).

45. Ibid. Alternatively, the specific heat at constant volume c_v would be given by 1/2 nR, where n is the total number of degrees of freedom (quadratic terms in the kinetic energy of a molecule).

46. Three angles, the Euler angles, are necessary to specify the rotation of a rigid body.

47. In this notation, c_v would be given by 1/2(m + e)R. Earlier, we wrote $\gamma = 1 + 2/n$, which is the same as $\gamma = (n + 2)/n$. A harmonic force is one in which the potential energy is given by $V = 1/2\ kx^2$. For small oscillations, this is a good approximation in a wide range of physical phenomena.

48. Maxwell, "Illustrations of the Dynamical Theory of Gases." The passage in question can be found in Niven, *Scientific Papers,* 2:409.
49. See Martin Klein, *Paul Ehrenfest,* (North-Holland, 1970), 1:110.
50. c_v = 5/2R can also be expressed as a specific heat of 4.96 calories/mole. Measured values include 4.84 for H_2, 4.94 for N_2, and 4.98 for O_2.
51. Lord Rayleigh, "The Law of Partition of Energy," *Phil. Mag.* 49 (1900): 98–118.
52. Clausius, "On the Conduction of Heat by Gases."
53. On this issue, see P.M.C. Dias, "Clausius and Maxwell, the Statistics of Molecular Collisions (1857–1862)," *Ann. Sci.* 51 (1994): 249–61.
54. Garber et al., *Maxwell on Molecules and Gases.*
55. Which to Brush represented "the introduction of continental statistical theory into British science" (Brush, *Statistical Physics,* 59). See Garber et al., *Maxwell on Molecules and Gases,* 9.
56. Niven, ed., *Scientific Papers,* 2:45–46; Garber et al., *Maxwell on Molecules of Gases.*
57. Letter from James Clerk Maxwell to Peter Guthrie Tait, dated 8 February 1872, Maxwell Collection, Cambridge University Library. Elsewhere, Maxwell complained to Tait of Clausius's self-importance. Maxwell did make changes in the second edition of *Theory of Heat* in response to Clausius comments in *Phil. Mag.* 43 (1872): 106.
58. See Garber, "Clausius and Maxwell's Kinetic Theory of Gases." Earlier, while both Boltzmann and Clausius attempted to prove the second law analytically from mechanical principles, Maxwell wrote to Tait, "It is rare sport to see those learned Germans contending for the priority of the discovery that the 2nd law of thermodynamics is the *Hamiltonsche Princip."* Letter from James Clerk Maxwell to Peter Guthrie Tait, 1 December 1873, Maxwell Collection, Cambridge University Library.
59. When Max Planck tried to contact Clausius during the course of his own work on thermodynamics, Clausius would not answer his letters. See Max Planck, *Scientific Autobiography and Other Papers,* trans. Franz Gaynor (New York: Philosophical Library, 1949), 19. This was pointed out by Martin Klein in "Gibbs on Clausius," *Hist. Stud. Phys. Sci.* 1 (1969): 127–49.
60. Letter from James Clerk Maxwell to Peter Guthrie Tait, 11 August 1873, in the Maxwell-Tait Correspondence, Maxwell Collection, Cambridge University Library.
61. J. W. Gibbs, "Rudolph Julius Emanuel Clausius," *Proceedings of the American Academy of Arts and Sciences,* new series, 16 (1889): 458–65.
62. It is interesting to compare the rather complicated academic careers of German physicists, who moved around frequently when someone retired or died, with the long and placid careers of English scientists such as Thomson and Maxwell. In *Ludwig Boltzmann* (Woodbridge, Conn.: Ox Bow Press, 1983), E. Broda remarks that, according to Flamm, Boltzmann left Vienna and then Leipzig because of the opposition to atomism by Mach and Ostwald, respectively.
63. L. Boltzmann, "Studien über das Gleichgewicht der lebehndige Kraft zwischen bewegten materiellen Punkten," *Sitzungsberichte, K. Akademie der Wissenschaften, Wien, Mathematisch-Naturwissenschaftliche Klasse* 58 (1868): 517–60.
64. L. Boltzmann, "Weitere Studien über das Wärmegleichgewicht unter Gasmolekulen," *Wiener Ber.* 66 (1872): 275–370, translated by Brush in *Kinetic Theory,* vol.

2. Boltzmann's collected works were edited by Friedrich Hasenöhrl as *Wissenschafliche Abhandlungen*, 3 vols. (Leipzig, 1909).

The function f that satisfied this collision equation, or transport equation, denoted by f, determined a quantity E = f log f, which is now universally represented by the letter H (after S. H. Burbury [1890]; see Brush, *Statistical Physics,* 63). The H-theorem states that H increases with time unless the velocity distribution is Maxwellian. Thus, H is, within a constant factor, the negative of the entropy. The H-theorem is applicable to nonequilibrium states and thus generalizes the thermodynamic concept of entropy. See Brush's article on Boltzmann in the *Dictionary of Scientific Biography.*

65. The integral is over phase space, and f is a function of x and t, representing the distribution of the particles in phase space as a function of time. See any text on the subject—for example, Richard Tolman's classic *The Principles of Statistical Mechanics* (Oxford: Oxford University Press, 1938; reprinted, New York: Dover, 1979).

66. L. Boltzmann, "A Word from Mathematics to Energism," quoted in Broda, *Ludwig Boltzmann,* 74; see also the quotations on 86–87, including one from a 1904 lecture given in St. Louis, Missouri.

67. William Thomson, "The Kinetic Theory of the Dissipation of Energy," *Proc. Roy. Soc. Edin.* 8 (1874): 325–34; J. Loschmidt, "Über den Zustand des Wärmegleich-gewichtes eines Systems von Korpern mit Rucksicht auf die Schwerkraft," *Sitzungberichte der Akademie der Wissenshaften in Wien, Math-natwiss* 73 (1876): 135, 366–72; 75 (1877): 67; 76 (1877): 209–25. See Klein, *Paul Ehrenfest,* vol. 1, *The Making of a Theoretical Physicist,* 102–3.

68. L. Boltzmann, "Über die Beziehung zwischen dem zweiten Hauptsatze der mechanischen Wärmetheorie und der Wahrscheinlichkeitsrechnung respective den Satzen uber das Wärmegleichgewicht," *Sitzungberichte der Akademie der Wissenshaften in Wien, Math-natwiss* 76 (1877): 373. Translated by Stephen Brush as "On the Relation of a General Mechanical Theorem to the Second Law of Thermodynamics," in Brush, *Kinetic Theory,* 2:189. See Klein, *Paul Ehrenfest,* 1:102–5, for other references. See also Brush, *Kinetic Theory,* and *Statistical Physics;* L. Boltzmann, *Lectures on Gas Theory,* trans. Stephen G. Brush (Berkeley: University of California Press, 1964); and Brush's article on Boltzmann in the *Dictionary of Scientific Biography.* On statistical mechanics, see the classic by Paul and Tatyana Ehrenfest, *The Conceptual Foundations of the Statistical Approach in Mechanics,* trans. M. J. Moravcsik (Ithaca, N.Y.: Cornell University Press, 1959).

69. L. Boltzmann, "Ueber die Beziehung zwischen dem Zweiten Hauptsatze der mechanischen Wärmtheorie und der Wahrscheinlichkeitsrechung, respective den Sätzen über das Wärmegleich gewicht," *Sitzungberichte der Akademie der Wissenshaften in Wien, Math-natwiss* 76 (1877): 373.

70. These ideas were generalized and clarified by Gibbs, who introduced the idea of an average over an ensemble of copies of the system, each representing the same thermodynamic state but a different microstate; see the discussion later in this chapter.

71. Boltzmann's use of them also reflected a philosophical bias; see Brush's article in the *Dictionary of Scientific Biography.* This approach is also used in the 1872 paper "Further Studies on the Thermal Equilibrium of Gas Molecules," which is translated by Brush in *Kinetic Theory,* vol. 2.

72. In fact, however, Boltzmann's transport equation is a special case of Maxwell's. Solutions to Boltzmann's and Maxwell's equations were obtained only in 1916 and 1917, respectively, by Chapman and Enskog. These papers are reprinted in Brush, *Kinetic Theory,* vol. 3. See also, for example, F. Mohrling, *Statistical Mechanics* (New York: Wiley, 1982), or Walter Grandy, *Foundations of Statistical Mechanics* (Boston: Reidel, 1987).

73. Quoted in Broda, *Ludwig Boltzmann,* 83.

74. The mathematical argument has to be formulated in a 6N-dimensional phase space, in which the position and momentum of each particle, in three dimensions, are represented by a single point. The trajectory of the system is a curve in this 6N-dimensional phase space, and Poincaré's ergodic theorem proved that eventually all points in the phase space would be visited by the system—in other words, that it would eventually return to the same phase space point. Poincaré's paper, "On the Three-Body Problem and the Equations of Dynamics," which was originally published in *Acta mathematica* 13 (1890): 1–270, is translated, in part, in Brush, *Kinetic Theory,* vol. 2.

75. L. Boltzmann, "Entgegung auf die Wärmtheoretischen Betrachtungen des Hrn. E. Zermelo," *Ann. d. Phys.* 57 (1896): 773–84.

76. This point was first made, apparently, by Maxwell. Boltzmann wrote: "The foregoing results are intimately connected with my interpretation of the second law of thermodynamics. . . . According to the molecular-kinetic view, this is merely a theorem of probability theory. According to this view, it cannot be proved from the equations of motion that all phenomena must evolve in a certain direction in time. . . . On the other hand, when the motion involves a very large number of small molecules, then there must be (aside from a small number of exceptional cases) a progression from less probable to more probable states, and therefore a continual change in a definite direction, such as, in a gas, the evolution toward a Maxwellian distribution" (from ibid.; translated by Brush, *Kinetic Theory,* 2:223–24, as "Reply to Zermelo's Remarks on the Theory of Heat").

77. Ehrenfest also died by his own hand, in 1933, in Amsterdam. One of Boltzmann's most outstanding students was Fritz Hasenöhrl, who also studied under Boltzmann's colleague Franz Exner, and was killed in World War I. Schrödinger did his thesis under Exner but was strongly influenced by Hasenöhrl, as was Hans Thirring. Both Exner and Boltzmann were students of Stefan. See Broda, *Ludwig Boltzmann.*

78. J. W. Gibbs, *Elementary Principles in Statistical Mechanics Developed with Special Reference to the Rational Foundations of Thermodynamics* (New York: Scribner's, 1902; reprint, Woodbridge, Conn.: Oxbow Press, 1981).

79. L. Boltzmann, *Vorlesungen über Gastheorie,* 2d ed. (Leipzig, 1898), quoted in M. Klein, "Planck and the Beginnings of the Quantum Theory," *Arch. Hist. Exact Sci.* 1 (1962): 459–479, 476. See Boltzmann's *Lectures on Gas Theory,* trans. Brush.

80. Quoted in Broda, *Ludwig Boltzmann,* 43.

81. Brownian motion, named after nineteenth-century English botanist Robert Brown, is the apparently random agitation of smoke particles, which both Smoluchowski and Einstein showed could be explained in terms of collisions between the particles and gas molecules.

82. After the Civil War, physics research in the United States began to approach the level of Europe's. Earlier, Joseph Henry, the first director of the Smithsonian Institution, made significant contributions to electrical science, especially induction.

Henry Rowland at Johns Hopkins (1848–1901) was a leading figure in American science and a consummate experimentalist, having worked on a variety of topics in magnetism and electricity. He was the first to establish that a moving charge produced a magnetic field and was renowned for his fine diffraction gratings. Rowland was the first president of the American Physical Society.

83. Among Gibbs's inventions was a governor for steam engines, which was never actually built. See Martin Klein's article on Gibbs in the *Dictionary of Scientific Biography.*

84. See Lynde Phelps Wheeler, *Josiah Willard Gibbs* (New Haven: Yale University Press, 1951). One might also consult Muriel Rukeyser, *Willard Gibbs* (Garden City, N.Y.: Doubleday, 1942).

85. See Klein's article on Gibbs in the *Dictionary of Scientific Biography.*

86. He received a salary only after 1880, when Johns Hopkins nearly wooed him away from Yale.

87. See Klein's article on Gibbs in the *Dictionary of Scientific Biography.*

88. Gibbs, *Elementary Principles in Statistical Mechanics*, 165–66. See also the *Collected Works of Josiah Willard Gibbs,* ed. W. R. Longley and R. G. Van Name, 2 vols. (New Haven: Yale, 1948).

89. See any standard treatise on statistical thermodynamics, such as Terrell L. Hill, *Introduction to Statistical Thermodynamics* (Reading, Mass.: Addison-Wesley, 1960), or J. Kestin and J. R. Dorfman, *A Course in Statistical Thermodynamics* (New York: Academic Press, 1971). Neither of these works offers any historical perspective.

90. Albert Einstein, "Zur allegemeinen molekularen Theorie der Wärme," *Ann. d. Phys.* 14 (1904): 354–62. See, for example, *The Collected Papers of Albert Einstein,* trans. Anna Beck, vols. 1 and 2 (Princeton, N.J.: Princeton University Press, 1989).

CHAPTER 8   *Fin de Siècle*

1. This is very different from saying that it was generally *felt* that they foreshadowed an impending revolution, which was certainly not the case.

2. Both (1) and (2) involve the question of the validity of the equipartition theorem.

3. With Thomson stood Charles Darwin, who died in 1882, three years after Maxwell. Thomson had outlived others who might have challenged him—for example, Maxwell, Helmholtz, and Clausius.

4. See Crosbie Smith and Norton Wise, *Energy and Empire* (Cambridge: Cambridge University Press, 1989); the opening page of chapter 13 gives an eloquent summary of Thomson's situation at the century's end.

5. J. J. Thomson, "Cathode Rays," *Phil. Mag.,* series 5, 44 (1897): 293–316. He announced his results at the Royal Institution on 29 April 1897. This paper is reprinted in Stephen Wright, *Classical Scientific Papers: Physics* (New York: Elsevier, 1964).

6. R. McCormmach, "Einstein, Lorentz, and the Electron Theory," *Hist. Stud. Phys. Sci.* 2 (1970): 61.

7. Blackbody radiation was also known as "complete radiation," "natural radiation" (Planck), and "cavity radiation." Many heated objects approximate the ideal blackbody, including (somewhat surprisingly, perhaps) the surfaces of stars and, much less perfectly, a warm body such as the Earth's that is radiating into space. If elec-

tromagnetic radiation in a cavity is allowed to come into equilibrium with the walls of the cavity, the spectrum of the radiation is that of a blackbody.

8. Indeed, for a long time the whole question of radiant heat was a puzzle that fell outside the problem of the emission and wave theories of light.

9. Josef Stefan, "Über die Beziehung zwischen der Wärmestrahlung und der Temperatur," *Wiener Ber.*, series 2, 79 (1879): 391–428. Stefan was born in 1835 and died in 1893. One of the most fundamental sets of measurements on which the T^4 law was based was due to John Tyndall.

10. See Hans Kangro's article on Wien in the *Dictionary of Scientific Biography*. $\rho(\lambda)d\lambda$ is the energy per unit volume in the wavelength range $d\lambda$. Alternatively, in terms of frequency ν, $\rho(\nu) = \nu^{-3} \phi(\nu/T)$.

11. See the discussion in Thomas Kuhn, *Blackbody Theory and the Quantum Discontinuity, 1894–1912* (Oxford: Clarendon, 1978), chap. 1.

12. This had a mean value of 5.66.

13. Wilhelm Wien, "Über die Energieverteilung im Emissionsspectrum eines schwarzen Körpers," *Ann. d. Phys.* 58 (1896): 662–69. The exponential form was actually proposed by Paschen, but Wien provided a theoretical basis for it and first published it. See Kuhn, *Blackbody Theory*, 10–11.

14. O. Lummer and E. Pringsheim, "Die Vertheilung der Energie im Spectrum des schwarzen Körpers," *Verh. d. D. Phys. Ges.* 1 (1899): 23–41, and Lummer and Pringsheim, "Die Vertheilung der Energie im Spectrum des schwarzen Körpers und des blanken Platins," *Verh. d. D. Phys. Ges.* 1 (1899): 215–35. See Kuhn, *Blackbody Theory*, chap. 4, and H. Kangro, *Vorgeschichte des Planckschen Strahlungsgesetzes* (Wiesbaden, 1970); trans. as *Early History of Planck's Radiation Law* (London: Taylor and Grancis, 1976).

15. O. Lummer and E. Pringsheim, "Ueber die Strahlung des schwarzen Körpers für lange Wellen," *Vehr. d. D. Phys. Ges.* 2 (1900): 163–80. See Kuhn, *Blackbody Theory*, 281, n. 10, for details.

16. $\exp(-b/\lambda T)$ tends to unity at high temperatures.

17. Lord Rayleigh, "The Law of Partition of Kinetic Energy," *Phil. Mag.*, series 5, 49 (1900): 98–118.

18. Lord Rayleigh, "Remarks on the Law of Complete Radiation," *Phil. Mag.*, series 5, 49 (1900): 539–40.

19. Alternatively, in terms of the wavelength λ, he had found, from a calculation of the number of modes of oscillation of the radiation field, that the energy in a wavelength range $d\lambda$ should be proportional to λ^{-4} and would become very large at short wavelengths (and infinite at $\lambda = 0$). $\rho(\nu)d\nu \approx \nu^2 T$ implies that $\rho(\lambda)d\lambda \approx \lambda^{-4}T$.

20. See Kuhn, *Blackbody Theory*, 144–47, for details. Rayleigh's initial work ("Remarks on the Law of Complete Radiation") was followed by another paper in which he "derived" this form by treating the state of the electromagnetic field in a cavity like the modes of vibration of an elastic solid. Lord Rayleigh, "The Law of Partition of Kinetic Energy," *Phil. Mag.*, series 5, 49 (1900): 98–118.

21. See especially Kuhn, *Blackbody Theory*, 36 and chap 3. The term *cavity radiation* refers to the fact that electromagnetic radiation in a cavity is a useful and effective model of a perfect blackbody.

22. Martin Klein, *Paul Ehrenfest* (Amsterdam: North-Holland, 1970), 1:140.

23. Ibid., 1:218.

24. Quoted in J. L. Heilbron, *The Dilemmas of an Upright Man* (Berkeley: University of California Press, 1986), 14. See Planck's comment on atomism in Max Planck, *Scientific Autobiography and Other Papers*, trans. F. Gaynor (New York: Philosophical Library, 1949).

25. Martin Klein, "Max Planck and the Beginnings of the Quantum Theory," *Arch. Hist. Exact Sci.* 1 (1962): 468.

26. See Kangro's article on Planck in the *Dictionary of Scientific Biography*. See also H. Kangro, *Early History of Planck's Radiation Law.*

27. Klein, "Max Planck," 459–79.

28. He used Boltzmann's approach with considerable reluctance. See Kuhn, *Blackbody Theory,* 76–77.

29. Rayleigh, "The Law of Partition of Kinetic Energy," "Remarks on the Law of Complete Radiation."

30. Klein, "Max Planck."

31. By this age most theoretical physicists have done their best work.

32. Klein, "Max Planck."

33. Here we have written ρ in terms of frequency instead of wavelength; the two are related through the speed of light c: $c = f\lambda$.

34. Or $\rho(\lambda,T) = a' \, \lambda^{-5} \exp(-b/\lambda T)$.

35. H. Rubens and F. Kurlbaum, "Anwendung der Methode der Reststrahlung zur Prüfung des Strahlungsgesetzes," *Ann. d. Phys.* 4 (1901): 649–66.

36. See Klein, "Max Planck," 462–65.

37. Its purpose was to bring the radiation into equilibrium with the cavity walls. See Thomas Kuhn, "Revisiting Planck," *Hist. Stud. Phys. Sci.* 14 (1984): 231–52.

38. Max Planck, "Über das Gesetz der Energieverteilung im Normalspectrum," *Ann. d. Phys.* 4 (1901):564–66.

39. From Planck's Nobel Prize lecture; quoted in Klein, "Max Planck," 468.

40. This entropy function had the form $S = -(E/bv)[\ln (E/av) - 1]$.

41. Planck's problem was to find the average energy per oscillator. But he was also constrained by what he had learned in his studies of the second law of thermodynamics. He had found, for example, that the second derivative of the entropy had to be a negative function of the energy u; he wrote $\partial^2 S/\partial u^2 = -a\, u$. Integrating once gave $\partial S/\partial u$ but, by the definition of entropy, was also equal to $1/T$. The result was $u(T)$—that is, the expression for the average energy per oscillator and hence the Wien law. Integrating twice gave $S(u)$, the formula for the oscillator entropy. For his new blackbody formula, he had to adopt a more general form for S, which would yield $u = bv \, /(e^{av/T} - 1)$. But how to justify the form for the oscillator entropy? To do so, Planck applied what he thought was Boltzmann's method, to justify the form of S on statistical grounds—that is, $S = k \ln W$, where W was a combinatorial factor. It is this statistical calculation that Kuhn shows to have been flawed. See Kuhn, *Blackbody Theory,* chap. 4, or Klein, "Max Planck."

42. See Kuhn, *Blackbody Theory,* 105–6.

43. Max Planck, "Über die Elementarquanta der Materie und der Elektricität," *Ann. d. Phys.* 4 (1901): 564–66.

44. See especially Kuhn, "Revisiting Planck," 241–42.

45. James Jeans, "A Comparison between Two Theories of Radiation," *Nature* 72 (1905): 293.

46. Planck found that the energy increment ε had to be proportional to the frequency

v: That is, he introduced the constant of proportionality h. Its value could be determined from Stefan's law for the total energy density of the radiation and from the Wien displacement law (see Klein, "Max Planck"). The constant soon became known as Planck's constant.

47. Jeans, "A Comparison between Two Theories of Radiation," 294.
48. In fact, it is not necessary, and indeed invalid, to take this limit. ε is irrelevant and can be eliminated from the expression for the entropy. See Kuhn, "Revisiting Planck," and his *Blackbody Theory.*
49. Klein, "Max Planck," 474.
50. A cell labeled k, for example, would specify the energy $k\varepsilon$, and w_k would be the number of oscillators with energy $E_k = k\varepsilon$. So as an oscillator gained or loss energy, it would move from cell k to another. See Kuhn, "Revisiting Planck."
51. Planck's constant is now regarded as one of the three fundamental constants of nature: h, c (velocity of light), and G, the gravitational constant.
52. See Kuhn, "Revisiting Planck," 235–36. In a personal comment, Stephen Brush pointed out that, in Planck's Nobel lecture, he gives the credit to Einstein for *physical* quantization. See Max Planck, *Vorlesungen über die Theorie der Wärmstrahlung* (Leipzig: Barth, 1906).
53. Nearly a half-century later, Einstein wrote: "It was as if the ground had been pulled out from under one, with no firm foundation to be seen anywhere upon which one could have built" (Albert Einstein, *Philosopher-Scientist,* ed. P. A. Schilpp [LaSalle, Ill.: Open Court, 1970], 45). See Kuhn, "Revisiting Planck," 237–38, and his *Blackbody Theory,* chap. 7 and p. 140.
54. Kuhn, *Blackbody Theory,* 138–39.
55. "Es is namlich zu bermerken, dass nach der Theorie von Planck die Resonatoren in ganz stetiger Weise (ohne dass von einem endlichen Elemetarquantum die Rede ist) Energie von dem Ather erhalten oder an ihn abgeben konnen." The letter from Lorentz to Wien is dated 6 June 1908. Although the original has been lost, it is quoted in full in Kuhn, *Blackbody Theory,* 302.
56. Albert Einstein, "Zur gegenwarten Stand des Strahlungsproblems," *Phys. ZS.* 10 (1909): 185–93; quoted in Kuhn, *Blackbody Theory,* 185. Kuhn also quotes Einstein's pointing out (in the same year) that the energy quantum ε must be small compared to the average energy per oscillator, which is not in general the case.
57. James Clerk Maxwell, "On the Dynamical Evidence of the Molecular Constitution of Bodies," *Nature* 11 (1875): 357–59, 374–77; reprinted in Niven, ed., *Scientific Papers.* Here Maxwell commented on the structure of the aether: "Among the properties of a gas, it will have that established by Dulong and Petit, so that the capacity for heat of unit of volume of the aether must be equal to that of unit of volume of any ordinary gas at the same pressure. Its presence, therefore, could not fail to be detected by our experiments on specific heats, and we may therefore assert that the constitution of the aether is not molecular" (377).
58. See, for example, F. K. Richtmyer and E.H. Kennard, *Introduction to Modern Physics,* 3d ed. (New York: McGraw-Hill, 1942), 457–68.
59. Albert Einstein, "Die Plancksche Theorie der Strahlung und die Theorie der spezifischen Wärme," *Ann. d. Phys.* 22 (1907): 180–190.
60. See William McGucken, *Nineteenth-Century Spectroscopy* (Baltimore: Johns Hopkins University Press, 1969).
61. There were numerous attempts, of course, including one by Baron Fabian Jacob

von Wrede, who explained solar absorption lines as interference phenomena. See ibid., 20.

62. J. Herschel, quoted by David Brewster in "Report on the Recent Progress of Optics," *Rep. Brit. Assoc.* 1 (1831–32): 308–22. See McGucken, *Nineteenth-Century Spectroscopy,* 15.

63. See David Wilson, ed., *The Correspondence of Sir George Gabriel Stokes and Sir William Thomson, Baron Kelvin of Largs* (Cambridge: Cambridge University Press, 1990), or G. G. Stokes, *Mathematical and Physical Papers* (Cambridge: Cambridge University Press, 1904).

64. Letter from William Thomson to Leo Königsberger, 26 September 1902, Cambridge University Library.

65. Gustav Kirchhoff, "Ueber das Sonnenspectrum," *Verhandlungen des naturhistorisch-medicinschen Vereins zu Heidelberg* (1857–59): 251–55. See especially the discussion in McGucken, *Nineteenth-Century Spectroscopy,* 30–34.

66. Balfour Stewart, "An Account of Some Experiments on Radiant Heat, Involving an Extension of Prevost's Law of Exchanges," *Trans. Roy. Soc. Edin.* 22 (1858): 1–20.

67. See references in McGucken, *Nineteenth-Century Spectroscopy,* 48–49. Note especially Gustav Kirchhoff, *Researches on the Solar Spectrum and the Spectra of the Chemical Elements,* trans. Henry Roscoe (London, 1862–33).

68. G. J. Stoney, "The Internal Motions of Gases Compared with the Motions of Waves of Light," *Phil. Mag.,* series 4, 36 (1868): 132–41.

69. See McGucken, *Nineteenth-Century Spectroscopy,* 73ff, for these discussions.

70. On the life of Jean Baptiste Perrin, who proposed a planetary model of the atom in 1901, see Mary Jo Nye, *Molecular Reality* (London: Macdonald, 1972). Hantaro Nagaoka's Saturnian model appeared in "Kinematics of a System of Particles Illustrating the Line and the Band Spectrum and the Phenomenon of Radioactivity," *Phil. Mag.,* series 6, 7 (1904): 445–55.

71. J. J. Thomson's model was first published in *Proceedings of the Cambridge Literary and Philosophical Society* 15, part 5 (1910).

72. See David Wilson, *Rutherford, Simple Genius* (Cambridge: MIT Press, 1983), 211.

73. The Zeeman effect is a semiclassical phenomenon, resulting from the interaction of the orbital magnetic moment of the electron with an external magnetic field. The anomalous Zeeman effect arises from the introduction of the electron spin. See, for example, Paul Forman, "Alfred Landé and the Anomalous Zeeman Effect, 1919–1921," *Hist. Stud. Phys. Sci.* 2 (1970): 153–261. On the discovery of the spin of the electron, see A. Pais, "George Uhlenbeck and the Discovery of Electron Spin," *Physics Today* 42 (1989): 34–40.

74. P. Lenard, "Über Kathodenstrahlen in Gasen von atmosphärischem Druck und in äussersten Vakuum," *Ann. d. Phys.* 51 (1894): 225–68.

75. H. Geiger and E. Marsden, "On a Diffuse Reflection of the α-Particles," *Proc. Roy. Soc.* A82 (1909): 495–500.

76. E. Rutherford, "The Scattering of α and ß Particles by Matter and the Structure of the Atom," *Phil. Mag.,* series 6, 21 (1911): 669–88. Yet in a paper published the year after the Geiger-Marsden paper (see previous note), Geiger wrote: "It does not appear to be profitable at present to discuss the assumption which might be made to account for the difference" ("The Scattering of the α-Particles by matter," *Proc. Roy. Soc.* A83 [1910]: 492–504).

77. A. van den Broek, *Phys. ZS.* 14 (1913): 32–41. Barkla had shown by 1911 that each atom has about A/2 electrons. Wilson says Bohr came to this conclusion at Manchester in 1912 (*Rutherford,* 327).

78. J. J. Thomson, "Cathode Rays."

79. In a discussion with Eugene Wigner in the late 1970s concerning space-time singularities, I expressed the view that "nature abhors singularities." He countered with "Is not the electron a singularity?" Indeed, it may be. On the discovery of X rays, see the discussion later in this chapter.

80. J. Stark, "Elementary Quantum of Energy, Model of Negative and Positive Electricity" and "Relationship of the Doppler Effect for Canal-Rays in the Planck Radiation-Theory," both in *Phys. ZS.* 8 (1907): 881–84; 913–919 (see Kuhn, *Blackbody Theory,* 344, for the original German titles); W. Wien, "Über eine Berechnung der Wellenlänge der Röntgenstrahlen aus dem Planckschen Energie-Element," *Nachrichten von der Gesellschaft der Wissenschaften zu Göttingen* (1907): 598–601; F. Hund, *The History of Quantum Theory* (New York: Barnes and Noble, 1974), 45–46.

81. N. Bohr, H. Kramers, and J. Slater, "The Quantum Theory of Radiation," *Phil. Mag.* 47 (1924): 785–802. See the discussion in A. Pais, *Subtle Is the Lord* (Oxford: Oxford University Press, 1982), chap. 22.

82. See Bruce Wheaton, "Philipp Lenard and the Photoelectric Effect, 1899–1911," *Hist. Stud. Phys. Sci.* 9 (1978): 299–313.

83. P. Lenard, "Über die lichtelektrische wirkung," *Ann. d. Phys.* 8 (1902): 149–98.

84. Albert Einstein, "Über einen die Erzeugung und Verwandlung des Lichtes betreffenden heruistischen Gesichtspunkt," *Ann. d. Phys.* 17 (1905): 132–48; translated in D. ter Haar, *The Old Quantum Theory* (Oxford: Pergamon, 1967).

85. Einstein references Planck twice, as well as Lenard and Stark. The Lenard reference is to his 1902 paper (see note 82). Both Lenard and Stark were to protest Einstein's "non-Aryan" science.

86. A. Einstein, "Über einen die Erzeugung und Verwandlung des Lichts betreffen neuristischen Gesichtspunkt"; translated and reprinted in ter Haar, *The Old Quantum Theory,* 102.

87. A. Einstein, "Über die Entwicklung unserer Anschauungen über das Wesen und die Konstitution der Strahlung," *Phys. ZS.* 10 (1909): 817–25.

88. A. Einstein, "Zur Quantentheorie der Strahlung," reprinted in *Phys. ZS.* 18 (1917): 121–28, translated in ter Haar, *The Old Quantum Theory,* 167–83.

89. A. Einstein, *Verh. d. D. Phys. Ges.* 18 (1916): 318.

90. See, for example, W. Robert Nitske, *Wilhelm Conrad Rontgen* (Tucson: University of Arizona, 1971). The event was described in the journal *Sitzungsberichte der Physikalisch-Medizinischen Gesellschaft.*

91. Letter from J. J. Thomson to William Thomson, 10 April 1896, Cambridge University Library.

92. Bruce Wheaton, *The Tiger and the Shark: Empirical Roots of Wave-Particle Dualism* (Cambridge: Cambridge University Press, 1983).

93. Ibid.

94. Ibid., 19

95. Ibid., 208, 210

96. Both Braggs were associated with the Royal Institution, and they shared the Nobel Prize in 1915.

97. Wheaton, *The Tiger and the Shark,* 208–12. Mosely was killed in Gallipoli in World War I.

98. Letter from William Thomson to Antoine Becquerel, 4 December 1903, Cambridge University Library.

99. Becquerel reported these and other results in a series of short papers in 1896, published by the Academy of Sciences, including "On the Invisible Radiations Emitted by Phosphorescent Substances." Apparently he simply wanted to test some old plates that had been set aside in a drawer, which also had a packet of uranium salt in it.

100. Wilson, *Rutherford,* 127–29.

101. Ibid., 127–28; E. Rutherford, *Radioactivity* (Cambridge: Cambridge University Press, 1904). See also James Chadwick, *The Collected Papers of Lord Rutherford,* 3 vols. (London: Allen and Unwin, 1965), and Wheaton, *The Tiger and the Shark,* 220f.

102. E. Rutherford, "A Radioactive Substance Emitted from Thorium Compounds," *Phil. Mag.,* series 5, 49 (1900): 1–14. Radon ("emanation"; ^{222}Rn) has a short half-life of 3.8 days, decaying by α emission. Thorium (^{232}Th), on the other hand, has a half-life of 14 billion years.

103. See Marie Curie, *Pierre Curie* (New York: Macmillan, 1955), and Robert Reid, *Marie Curie* (London: Collins, 1974).

104. E. Rutherford and F. Soddy, "Condensation of the Radioactive Emanations," *Phil. Mag.* 5 (1903): 561–76. Three other papers by Rutherford and Soddy on uranium, radium, and thorium appear in that volume of *Philosophical Magazine.* See, for example, Wright, *Classical Scientific Papers.*

105. Wilson, *Rutherford,* 163.

106. A. A. Michelson, "The Relative Motion of the Earth and the Luminiferous Ether," *Am. J. Sci.* 22 (1881): 120–29; A. A. Michelson and E. W. Morley, "On the Relative Motion of the Earth and the Luminiferous Ether," *Am. J. Sci.* 34 (1887): 333–45.

107. E. Whittaker, *A History of the Theories of Aether and Electricity,* (London: Nelson, 1951–53), vol. 2, chap 2. See also G. H. Keswani, "Origin and Concept of Relativity," *Brit. J. Sci.* 15 (1965): 286–306; 16 (1965): 19–32.

108. H. A. Lorentz, "Electromagnetic Phenomena in a System Moving with Any Velocity Less Than That of Light," *Proceedings of the Royal Academy of Amsterdam* 6 (1904): 809–31; H. Poincaré, "Sur la dynamique de l'electron," in *Ouevres de Henri Poincaré,* 11 vols. (Paris: Gauthier-Villars, 1934–54), 9:494–550. Poincaré, who inherited the mantle of Gauss, was the greatest mathematician of his time. His contributions to physics are rather more equivocal.

109. E. Mach, *The Science of Mechanics: A Critical and Historical Account of Its Development,* trans. T. J. McCormack (LaSalle, Ill.: Open Court, 1960); M. Abraham, *Theorie der Elektrizität* (Leipzig: Teubner, 1904); H. Poincaré, *Science and Hypothesis,* trans. W. J. Greenstreet (New York: Dover, 1952).

110. See Gerald Holton, "On the Origins of the Special Theory of Relativity," *Am. J. Phys.* 28 (1960): 627–36.

111. The postulates are not set apart in Einstein's text, but we may state them as follows: (1) According to the principle of relativity, the laws of physics are the same for all unaccelerated observers; and (2) the speed of light is a constant, c, independent of the motion of source of observer. See John Stachel, "History of Relativ-

ity," in *Twentieth Century Physics,* ed. Laurie Brown, Abraham Pais, and Sir Brian Pippard, 3 vols. (New York: American Institute of Physics, 1995), vol. 1, chap 4.

112. Newton says: "The motions of bodies included in a given space are the same among themselves, whether that space is at rest or moves uniformly forward in a right line without any circular motion." See *Sir Isaac Newton's Mathematical Principles of Natural Philosophy,* 2 vols, trans. Andrew Motte (1729), revised by F. Cajori (Berkeley: University of California Press, 1934), 20.

113. For example, see A. I. Miller, *Frontiers of Physics, 1900–1911: Selected Essays* (Boston: Birkhäuser, 1986).

114. A. I. Miller, *Albert Einstein's Special Theory of Relativity: Emergence (1905) and Early Interpretation (1905–1911)* (Reading, Mass.: Addison-Wesley, 1981).

115. N. Bohr, *On the Constitution of Atoms and Molecules,* intro. by L. Rosenfeld (Copenhagen: Munksgaard, 1963), 5.

116. J. Franck and G. Hertz, "Über Zusammenhang zwischen Gasmolekülen und langsamen Elecktronen," *Verh. d. D. Phys. Ges.* 15 (1913): 34.

CHAPTER 9   *Epilogue*

1. John Dalton, *A New System of Chemical Philosophy* (London, 1808).
2. Here I simply mean that the vacuum of modern field theory is a much more complex thing than mere absence of matter.
3. See Y. Elkana, *The Discovery of the Conservation of Energy* (Cambridge: Harvard University Press, 1974).
4. This demise was based on the experiments of Davy and Thompson, the ambivalence of Carnot; the experiments of Joule; and the kinetic theory of Herapath, Clausius, and Maxwell.
5. See chapter 5.
6. The concept of transformation or revolution in physics has generated an enormous literature in recent years, led by Thomas Kuhn, Bernard Cohen, and Paul Feyerabend. See Kuhn's *The Structure of Scientific Revolutions,* 2d ed. (Chicago: University of Chicago Press, 1970), and perhaps Cohen's *Newtonian Revolution* (Norwalk, Conn.: Burndy Library, 1987), and Feyerabend's *Against Method,* rev. ed. (London: Verso, 1988).
7. One might mention Leon Lederman, an important particle experimentalist, but he is known to the public more for his polemical and popular writing than for his scientific work.
8. Kapitza called Rutherford "the Faraday of our time."
9. See especially Allan Franklin, *The Neglect of Experiment* (Cambridge: Cambridge University Press, 1986), and David Gooding, Trevor Pinch, and Simon Schaffer, eds., *The Uses of Experiment* (Cambridge: Cambridge University Press, 1989). In emphasizing the role of experiment, we do not mean to dismiss the serious epistemological problems that arise in the context of experimental verification or falsification.
10. See Carl Schorske, *Fin de Siècle Vienna: Politics and Culture* (New York: Knopf, 1979). Note especially the connections between the painter Oskar Kokoshka and the composer Arnold Schoenberg.

INDEX

{Page numbers in italics indicate illustrations}

Abraham, Max, 167
absolute temperature scale, 84
absolute zero, 84
Academy of the Sciences (France), 10, 11, 12, 18
acoustics, 4–5, 149, 170, 172
action at a distance, 20, 61, 70–73; Faraday on, 48, 52–55
active powers, 7, 20, 22, 25–26
adiabatic processes, 86
Aepinus, Ulrich Theodor, 34
aether, 6, 24, 39, 54, 62, 102, 128, 149– 150, 161–162, 164, 166–167, 169– 170; elastic solid, 39; luminiferous, 24, 33, 53, 74
air engine, 81, 96
Airy, George Biddle, 12
alpha rays, scattering, 161–162, 165
American science, 6, 7, 19, 178n10
American Philosophical Society, *Transactions*, 19
Amontons, Guillaume, 76–77
Ampère, André Marie, 20, 37–38, 41, 43– 45, 56, 61, 79, 86, 122, 125, 128, 130; on aether, 47, 61; on electrical fluids, 47; on interaction of currents, 44, 45, 47; on magnetic induction, 51; on molecular currents, 45–47
Alembert, Jean le Rond d', xiv, 23, 117

analogy, 6, 27–29; mathematical, 29; physical, 28, 68–69
analysis, 25
Analytical Society, 128
Ångström, Anders Jonah, 160
animal heat, 72, 107–108
Annalen der Physik, 18–19, 110
Annales de Chimie et de Physique, 19
Aquinas, Thomas, 23
Arago, François, 21, 39, 41, 43, 56, 79, 86
Arcueil, Society of, 11, 22, 78, 79
Arrhenius, Svante, 18, 130, 142
arrow of time, ix, 143
Association for the Advancement of Science (Germany), 16
Astronomical Society of London, 19
astronomy, xv, 5–6, 178n8
astrophysics, 5–6
atomic nucleus, 161, 165
atomic physics, 5
atomic spectra, 122, 131, 149, 159, 161, 171; pressure and temperature dependence of, 161
atomic theory, 20, 153–154, 159
atomic weight, 85
atomism, 21, 30, 75–76, 85, 102, 116, 120, 132, 145, 147, 150, 169; Boscovichean, 115–117, 123–125; chemical, 114, 116, 118–119, 127, 130–131; Greek, 75, 113–114, 130– 131; physical, 113–114, 116, 128, 130–131; and quantum theory, 131

atomistic philosophy, 22, 26
atoms: diameters of, 128; reality of, 113, 122, 128–129, 131
automobile, 3
avant garde, 173
Avogodro, Amedeo, 116, 120, 161

Babbage, Charles, 12, 13, 23, 56, 128
Back, Ernst, 161
Bacon, Francis, 20, 75
Banks, Joseph, 118
Barkla, Charles, 131, 164
Bartholin, Erasmus, 39
battery, *see* voltaic pile
Bequerel, Henri, 164–165; and discovery of radioactivity, 164
Berard, E., 133
Berkson, William, ix
Berlin Academy of Sciences, 10
Bernoulli, Daniel, 77, 114–116, 133; *Hydrodynamica*, 133, *134*
Bernoulli, John, 39, 103
Berthollet, Claude-Louis, 11, 77–78, 84, 125
Berzelius, Jacob, 32, 37, 41, 79, 123–124, 126–127; on Dalton and Davy, 126; on nomenclature, 126
beta rays, 165
Biot, Jean-Baptiste, 38–39, 41, 43, 71, 79, 125
Biot-Savart Law, 41, 44, 189n40
birefringence, 39, 72
Black, Joseph, 75, 78–80, 117
blackbody: radiation, 151, 158, 168; spectrum, 151–156, 171
Boerhaave, Hermann, 83, 205n43
Bohr, Neils, 7, 158, 173; and theory of atom, 161–162, 168
Boltzmann, Ludwig, xiv, 17–18, 30, 32, 63, 67, 101, 138, 140, 146–147, 150, 152, 154, 156, 169, 171; on atoms, 131; on collision equation, 141–142; on H-theorem, 142–143, 228n64; on irreversibility, 141–142; on recurrence paradox, 144; on specific heats, 157; on statistical interpretation of Second Law, 141–145, 153; on statistical mechanics, 142–144; on transport

equation, 144; on X rays, 164
Born, Max, xiv
Bosanquet, R.H.M., 140
Boscovich, Roger, 22, 48, 54, 115–117, 122–125
Boulton, Matthew, 13
Boyle, Robert, 32, 33, 39, 75–77, 81, 84, 93, 114, 116, 118, 133
Bragg, William Henry, 13, 15, 164
Bragg, William Lawrence, 13, 15, 164
British Association for the Advancement of Science (BAAS), 10, 12, 15, 56, 58, 62, 109–110; and Thomson, 56, 58
Brownian motion, 130–131, 145
Brugmans, Anton, 34
Buffon, Georges-Louis, 116
Bunsen, Robert, 18, 146, 160
Buys-Ballot, Christoph, 138

caloric theory, 75–79, 81, 83–84, 91–92, 105, 119–120, 124, 128, 130, 170
Cambridge Mathematical Journal, 19, 56
Cambridge Mathematical Tripos, *see* Tripos
Cambridge Philosophical Society, 64–65
Carlisle, A., 37
Carnap, Rudolf, 27
Carnot, Lazare, 88, 105
Carnot, Sadi, xv, 80, 84–86, *87*, 88, 91–94, 96, 98, 106–110, 125, 136, 170
Carnot cycle, 80, 88, *89*, 93, 95, 100
cathode rays, 130–131, 150, 161–165
Cauchy, Augustin-Louis, Baron, 20, 23, 57, 72, 125, 145
Cavendish, Henry, 34, 38, 70, 117, 119, 127
Cavendish, William, 70
Cavendish Laboratory (Cambridge), 13, 70, 150, 162
Cayley, Arthur, 14
charge: conservation, 33; electron, 162; nuclear, 162
Charles, J.A.C., 77, 81, 133
Chasles, Michel, 57, 145
chemical equivalents, *see* combining proportions
Chemical Society, 127, 130
chemistry, 118, 132; organic, 126;

pneumatic, 117
City Philosophical Society (London), 47
Clapeyron, Emile, 88, 91, 93, 96
Clarendon Laboratory (Oxford), 13
Clausius, Rudolf, xv, 16–17, 32, 73, 92, 93, 129, *137*, 153, 171; on disgregation, 95; on internal energy, 95; on kinetic theory, 135–136, 138–139, 146–147, 169; on mean free path, 138; on Second Law, 92, 94–96, 98–100, 170; on specific heats of gases, 136, 157; and Thomson, 93
Clifton, Robert, 160
Colding, Ludwig, 92, 104, 107–109, 111
Coleridge, Samuel Taylor, 7, 24, 27–28, 123
Collège de France, 11, 43
combining proportions, 116, 120, 122–126, 129–130
commonsense philosophy, Scottish, 21, 63
Comptes Rendus, 18
Compton effect, 162–163
Comte, Auguste, 7, 21
conductivity, electrical, 123, 125
conservation of energy, *see* energy: conservation of
continuum mechanics, 162–163, 169
conservation of force, *see* force
continuum theories, vs. particle models, 30, 169
corpuscular philosophy, 33, 39
Coulomb, Charles Augustin, 34, *36*, 37–38, 41, 45, 57–58
Crawford, Adair, 133
cryogenics, 172
Curie, Marie, 165
Curie, Pierre, 165

Dalton, John, 20–21, 26, 77, 85, 90, 113, 117–120, *121*, 122–123, 131–132, 169; and law of partial pressures, 119
Darwin, C. G., 164
Darwin, Charles, 3, 7, 13
Darwin, Erasmus, 13
Davisson, C. J., 163
Davy, Humphrey, 7, 13, 15, 20, 25, 27, 37, 43, 46, 50, 79, 83, 108; on atoms, 120, 122–124; on dynamical theory of heat,

76; and Faraday, 47, 48; ice experiment, 82, 84, 105; and Royal Institution, 83; and Rumford, 84
Davy, John, 123
Debierne, A., 165
Delaroche, François, 79, 133
Desaguliers, Jean Theophilus, 33
Descartes, René, 75, 103
diamagnetism, 52–53
Diderot, Denis, 43
dielectrics, 53, 65
differential equations, 21; partial, 30, 38, 72, 115, 169
diffraction, 73; of electrons, 163
diffusion, 139–140
Dirac, P.A.M, xiv
Dirichlet, Gustav, 72
displacement current, 70, 200n195
dissipation, 96, 99–100
du Fay, Charles-François, 33
Duhamel, J.-M.-C., 145
Duhem, Pierre, 7, 11, 19–21, 44; on atomism, 169
Dulong, Pierre Louis, 21, 41, 79, 85, 125–126, 133, 136
Dulong-Petit Law, 126
dynamical philosophy, 7, 20–23, 25, 29, 46, 56, 115–116
dynamical theory of heat, *see* heat
dynamistic tradition, 20–22

Earth: age of, 100–101; cooling of, 101
Ecole Militaire, 11
Ecole Normale, 11, 43
Ecole Polytechnique, 79, 85, 88
Eddington, Sir Arthur, xiv, 18
educational systems, 6, 12, 17–18, 31
Ehrenfest, Paul, 18, 142, 145; on Planck, 157, 161
Ehrenfest, Tatyana, 145
Ehrenraft, Felix, 162
Einstein, Albert, xiv, 29, 30, 63, 73, 76, 101, 162, 166, 173; on blackbody radiation, 147; on Brownian motion, 131, 145, 147, 169; on cosmology, 147; on general relativity, 147; on light quantum, 147, 150, 162, 168; on photoelectric effect, 163; on photon

Einstein, Albert, (*continued*)
  momentum, 163; on Planck, 157; on
  quantum, 144; on special relativity,
  147, 166; on specific heats of solids,
  158, 168; on statistical mechanics,
  147; on stimulated emission, 164
elasticity, xv, 4–5, 170, 172
electricity, two-fluid model, 34, 37
electrochemistry, 37, 50, 52, 124–125, 130
electromagnetic waves, 72, 74, 150
electromagnetism, 32–74, 170, 172;
  discovery of, 40; and light, 38
electron, 130–131; charge, 162; spin, 161
electronics, 169, 172–173
electrostatics, 33, 37, 38, 71
electrotonic state, 66
elements, 118; nomenclature of, 126;
  periodic table of, 126
Eliot, George, 3–4
emission theory of light, 39
energeticism, 150
energy, 25, 66, 102–112; conservation of,
  38, 68, 72, 73, 76, 83, 85, 89, 91, 94,
  103–112, 170, 213n2
Engels, Friedrich, 7
Enlightenment, Scottish, 9, 13
entropy, 85–86, 95, 100, 142, 146, 154, 156
equation of state, 85, 132, 221n78
equipartition theorem, 135–136, 139–140,
  152, 154–155, 158
equivalent weights, *see* combining
  proportions
Euler, Leonard, xiv, 39, 105, 115
Ewart, Peter, 89, 91
Exner, Franz, 18, 145
experiment and theory, xiii, xiv, 18, 29–
  30, 171, 176n10

Faculty of Science (France), 11
Faraday, Michael, *ii*, xv, xvi, 7, 13–16, 18,
  23, 25, 28, 29, 32, 38, 43–44, 46, *51*,
  57, 63, 65, 71–72, 74, 88, 107–108,
  110, 116, 122–123, 126, 130, 169,
  172; on action at a distance, 52–53; on
  aether, 53–54; and Ampère, 47, 50; on
  analogy, 48; on analogy between heat
  flow and magnetism, 52; on benzene,
  50; on Boscovichean atoms, 48, 52–
  54, 124; on conductors and insulators,
  125; on contiguous particles, 125; on
  diamagnetism, 52–53, 193n97; on
  dielectrics, 53; on effect of magnetism
  on polarization of light, 52; on
  electrochemistry, 37, 50, 52, 124–125;
  on electrotonic state, 49; on energy
  conservation, 104; on experimentation,
  171; on field concept, 52–54, 170;
  health of, 52, 194n112; on induction in
  curved lines, 52; on lines of force, 52–
  55, 125; on magnetic induction, 50–
  51, 54; on magnetic rotation, *49*, 50;
  on magneto-optic effect, 40, 52; on
  mathematics, 55; on nature of
  electricity, 52; on paramagnetism,
  193n97; and Royal Institution, 48, 50;
  and Royal Society, 48, 51; and
  Rumford, 84; and Sandemanians, 48–
  49, 191n76; on symmetry, 51, 55; as
  theorist, 48; and Thomson, 52, 58
Fichte, Johann Gottlieb, 108
field concept, 52–54
Fischer, E. G., 39, 116
Fitzgerald, George, 62
Fizeau, A.-H.-L., 39, 166
fluid dynamics, *see* hydrodynamics
fluids: electrical, 33, 34; imponderable, 7,
  21, 25, 33, 34, 37, 53, 75–76, 78, 85,
  169; magnetic, 37
Forbes, James, 13, 14, 29, 55, 63, 67, 141
force, 26, 214n16; conservation of, 103–
  104, 107; dead, 103–104; living, *see*
  vis viva; of motion, 103–104
Fortschritte der Physik, 19
Foucault, Jean Bernard, 39
Fourcroy, Antoine, 116, 125
Fourier, Joseph, 21–22, 32, 55, 58, 71, 81,
  85, 96, 125, 169–170; Fourier series,
  85
Franck-Hertz experiment, 168
Franklin, Benjamin, 6, 11, 14, 17, 32, 33,
  35, 38, 145
Fraunhofer, Joseph von, 159
French Revolution, ix, xv, 7, 9, 41, 117–
  118
Fresnel, Augustin Jean, 20–21, 32–33, 39,
  45–46, 56, 72–74, 79, 125

Galileo, xiv, 20, 75–76, 145, 167
gamma rays, 165
Gassendi, Pierre, 75, 116
Gauss, Carl Friedrich, xiv, 16, 38, 57–58, 61, 71–72, 128, 140
Gay-Lussac, Joseph Louis, 11, 76, 77, 81, 88, 93, 133; on caloric theory, 78, 120, 125, 130
Gehrcke, Ernst, 163
German Physical Society, 10
Germer, L. H., 163
Gibbs, Josiah Willard, 138, 145–146; on chemical potential, 146; and Clausius, 141; on ensembles, 146; and Maxwell, 141; on statistical mechanics, 146–147
Gilbert, Davies, 50
Goethe, Johann Wolfgang von, 26
Goudsmit, Samuel, 161
Gough, John, 118
gravitation, 37, 54, 71, 100, 124
Gray, Joseph, 165
Gray, Stephen, 33
Green, George, 21, 23, 25, 37–38, 56–58, 61, 71, 128, 186n21; discovery by Thomson, 56
Grimaldi, Francesco, 39
Grove, William, 107–108, 111
Guericke, Otto von, 76
Guyton (de Morveau), Louis Bernard, 116

Hacking, Ian, 172
Hales, Stephen, 33, 63, 117
Hamilton, William Rowan, 13, 29, 71
Harris, Snow, 58
Hasenöhrl, Friedrich, 18, 144
Hatchette, J.N.P., 91
Hauksbee, Francis, 116
heat: caloric theory of, *see* caloric theory; conduction, 85, 96, 139; conservation of, 86; dynamical theory of, 25, 68, 76, 78–79, 81, 83–84, 86, 89, 92, 105, 119, 129; nature of, 78, 91; seventeenth-century theory of, 75; theory of, 2, 6, 169–170; wave theory of, 53, 79, 83
heat death, 100
Heaviside, Oliver, 73, 154
Hegel, Georg Wilhelm Friedrich, 26

Helmholtz, Hermann von, xiv, xv, 15–18, 21, 26–27, 38, 49, 70–73, 100–101, 104, 108, 110, *111*, 142, 146, 150, 154, 160, 170–171; on acoustics, 72; on animal heat, 72; on atoms, 130; on electricity, 73; on electromagnetism, 73; on energy conservation, 7, 89, 92, 104–105, 109, 112; on free energy, 146; on gravitation, 108; on Joule, 73, 110; on molecular vortices, 67–68, 128–129, 72; on optics, 72; on physiology, 72; and Thomson, 73; on X rays, 164
Henslow, John, 13
Herapath, John, 84; on absolute zero, 84; on kinetic theory, 84, 91, 129
Herschel, John, 8, 12, 13, 23, 56, 128, 133–135, 141; and Seguin, 108; on solar spectrum, 159
Herschel, William, 79, 116
Hertz, Heinrich, 17, 73–74, 154; on cathode rays, 162; on electromagnetic waves, 150; on photoelectric effect, 162
Hirn, Gustav Adolph, 107–108
Hittorf, J. W., 160
Holtzmann, Karl, 88, 108, 111
Hooke, Robert, xiv, 77, 81
Hopkins, William, 14, 18, 64
Humboldt, Alexander von, 16
Hume, David, 78
Huxley, T. H., 100
Huygens, Christian, xiv, 32, 39, 73, 75, 107
hydrodynamics, xv, 4–5, 61, 62, 115, 170, 172
hypothetical entities, 21–22, 25

ideal gas law, 133
idealism, 7, 27; transcendental, 17
Il Nuovo Cimento, 19
images, method of, 62
imponderable fluids, *see* fluids
indicator diagram, *see* p-V diagram
induction, magnetic, 50–51, 54, 68, 173
Industrial Revolution, ix, 2, 8, 9, 16, 31; and capitalism, 2
Institut de France, 11

Institut Imperial, 43
institutionalization, of science, 10–11, 13, 15
institutions, scientific, 6
instrumentalism, 19
instrumentation, 171
interconvertibility of heat and work, 88–89, 91–92, 96, 104, 109, 170
internal combustion engine, 101, 172–173
inverse-square law, 34, 55, 71; and Coulomb, 37; and electrostatics, 117; in magnetism, 34, 41
irreversibility, 143
Irvine, William, 133

Jacobi, Carl Gustav, 16
Jeans, Sir James, 13, 18, 101, 150; on Planck, 156
Joule, James Prescott, xv, 14, 18, 32, 73, 76, 88, 90, 93, 96, 104–105, 108, 110, 120, 133–135, 170–171; on dynamical theory of heat, 89; on energy conservation, 73, 89, 91, 92, 112; and Herapath, 91; on Joule heating, 90; on kinetic theory, 91, 129; on mechanical equivalent of heat, 91, 107, 109, 111; St. Ann's Reading Room lecture, 109; on speed of hydrogen atoms, 129; and Thomson, 98–99
Journal de Physique, 19
Journal der Physik, 19
journals, 18–19

Kant, Immanuel, 10, 16, 20, 22–24, 26–27, 30, 40, 46, 54, 63, 115
Kastner, A. G., 27
Kelland, Phillip, 63
Kelvin, Lord, *see* Thomson, William
Kepler, Johannes, 20
Kier, James, 14
Kierkegaard, Søren, 7
kinetic theory, 63, 86, 132–147, 170
Kirchoff, Gustav, 17–18, 38, 72–73, 101, 142, 146, 150–151, 154, 160, 171, 173; on blackbody radiation, 151; on diffraction, 73; on sodium D-line, 160
Kirwan, Richard, 116
Königsberger, Leo, 18

Kramers, H. A., 162
Krönig, August Karl, 77, 129, 135–136, 138
Krönig, R., 161
Kuhn, Thomas, 156–157
Kummer, E. E., 145
Kundt, August, 145
Kurlbaum, Ferdinand, 152, 155

Lagrange, Joseph Louis, 11, 20, 22–23, 26, 37–38, 46, 84–85, 103, 145, 183n76
Lambert, J. H., 34
Langley, Samuel P., 152
Laplace, P. S., marquis de, xiv, 11, 20, 22–23, 26, 32, 37–39, 41, 46, 77, 84, 103, 105, 133, 140, 145; on atoms, 116, 127; on caloric theory, 75, 78–79; on potential, 37; on speed of sound, 86; on statistics, 141
Larmour, Joseph, 13
latent heat, 75, 78, 81, 96, 202n7
Laue, Max von, 164
Lavoisier, Anton-Laurent, 20, 26, 75, 77–78, 83, 116–118, 125
Legendre, Adrien-Marie, 37
Leibniz, Gottfried Wilhelm, 20, 103, 115
Lenard, Phillip von, xiv, 131; on cathode rays, 162–163
Leyden jar, 32–34
Liebig, Justus von, 16, 107–108
light: diffraction of, 39; dual nature of, 74; corpuscular theory of, 79, 128, emission theory of, 74; effect of magnetism on, 68; and interference, 39, 163; polarization of, 39; and quantum, 162–163; transverse nature of, 39, 68–69; wave theory of, 5, 21, 30, 33, 32, 39, 68, 79, 128
lines of force, 53–54, 58, 65, 67–69
Liouville, Joseph, 38, 57, 145
Locke, John, 20, 26, 75
Lodge, Oliver, 164
London Institution, 108
Lorentz, Hendrik Antoon, xiv, 13; on electron, 130, 149–150, 161; and Planck, 157; transformation, 167
Lorentz-Fitzgerald contraction, 167

Loschmidt, Johann Joseph, 28, 128, 143
Lummer, Otto, 152, 154–155, 163
Lunar Society of Birmingham, 13, 117

Mach, Ernst, 7, 17, 27, 130, 145, 167; on
    atomism, 169; on X rays, 161, 167
machine tools, 172
magnetic induction, *see* induction,
    magnetic
magnetism, 32, 34, 37, 38, 68; effect on
    light of, 68; and molecular currents,
    32; and scalar potential, 37; two-fluid
    model of, 34
Magni, Valeriano, 76
Magnus, H. G., 73, 145
Malus, Etienne, 39
Manchester Literary and Philosophical
    Society, 14, 18, 91, 119
Marcet, Jane, 47
Mariotte, Edmé, 76–77, 81, 93
Marsden, Ernest, 161
Martine, Georges, 78
Marx, Karl, 2
materialism, 26
mathematics: continental, 23; reform in
    Britain, 56
Maupertuis, Pierre Louis, 23, 38
Maxwell, James Clerk, xiv, xv, 5, 13–14,
    18, 21, 23, 25, 28–29, 31, 32, 38, 40,
    49, 56, 62, *64*, *69*, 74, 76, 110, 129,
    141, 146, 149–150, 153–154, 171; on
    aether, 69–70; on Ampère, 46–47, 70;
    on analogy, 61, 65, 67–68; and
    Cavendish Laboratory, 70; and
    Cavendish papers, 34, 70; on diffusion,
    139–140; on displacement current, 70;
    on electromagnetic theory of light, 7,
    79; on electromagnetic waves, 69, 72–
    73; on electrotonic state, 66; on
    energy, 66, 70–71; on equipartition,
    136, 139–140; and Faraday, 48, 57, 67;
    and Forbes, 67; and Glenlair, 67, 70,
    71; on images, 66; on kinetic theory,
    77, 79, 81, 101, 135, 139, 147, 169; on
    magneto-optic effect, 68; on mechani-
    cal models, 67–68, 70; on molecular
    vortices, 67–70, 128; on molecules,
    130; on probability, 63; reception of
    ideas of, 73; on specific heats, 139–
    140, 157–158; and Stokes, 64; on
    symmetry; 66; and Tait, 63, 67; on
    thermal conductivity, 139–140; *Theory
    of Heat*, 100; on transport phenomena,
    138, 140, 144; *Treatise*, 63, 65, 67, 71,
    141, 146, 170; on viscosity, 129, 139–
    140; and Weber, 66, 70; and Whewell,
    63–64
Maxwell-Boltzmann distribution, 138,
    140, 142–144, 155
Maxwell's equations, 67, 73, 166
Mayer, Julius Robert, 92, 104, 107–111,
    170
Mayer, Tobias, 34
mean free path, 138
mechanical equivalent of heat, 88, 91,
    107–108, 111, 208n79, 216n40
mechanical models, 21, 22, 28, 56, 67–68,
    70
mechanical philosophy, 8
mechanics, 4
mechanistic philosophy, 7, 20, 25, 34, 46,
    56
medicine, 3
Meikleham, William, 56, 58
Meitner, Lise, 18, 142
Mendeleev, D. I., 126
methodology, scientific, 20
Meyer, Julius L., 126
Michell, John, 41, 141, 185n8
Michelson, W. A., 62, 165
Michelson-Morley experiment, 166–167
Millikan, R. A., 162
Minkowski, Hermann, 167
models, mechanical, 21, 22
modern physics, and nineteenth century, 1
Mohr, Friedrich, 107, 110–111
molecular sizes, 126, 128, 140
molecular vortices, 22, 28, 69, 70
momentum, 103–104
Montgolfier, Joseph, 108
Morley, Edward, 62, 166–167
Moseley, Henry, 164
Müller, Johannes, 72
Muncke, Georg, 22
Musschenbroek, Petrus van, 33

Nagaoka, Hantaro, 161
nationalism, 31
Naturphilosphie, 7, 10, 16, 20, 22, 26, 40, 47, 107
Navier, C.-L.-M.-H., 72, 170
Nernst, Hermann, 18, 142
Neumann, Franz, 72
Newcastle Circle, 13
Newcomen, Thomas, 80, *81*, 105
Newlands, John, 126
Newton, Isaac, 33, 38–39, 44, 46, 48, 55, 63, 78, 103, 114–116, 118, 163, 173; on atoms, 75, 202n3, 217n6; on light, 79; *Opticks*, 38, 75, 122; *Principia*, 38, 103, 122, 127, 167; on speed of sound, 86
Nichol, John Pringle, 85
Nicholson, William, 37
Northumberland Circle, 13
nuclear atom, 161, 165

Oersted, Hans Christian, 26, 32, 37–38, 40–41, *42*, 43–45, 50, 108
Ohm, George Simon, 44, 71
Oken, Lorenz, 16, 26
optics, 149, 170
Ostwald, Friedrich Wilhelm, 130, 143, 146, 150

p-V diagram, 80, 88, 93–94
paramagnetism, 65
Parrot, G. F., 22
partial differential equations, 30, 38, 72, 115, 169
partial pressures, law of, 119
particle theories, *see* corpuscular philosophy
Pascal, Blaise, 76
Paschen, Frederick, 152, 154, 161
Paschen-Back effect, 171
Peacock, George, 12–13, 23
periodic table, 126, 161
Perrin, J. B., 161
Petit, Alexis Thérèse, 21, 79, 85, 125, 135
Philosophical Magazine, 10, 18
Philosophical Society of Edinburgh, 15
Philosophical Society of London, 10, 18
Philosophical Transactions, see Royal

Society of London
philosophy: moral, 5; natural, 5; of nature, 7
phlogiston, 77, 117–118
photoelectric effect, 149, 162–163, 171
photography, 172
physics: American, 6; Italian, 6; solid-state, 5, 173
Physikalische Zeitschrift, 19
Pictet, Marc-Auguste, 79
Planck, Max, xiii, xiv, 17–18, 30, 73, 144–145, 147, 152, 157, 162; on atomic theory, 153–154; on blackbody formula, 155–156, 164; and Boltz-mann, 153; on entropy, 155–156; on quantum, 162; on Second Law, 153, 155
Planck's constant, 232n46
Playfair, John, 21, 28
Poggendorff, Johann Christian, 19
Poincaré, Henri, 21, 32, 37–38, 56–57, 71–72, 145, 162, 166–167; on recurrence paradox, 144
Poisson, Simeon-Denis, xiv, 84–85, 88, 125, 145; on caloric theory, 78–79
polarization, of light, 29, 65
polonium, 165
Popper, Karl, 19, 172
positivism, 7, 19–20, 22, 26–27, 145, 182n60
potential theory, 37–38, 72; gravitational, 37; and potential energy, 103
Power, Henry, 77, 81
Prechtl, J. J., 41, 43, 47
Priestley, Joseph, 14, 22, 34, 38, 116–118, 133; and discovery of oxygen, 117
Pringsheim, Ernst, 152, 154–155
probability, 63
professional societies, 10, 19
professionalism, 5, 9–13
Proust, Louis, 116
Prout, William, 126
Prussian Academy of Sciences, 10, 19

quantum, 144, 156, 166, 168; and discontinuity, xiii; and electrodynam-ics, 5; and ontology, 7; revolution, xv; theory, xiv, 4–5, 7, 20, 63, 86, 101,

131, 140, 147, 149, 154, 158, 166, 168, 171, 173
Quetelet, L.-A.-J., 141
Qunicke, Georg Hermann, 18

radiation, 85; blackbody, 149, 151, 158, 168; discrete, 101; thermal, 7, 152
radicalism, 173
radioactivity, 130, 171
radium, 165
radon, 165
Ramsay, William, 165
Rankine, William, 21, 92, 96, *99*, 135; on molecular vortices, 67, 96
Rayleigh, Lord (William Strutt), 4, 13, 15, 18, 62, 101, 135, 140, 150, 152; on equipartition, 153–155, 158
realism, xi, 7, 22
Regnault, Victor, 57, 62, 96, 133
Reichenbach, Hans, 27
relativity, 4, 171; general, 5–6; special, 5, 62–63, 166–167
religion, 3, 7, 9, 48–49; dissenters, 117–118
research ethos, 6, 16
revolutions of 1848, 7
revolutions, scientific, twentieth-century, 8, 173
Richter, Jeremias Benjamin, 116
Riemann, Bernhard, 72; and EM waves, 72
Ritter, Wilhelm, 26, 40
Rive, August de la, 41, 51
Roberval, Gilles, 76
Robison, John, 21, 34, 80
romanticism, 10, 16
Römer, Ole, 39
Röntgen, Wilhelm Conrad, 17, 164; on X rays, 163
Routh, E. J., 13–15, 18, 64, 83
Rowland, Henry, 62
Royal Institution (of Great Britain), 10, 13–15, 18, 44, 47–48, 120, 123–124
Royal Society of Edinburgh, 10, 15
Royal Society of London, xii, 10, 12, 18, 48, 65, 67, 70, 83, 135, 163, 178n3, 179n15
Rubens, Heinrich, 152, 155
Rumford, Count, *see* Thompson,

Benjamin
Rupp, E., 163
Rutherford, Ernest, xvi, 62, 162; on alpha rays, 161; and Cavendish Laboratory, 150; on emanation, 165; on experimentation, 171; on gamma rays, 165; on nuclear atom, 161

Saussure, H. B. de, 79
Savart, Felix, 41
Savery, Thomas, 80
saving the appearances, 21, 79
Scheele, Carl Wilhelm, 79, 117
Schelling, F.W.J., 10, 16, 26–27
Schödinger, Erwin, 18
Schweigger, Johann, 37, 117
Scottish Enlightenment, 13, 14, 63
Second Law of Thermodynamics, *see* thermodynamics
Sedgewick, Adam, 13
Seebeck, Thomas, 43
Seguin, Marc, 104, 107–108
Select Society of Edinburgh, 15
Slater, John, 162
Smith, Adam, 14, 78
Smoluchowski, Marian, 145
Snell, W., 39, 79
Société Française de Physique, 11
Societé Philomatique, 10–11
Soddy, Frederick, 165
solar spectrum, 158–159
solid-state physics, 173
Sommerfeld, Arnold, 164
specific heats, 81, 85–86, 126, 129, 132–133, 136, 140, 149, 158, 168, 171–172, 206n61; and solids, 133, 149, 158, 168, 171–172; temperature dependence, 132, 140, 158, *159*
spectra: atomic, 149, 158, 160–161; emission, 160; molecular, 149, 158, 160–161
Stark, Johannes, xiv, 131, 162, 164
statistical mechanics, 139, 141, 143–144, 146
steam engine, 15, 76, 80–81, 105, 107
Stefan, Joseph, 18, 101, 142, 145, 152
Stewart, Balfour, 151, 160; on blackbody radiation, 151

Stokes, George Gabriel, xv, 13–14, 18, 23, 25, 56, 58, *60*, 101, 170; on analogy between heat and fluids, 61; on sodium D-line; and Thomson, 58, 61
Stoney, Johnstone, 128; on atomic and molecular spectra, 160; on "electron," 130, 150
Strutt, William, *see* Rayleigh, Lord
Sturgeon, William, 37, 186n15
Swan, William, 160
Symmer, Robert, 34
symmetry, 51, 54–55, 66

Tait, Peter Guthrie, 13–14, 23, 55–56, 63, 67, 100, 112, 146
Talbot, Fox, 159
Taylor's Scientific Memoirs, 19
technology, xiv, 3–4, 8, 101, 171–172
temperature scale, 77; absolute, 96
Thales, 32
Theophrastus, 57
theoretical physics, xiii, 154
thermodynamics, 4–5, 30, 76; First Law, 88, 94, 170; irreversible, 149, 153–155; rational, 76; Second Law, 86, 88, 92, 94, 96, 100, 153, 155
thermometry, 75, 77
Thompson, Benjamin (Count Rumford), 6–7, 15, 82, 84, 106, 108, 180n36; and cannon boring experiment, 79, 84, 105; on dynamical theory of heat, 79, 83; and Royal Institution, 83
Thomsen, Hans Peter, 126
Thomson, James (father), 56
Thomson, James (brother), 56, 96, 98
Thomson, J. J., 162; on atomic model, 131, 161; and Cavendish Laboratory, 150; on electron, 130, 150; on electron charge, 161–162; on X rays, 164
Thomson, G. P., 163
Thomson, Thomas, 122–123
Thomson, William (Lord Kelvin), 5–6, 14–15, 18–19, 21, 23, 25, 28–29, 31, 32, 38, 49, 55, *59*, 71, *97*, 104, 110, 141, 148, 165, 168, 171; on absolute temperature scale, 92; on aether, 58–60; on age of Earth, 62, 100; and

Ampère, 61; on analogy between electricity and heat, 57–58; and Atlantic cable, 62; on atoms, 132; and BAAS, 56, 58; on British mathematics, 57; and Cambridge, 56–58; and *Cambridge Mathematical Journal*, 56; on caloric theory, 84; and Carnot, 62, 88, 92–93, 100, 109; and Clausius, 99, 136; on dissipation, 99–100; on dynamical theory of heat, 92; on electrical fluid, 56; on equipartition, 158; and Faraday, 48; and Forbes, 56; and Fourier, 58, 85; and Hopkins, 56; and Green, 38, 57; on intrinsic energy, 95; and Joule, 91, 96, 100, 109; and Laplace, 56; on magnetism and light, 61; on materiality of heat, 56; on mathematical analogy, 61; and Maxwell, 62; on method of images, 58; on model of atom, 161; and Nichol, 55; on physical analogy, 57–58, 61; on Second Law, 92, 95–100, 170; on sizes of atoms, 128–129; and Stokes, 160; on terrestrial magnetism, 62; on theory of heat, 56, 61; Thomson-Stokes correspondence, 58; *Treatise*, 112; on vortex rings, 129; on X rays, 164
thorium, 165
Torricelli, Evangelista, 76
torsion balance, 34
Towneley, Richard, 77
Townsend, J. S., 162
transcendental idealism, 7, 10
transistor, 173
transmutation, in radioactive decay, 165
transport phenomena, 132, 138, 140, 144
Trevithick, Richard, 80
Tripos, mathematical, 6, 14, 56, 63, 180nn29, 31
Tyndall, John, xv, 13, 44
Tytler, James, 47

Uhlenbeck, George, 161
ultraviolet catastrophe, 153
unification, unity, 5, 28–29, 170; of electricity and magnetism, 46, 55, 74

universities: British, 12; German, 16–18; and research ethos, 6
uranium, 165

vacuum technology, 172
valence, 127
van't Hoff, J. H., 154
Vienna Circle, 7, 27
Villard, Paul, 165
vis viva, 78, 170; conservation of, 102, 105, 108, 110, 136; loss of, 103
viscosity, 139–140
Volta, Alessandro, 34, 37–38
voltaic pile, 34, 37, 41, 44, 173
vortex hypothesis, 54, 67, 128

Wallis, John, 103
Waterston J. J., 129, 135–136, 224n19
Watson, William, 33
Watt, James, 13, 15, 80, *82*, 117
wave-particle duality, 163
wave theory, of heat, *see* light: wave theory of
Weber, Wilhelm, 16, 70–72
Wedgwood, Josiah, 14
Weierstrass, K.T.W., 145
Wenzel, Karl Friedrich, 116
Wheatstone, Charles, 159

Whewell, William, 12–14, 23, *24*, 25, 46, 63–64, 91; on atoms, 126–127
"Whig history," 176n4
Whitehead, John, 117
Wien, Wilhelm, 101, 152, 157; on displacement law, 152; on distribution, 153–155, 163, 171; on X rays, 162
Wilcke, Johann Carl, 34, 78
Williamson, Alexander, 127
Wittgenstein, Ludwig, 7
Wollaston, William Hyde, 37, 48, 120, 122; on solar spectrum, 159
work, 104–105
Wren, Christopher, 105
Wullner, Adolph, 160

X rays, 74, 149, 162, 164–165, 168, 172; continuous, 164; discrete, 164; scattering, 149

Young, Thomas, 15, 21, 33, 39, 74; on "energy," 104; on molecular sizes, 126

Zeeman, Pieter, 161
Zeeman effect, 150, 161, 171; anomalous, 161, 171
Zeitschrift für Mathematik und Physik, 19
Zermel, Ernst, 144–145, 154

Robert D. Purrington is a professor of physics at Tulane University. In addition to papers in experimental physics, he is co-author of *Frame of the Universe* (1983), a highly acclaimed history of Western astronomical and cosmological thought.